T0177532

# The Great Silence

# THE GREAT SILENCE

*The Science and Philosophy of Fermi's Paradox*

MILAN M. ĆIRKOVIĆ

OXFORD
UNIVERSITY PRESS

OXFORD
UNIVERSITY PRESS

Great Clarendon Street, Oxford, OX2 6DP,
United Kingdom

Oxford University Press is a department of the University of Oxford.
It furthers the University's objective of excellence in research, scholarship,
and education by publishing worldwide. Oxford is a registered trade mark of
Oxford University Press in the UK and in certain other countries



First Edition published in 2018



Published in the United States of America by Oxford University Press
198 Madison Avenue, New York, NY 10016, United States of America

British Library Cataloguing in Publication Data
Data available

Library of Congress Control Number: 2017954178

ISBN 978–0–19–964630–2

Printed and bound by
CPI Group (UK) Ltd, Croydon, CR0 4YY

To Jelena Dimitrijević,
Who reads by flashlight and time-travels in libraries, with kittens

How many kingdoms know us not!

**Blaise Pascal**

The void is the only great wonder of the world.

**René Magritte**

Man is a very small thing, and the night is very large and full of wonders.

**Lord Dunsany**

Every classification throws light on something.

**Sir Isaiah Berlin**

# CONTENTS

# INTRODUCTORY NOTE

The perennial question *Are we alone?* is one of those crucial watersheds in human quest for understanding the universe which defy both simple digests and simple answers. One of its deepest components is the puzzle almost universally—though imprecisely—known as Fermi's paradox, the absence of extraterrestrial life and its manifestations from our vicinity in both space and time. Originating in lunchtime remarks of the great physicist Enrico Fermi in 1950 (although some precursors have been noted), the chestnut has proven to be one of the deepest, subtlest, and most persistent challenges in the history of science. This book is, at least in part, devoted to justifying such a sweeping claim. But there is much more, even beyond the borders of science: Fermi's problem not only shaped innumerable science-fiction visions, but also profoundly influenced all our thinking about the relationship of the cosmos and the mind, as well as about the future of humanity and cognition. Encapsulated in the legendary question *Where is everybody?*, the puzzle turned out to be a tough nut to crack. Even a cursory look at the literature will show that, since about 1975, the number of publications containing references to Fermi's paradox is well in the hundreds. Still, the problem remains completely and irritatingly unsolved.

The present book grew from a profound dissatisfaction I have felt for quite some time that people do not take Fermi's paradox seriously enough, in spite of the multiple recent discoveries and trends which make the problem objectively more disturbing than ever. As one example among many, in Fermi's time, no data on the age distribution of habitable planets and their parent stars were available, and one could plausibly argue—as with the 'rare-Earth' claims of some modern researchers—that Earth is in the extreme early tail of this age distribution, being among the oldest, if not *the oldest* habitable planet in the Milky Way. This would, in turn, mean that all other biospheres are younger than the terrestrial one and, as long as the emergence of intelligent beings and their civilizations follows the terrestrial chronology even roughly, that we must be the oldest civilized species in the Galaxy. Fermi's paradox is then, obviously, resolved, since nobody is around *yet*. The contemporary 'astrobiological revolution' brought about a revolutionary change in this respect: since 2001 and the seminal results of Charles Lineweaver, we are aware that Earth is, in fact, a latecomer on the Galactic scene: the formation of Earth-like planets started more than 9 Ga (a giga-annus = 1 billion years = $10^9$ years) before the birth of Earth and the Solar System, and reached its peak at −6.2 ± 0.7) Ga, still some 1.7 ± 0.7 Ga before our planet completed its original accretion.[1] That, obviously, means that most of the habitable sites in the Milky Way are billions of years older and, if we accept the straightforward Copernican assumption, so must be intelligent beings which have evolved on these sites. So, indeed, *where is everybody?*

This book is about many facets of one particular problem belonging to astrobiology and, more specifically, to its subfield dealing with the Search for Extra-Terrestrial Intelligence (SETI). SETI is, contrary to the popular impression, a very large field when the size is gauged not by the funds invested or the number of people involved, but by the scope of the research topics and the relevance of its questions. Thus, the reader will find the amount of general SETI content somewhat disappointing; fortunately, there are multiple high-quality introductions and reviews of the entire field, some of which are cited in the endnotes.[2] So, although I may mention some other important problems of SETI studies (e.g. target selection, wavelength choice, or even whether it is safe to transmit *our* messages) in passing, there is no way justice could be given to them in such a brief format. The Drake equation will not be discussed in detail either, but for an entirely different reason, explained in Chapter 3: while its historical role has been undisputed, the continuation of its use in the present epoch of the astrobiological revolution is, in my view, not only outdated, but can be counterproductive and supply ammunition to the 'fundamentalist' opponents of the whole endeavour.[3]

Fermi's paradox touches upon many issues belonging to various traditional disciplines; it is enough to look into items commonly used in formulations (such as those given in Section 1.2) of the paradox: the Galaxy, intelligence, civilization, technology, evolution, interstellar travel, ages, manifestations, detection, naturalism, cosmological horizon, and so on. The relevant ideas touch upon the key concepts in Galactic and planetary astronomy, evolutionary biology, social sciences, communication technology, astronautics, and cognitive and computer science. However, even that actually underestimates the broadness of the problem, since there are numerous philosophical assumptions underlying the individual formulations of both the problem and the multiple solutions offered in the literature. So, there is an additional dimension belonging to logic, epistemology, and even metaphysics. As we shall see in Chapters 4–7, the very richness of the multidisciplinary and multicultural resources required by individual explanatory hypotheses enables us to claim that it is *the* most complex multidisciplinary problem in contemporary science.

It is a *paradox* as well. Paradoxes are both fun and serious. Paradoxes are usually understood as *apparently* unacceptable conclusions derived by *apparently* acceptable reasoning from *apparently* acceptable premises. With a few exceptions in the arcane philosophical realm (a strange school of thought known as dialetheism on one side, and some crazier strands of postmodern social constructivism on the other), the consensus is that there must be an error somewhere, since acceptable does not lead by acceptable steps to the unacceptable. Hence, an investigation into both the premises and the methods of our reasoning is certainly warranted. Paradoxes raise problems which have been often associated with the great crises of thought and conceptual revolutions in science. This dual role of both destruction and construction constitutes part of their appeal.

Responses to paradoxes carve tunnels into deep, foundational issues—such were the historically well-understood cases of Russell's paradox in the set theory, and the 'ultraviolet catastrophe' paradox in early atomic and quantum theory. Even paradoxes which have been successfully resolved within the prevailing theory—like the Twin Paradox in special relativity—have played an important role in the clarification of foundational issues and continue to hold an honourable place in science education to this day. This indicates an evolution of paradoxes with time.

Fermi's paradox is still far from that stage—mainly because a *general astrobiological/SETI theory does not exist yet.* This is the main constraining circumstance which, unfortunately, has not been widely recognized. As I have tried to argue in *The Astrobiological Landscape*, astrobiology is still in a balkanized and sundered state, while it has many potential opportunities for becoming a unified, deeply interconnected field.[4] In that study, I have suggested the eponymous concept of an astrobiological landscape as a working concept capable of organizing current and future research, not just a metaphor. That particular is of minor importance, however, compared to the *general need* for more theoretical work which could forge unity among disparate astrobiological strands of thought and activity. Some of that need might be provoked by Fermi's paradox—and elucidating how such a thing may come to pass is an additional goal of the present book.

Paradoxes are related to the controversial scientific method of *thought experiment*. Since the beginning of modern science with Galileo, the thought experiment has been the foundation of much of contemporary physics and has made many inroads into other scientific disciplines. Without delving into the difficult epistemological questions of where the insights in thought experiments come from,[5] it is safe to assert that thought experiments are highly relevant in astrobiology, and in SETI studies as well. Some examples on various levels will be demonstrated in the course of this book, as we traverse the long list of explanatory hypotheses.

The very existence of so many hypotheses proposed for the resolution of Fermi's paradox in the last seven decades confuses the reader and makes serious research much more difficult. This creates the need for better organization of the ideas involved. *Darwin would not have been possible if he had not been preceded by Linnaeus*, famously wrote Claude Lévy-Strauss. The dictum applies here with full force: taxonomical work necessarily precedes the solution of a problem as complex as Fermi's paradox certainly is. A variety of hypotheses presented for the resolution of the puzzle often hides the fact that some hypotheses are just versions of the same underlying idea, being in the same time very different from other species of the same fauna. Obviously, we need to introduce some order into that chaos, and one of the main points of the present book is to argue that such classification is, in fact, much simpler than a cursory look would suggest. The specific taxonomic scheme presented here is clearly one of many possibilities in this respect; but it is of paramount importance that we embark on that journey in the first place. Nobody put it better than Steven Jay Gould:[6]

> Classifications are not passive ordering devices in a world objectively divided into obvious categories. Taxonomies are human decisions imposed upon nature—theories about the causes of nature's order. The chronicle of historical changes in classification provides our finest insight into conceptual revolutions in human thought. Objective nature does exist, but we can converse with her only through the structure of our taxonomic systems.

Of course, there will always be a kind of sneering criticism, of the 'oh, just another fancy way to organize our ignorance' kind. It was raised against Linnaeus, against the 'eightfold way' and the quark model in particle physics in the 1960s, against Carl Woese's restructuring of the basic branches of the "tree of life" in the 1980s, against Mendeleev's period table of elements in the nineteenth century, and in many other instances in the history of science. Organization of ignorance is as important to the real science as is the organization of knowledge—if not *more* important, since organization of knowledge, while crucial for deeper and wider theoretical syntheses, as well as for educational and promotional activities, still presupposes that the knowledge *is* at hand, after all. In contrast, the organization of ignorance emphasizes and points out the direction in which we must *seek* knowledge which is still not ours. While a cartographer might not had that in mind, the value of his maps for an explorer lies exactly in showing where exactly the blank patches—or dragons—of our ignorance are located.

In addition, the work on Fermi's paradox could be compared to a famous class of mathematical problems popularized by Martin Gardner and often known as 'impossible puzzles', where players have insufficient information to produce a solution, but gain additional knowledge from the fact that the other players cannot solve it either.[7] This delightful irony—that ignorance, in a sophisticated context, produces knowledge—is further compounded in the present case by the curious coincidence that the archetype of this kind of problem, the 'sum and product puzzle' in which the first player knows only the product of two numbers, and second player only the sum of the same numbers, was invented by none else than Hans Freudenthal. He was the Dutch mathematician who became well known in SETI circles for his proposal, which was made in 1960, of *LINCOS*: an artificial language intended for communication with extraterrestrial intelligence.[8] While the 'impossible puzzles' are, in general, quite difficult and touch upon arcane issues in number theory, there is another useful metaphor which one can use to understand why well-structured ignorance is important in our research efforts: the famous game *Battleship*, in which the players try to sink the opponent's flotilla on a grid, by improving their guesswork knowledge over time (many versions exist, including the solitaire ones).[9] From the perspective of a player, the ground rules of the game contain information on the sizes of the opponent's ships, although they are invisible; being able to formulate such 'ground rules' in the search for alien intelligence is the task at hand.

There is also the matter of full disclosure, which is necessary in a work such as this, where philosophical and aesthetical preferences still play an important role—in contrast to well-established and mature fields. Historians of science, such as John North, Helge Kragh, and Steven J. Dick have repeatedly pointed out how today's mature disciplines, such as physical cosmology, were heavily influenced by philosophical views in their early days. The same lesson is, of necessity, true of astrobiology as well. Following the dictum of Herbert Butterfield that biases are not problematic (since it is impossible to get rid of them completely), but avoiding to explicate them is, I admit to a moderate pro-SETI bias: I find that we have reasons to rationally believe in the existence of extraterrestrial intelligence elsewhere in the Galaxy (and, a fortiori, the wider universe); and I sharply disagree with both the traditional sceptics, such as Ernst Mayr and Frank Tipler, and more modern 'rare-Earth' theorists about the fundamental lack of plausible SETI targets in the Milky Way and the wider universe. I shall try to justify some of the reasons for such a belief in the later text, since they are relevant for at least some of the hypotheses historically suggested to account for Fermi's paradox. I have tried to present the adverse views as fairly as possible, but the degree of success in this has to be judged by the reader.

Therefore, I use this opportunity to state that I am an optimist regarding the existence of extraterrestrial intelligence, and hence meaningful SETI targets, mainly because I hold a deflationary view of the crucial attribute 'extraterrestrial': I do believe we neither can nor should draw a sharp distinction between the terrestrial and the extraterrestrial realms, and I believe that there is nothing both crucial and special about Earth, its physical, chemical, and so on, properties, or its location in space and time, and that this is true as well for billions of other planets in the Milky Way and beyond. If you look through the window of your home and see that it is raining outside, you will conclude that it is raining in your town; unless you are a radical sceptic, you will not limit your conclusion to the dozen or so of metres which you *directly observe*. Accordingly, you will decide pragmatically to take your umbrella even if you are going much farther than the tiny part of the town you actually see from your window. (Notice that you *might be wrong*: the rain might stop by the time you exit your flat or there might be local weather anomalies causing truly nice weather near your office. This is not *likely*, however, and it is exactly likelihood we are interested in as far as the real science of astrobiology is concerned.) Obviously, there is no real, physical boundary between the tiny part you observe and the rest of your town; neither will you conclude that your house or flat is located in such a way that your observation of the weather will be strongly biased. Your pragmatic decision to take your umbrella on the basis of a single local observation is justified by the intuitive—unless we are meteorologists, in which case it is much more than intuitive—understanding that weather patterns are variable on scales of space and time generally larger than the distance or travel time to our workplaces. We might say that weather patterns are determined by some *wider process*. This wider process (which we do not really need to understand in any detail!) makes the hypothesis about yet-unobserved raining around workplace the *default* hypothesis. And so it is with the cosmic life: we observe life and intelligence in a particular unremarkable cosmic locale—therefore, we pragmatically conclude that it is a manifestation of a wider process, *cosmic evolution*. This is a simple generalization of what the titans of biological thought, people like Alexander Oparin, John B. S. Haldane, Sidney Fox, and Joshua Lederberg thought about the origin of life on Earth: it is a lawful process occurring throughout the universe whenever necessary physical, chemical, and geological preconditions are satisfied. Therefore, the existence of life elsewhere—where the preconditions are satisfied—is the default hypothesis in astrobiological search. I shall elaborate on this position in Chapters 3 and 5.[10]

There are further arguments which have convinced me regarding the *existence* of viable SETI targets (and therefore, at least potential sources of Fermi's paradox). On the other hand, there simply isn't much around to convince me of the prospects of *communication* with such targets (what was once labelled CETI). On that issue, I profess agnosticism, since, contrary to many views present in pop-cultural discourse, the arguments of sceptics are much stronger here, and I do perceive very big difficulties in any attempt of even conceptualizing such communication. For one, many extraterrestrial intelligent communities in our Galaxy might be extinct by now. As we shall see in Chapters 1 and 6, such a scenario would not completely resolve Fermi's paradox, but will go some way towards its resolution, while certainly annihilating any chances for CETI. Even if such communities are not dead, there could be several kinds of reasons, including technical, economical, or safety-related obstacles, which could make SETI targets incapable or unwilling to engage in the CETI game. Finally, even if they are capable and willing, it still remains highly doubtful, to my best reading of arguments available in the literature, that attempts at such communication will *succeed*, as the cognitive and biosemantic differences might be too large for meaningful dialogue to be established. On top of all that, the question of *value* generated by such communication should not be discounted as obvious. Some of the people who thought hardest about this issue (like the great Polish novelist and philosopher Stanislaw Lem) warned us about scenarios where it seems that successful communication has occurred, but no real, tangible value—material, intellectual, or spiritual—has been generated in the process.[11]

A corollary is that the historical SETI *practice* might be misleading or wrongheaded, entirely independently of the real astrobiological issue of the density distribution of observers in the real universe. In the discussion pro et contra actual SETI methods employed since 1960, I admit to a moderate bias on the contra side of the debate. But it is important to understand that this debate has mostly been led on grounds firmly outside astrobiology, and occasionally outside science at all. Here, I aim at questions like *To what extent are our SETI methods grounded in specific human history and culture?*, *How confident one can be in identifying an artificial signal in the noise of natural sources?*, *Is it a cost-effective way of spending sparse funding for science?*, and so on. Some of the issues are very obviously relevant to Fermi's paradox (e.g. *Should we expect an extraterrestrial civilization to self-destruct by the time its message reaches Earth?*, expressing concerns which have bothered more than one 'disgruntled' founding father of SETI!) and I shall discuss them in the course of this book. Still, I have no doubt that empirical SETI should proceed, expand (although, in the present funding climate all over the world, it does not seem very likely, sadly enough), and incorporate alternative methods and approaches as much as constrains allow. While part of the reason for it is extrascientific—SETI, after all, is quite cheap as empirical sciences go and it offers interesting side benefits—the other part is very important from a heuristic point of view. Even if unsuccessful in terms of actually detecting a signal of undoubted extraterrestrial intelligent origin, each and every SETI project contributes to the reduction of the overall parameter space for cosmic civilizations in general. It enables a bit better focus on those regions of the astrobiological landscape which hold all trajectories leading to intelligent observers *compatible with all the evidence*. Even purely parasitic or archival SETI projects play this important heuristic role: even negative evidence does not really exist from the viewpoint of science, until it is *recognized as evidence*.[12]

Usually, it is not easy to persuade intelligent people—even astronomers!—that Fermi's paradox is a serious problem, and one which has become even more serious of late. In the course of the book, I shall try to peel that initial scepticism (not towards extraterrestrial intelligence as such, but towards appreciation of Fermi's paradox) layer by layer, as we traverse different aspects of this wide-ranging cluster of issues. This task is not simple because, among other reasons, the problem itself presents us with a mirror: some of the most deeply ingrained prejudices and dogmas about the universe and the place of life in it are clearly manifested in the reactions of both scientists and laypeople at the initial presentation of the paradox. Some prejudices and dogmas are, as we shall see, motivated by extrascientific items, such as religious views; others are particulars of our modern culture; still others might have evolutionary origins in the deep past of our species. Also, several common logical errors surface time and again in the replies of people finding, for example, Fermi's paradox 'an easy problem', usually via mistaking a part for the whole. All this offers a lesson of much wider scope than 'merely' SETI and astrobiology studies: the solution of this, as well as many other 'Big Questions' is not just purely *out there*; a part always lies within us.

In the main part of this book, I shall survey the already voluminous literature dealing with Fermi's paradox, with an eye on the classification scheme which could help in understanding many hypotheses posed in this regard. The problem is fundamentally intertwined with so many different disciplines and areas of human knowledge that it is difficult to give more than a very brief sketch in the present format. It should be noted straight at the beginning that it is not entirely surprising that several scientific hypotheses resolving the paradox have been formulated, in a qualitative manner, in the recreational context of a piece of science-fiction art; astrobiology is perhaps uniquely positioned to exert such influence upon human minds of various bents. After all, much of the scientific interest in questions of life beyond Earth in the twentieth century was generated by epoch-defining works such as Herbert G. Wells's *War of the Worlds*, Sir Arthur Clarke's *2001: Space Odyssey*, and Sir Fred Hoyle's *The Black Cloud*.

After presenting the main families of solutions, I shall compare them critically, in view of the recent astrobiological results and methodological principles. It is important to identify the most promising hypotheses as targets for future research for several reasons. First of all, the tremendous expansion of astrobiology since 1995 offers a prospect of tackling many difficult issues which were considered entirely speculative and unapproachable only a couple of decades ago. Furthermore, some of the promising research strategies involve detailed numerical modelling, which needs to be performed according to clear theoretical guidelines, which in the case of hypotheses for answering Fermi's question are still lacking. Finally, we face the pertinent problem of making quantitative assessments of the prospects for future of humanity—something which is, as we shall amply see in Chapters 8 and 9, tightly connected to our insights into the generic families of solutions.

The plan of the exposition is as follows. After introducing the paradox and its various versions in Chapter 1, I discuss two key pieces of background knowledge, cosmological/astrophysical in Chapter 2, and philosophical in Chapter 3. In these two chapters, I identify a number of tenets which lead, when strictly applied, to the paradoxical conclusion. It turns out, naturally, that by abandoning some of these tenets we can generate four families of solutions to the paradox. These families are discussed in some detail in Chapters 4–7. A comparative analysis of the solutions offered thus far is given in Chapter 8, and some paths for advancing the discussion and study of Fermi's paradox are sketched in the concluding chapter, Chapter 9. It is also an exhortation and a call for more work in this exciting and far-reaching area.

Finally, a few words about formatting, technical details, referencing, and sources are in order. Occasionally, a shaded box with more advanced material will appear in the midst of a chapter. No reason for panic! Most of these contain only slightly more complex material, which is easily followed with the help of supplementary literature recommended in each case. Even if not, it is a perennial hope that an interested reader might return to those parts after subsequent thinking and consideration.

As far as general number of sources is concerned, Fermi's paradox has persistently been an intriguing topic, if not since Fermi's original lunch, then from the start of the SETI era (1959–60), which began with the study by Cocconi and Morrison, as well as Project Ozma). A simple Google search for 'Fermi's paradox' gives a large number of hits, about 430,000—still much less than 40+ million hits for Britney Spears, but sizeable for any even remotely scientific topic. As usual, most of the information available on the web ranges between imprecise, at best, and utterly misleading, at worst. Some resources can be useful as points of entry in the literature, while others contain exhaustive discussions of individual hypotheses. Whenever it was practical, those resources are listed in the endnotes.

The published literature on Fermi's paradox has grown substantially in the last half-century, although it is of highly uneven quality. Brin's 1983 review is more than three decades old, but it still contains the best overall review of the hypotheses and their theoretical background. It was the first attempt to bring some order into the pretty chaotic subfield of study and it has been largely successful. An important virtue of Brin's work has been his attempt to classify various hypotheses according to the presence or absence of some general feature, notably the desiderata as far as epistemology and methodology are concerned. A bonus is that Brin (as an astrophysicist-turned-author) has included some ideas which had appeared only in fictional context, but whose internal merit is such that they deserve serious scrutiny. Unfortunately, this is a road few decided to follow, and this has had adverse consequences for the field in general, as I shall try to show in some examples. Of course, many hypotheses were suggested only after Brin's review and, even more importantly, our entire outlook on the questions of cosmic life has been profoundly transformed by the advent of the *astrobiological revolution* since *c*.1995, which is why the update of Brin's work has been sorely needed. This book is an attempt to fill that lacuna in.

Stephen Webb's book *Where is Everybody? Fifty (/Seventy-Five) Solutions to the Fermi's Paradox* is a worthy contribution to popular astrobiological literature and an excellent introduction to some of the major themes which have arisen in the half-century following Fermi's lunchtime remarks. Appearing originally in 2002, it has incorporated much of what has happened after the seminal work of Brin, in a relaxed and good-humoured way. It has been unavoidably subject to some constraints of the genre: for instance, the fact of the very subtitle containing a finely rounded integer like 50 imposed obvious limitations; the second edition, published in 2015, increased the tally by 50%! The taxonomy of the solutions left much to be desired, and it has obviously been aimed at the very general public, who have little understanding of astronomy and other relevant scientific specialties. As a popular introduction, Webb's book has performed its task splendidly; still, it does not do justice to the wider astrobiological, philosophical, and future studies-related aspects of the problem. It is my modest hope that the present book will help fill the niche.

Paul Davies's book *The Eerie Silence: Renewing Our Search for Alien Intelligence* is an important newer (2010) contribution to the serious discussion of issues surrounding the problem of extraterrestrial intelligence and its apparent absence. It does touch upon some issues related to Fermi's paradox, although this subject is not, in spite of the main title, the focus of the narrative. Davies's book is important for another reason, which shall be discussed in more detail in Chapter 9: it advocates 'a massive expansion in SETI, not by doing more of the same (though that is good too) but by shifting the focus towards the search for general signatures of intelligence.'[13] It is impossible to properly assess the importance and timeliness of this wake-up call before we fully take into account the seriousness of Fermi's paradox in its modern versions. Therefore, it is a good introduction to many issues to be treated in the present book in more detail.

# ACKNOWLEDGEMENTS

No man is an island; no book is either. Many people have in one way or another contributed to this project during its almost 8-years-long gestation period. While it is certainly overambitious to offer an exhaustive list, mentioning those whose contributions have been the largest and most obvious is the least I could do repay the intellectual and spiritual debts which cannot ever be truly repaid. The foremost mention is due to my father, Milivoje Ćirković, whose aid and support has been essential in all respects.

My long-term editor at Oxford, Keith Mansfield, has been the *spiritus movens* of the whole enterprise; this book would have been flat-out impossible without his gentle but firm support. The rest of the crew at OUP, headed by Dan Taber, Lydia Shinoj, and Clare Charles, are warmly acknowledged as well. Stephen Webb not only wrote the most comprehensive book-length preview of the topic and kindly provided the expanded edition, but has strongly encouraged me to pursue this project in all its various stages.

Slobodan Popović Bagi, a legendary figure in better-informed Belgrade art circles, has invested much of his time and energy in making the manuscript look better, and has been a constant source of encouragement and support during the long and difficult years of its emergence. Time and again, it was only natural to rely on his kindest help in overcoming various obstacles and uncertainties, big and small.

Feedback from the first readers has been essential for the present manuscript. Zoran Knežević, Alastair Nunn, Srdja Janković, Milan Stojanović, Momčilo Jovanović, and Robin Mackay offered their invaluable comments, which greatly improved the quality of various parts of the text. Branislav Vukotić, a highly talented astrophysicist and astrobiologist, as well as a friend, contributed some of the most insightful ideas and suggestions. Zoran Živković, splendid author and editor, has indebted me in too many ways ever to hope to list here; perhaps crucial was his wise encouragement in all-to-frequent moments of doubt and uncertainty. Slobodan Perović at the Philosophy Faculty of the University of Belgrade has kindly given help over the years of working on this book, and provided for an inspirational collaborative environment.

It is also a great pleasure to express sincere gratitude to my Oxford friends and collaborators Nick Bostrom, Anders Sandberg, and Stuart Armstrong for their many pleasant—and challenging—provocations, as well as generous technical help on many issues. Also, helpful discussions with Paul Gilster, Vojin Rakić, Karl Schroeder, Petar Grujić, Biljana Stojković, Eva Kamerer, Dejan Urošević, George Dvorsky, Ivan Almár, John M. Smart, Larry Klaes, Mark Walker, Radomir Đorđević, George Musser, Steven J. Dick, Zoran Stokić, Viorel Badescu, Seth Baum, Jacob Haqq-Misra, Živan Lazović, Gregory Matloff, Slobodan Ninković, Aleksandar Palavestra, Aleksandar Bogojević, Guy Consolmagno, and Bojan Stojanović immensely contributed to the sharpening of some of the views presented in this book. Dušan Pavlović played an important role in finding some of the less accessible references, as well as in sustaining optimism regarding the project.

Words are not sufficient to describe the magnitude of debt to my dear friends and family, people who has offered incredible kindness and support at one or another moment of time, notably Zora Ćirković, Mirko Ćetković, Damir Jelisavčić, Aleksandar Obradović, Suzana Cvetićanin, Dejan (Ilić) Rajković, Sandra Đorić Jelisavčić, Branislav K. Nikolić, Predrag Ivanović, Zona Kostić, Dušan Inđić, Edi Bon, Goran Milovanović, Aleksandar Jaćimovski, Jelena Mirković, Miloš Aćimović, Edvard Nalbantijan, Vladimir Ljubinković, Ivana Kojadinović, Ana Eraković, Irena Diklić, Dragoljub Igrošanac, Tanja Milovanović, Duška Kuhlmann, Jelena Andrejić, Karla Ilić Đurđić, Andrea Subotić, Nikola Božić, Vigor Majić, Srdjan Samurović, Milena Jovanović, Sunčica Zdravković, Marija Rajičić, Miroljub Dugić, Ana Parabucki, Tanja Vukadinović, Marko Stalevski, Maša Lakićević, Miroslav Mićić, Ana Vudragović, Aleksej Tarasjev, Ivan Halupka, Marija Nikolić, Srđan Cvetković, Jana Ristić, Miško Bilbija, Miloš Nikolić, Đorđe Trikoš, Katarina Atanacković, Sonja Kukić, Bojana Pavlović, Mihailo Gajić, and Branko Pavić. Every journey has its losses; this book owes much to my mother, Danica Ćirković (1935–2016), as well as the recently deceased colleagues and friends Robert J. Bradbury, Branislav Šimpraga, and Ljubomir Aćimović, who will always be fondly remembered for their kindness, support, and companionship in days both dark and sunny.

Important Internet services, such as the NASA Astrophysics Data System, ArXiv, and the KoBSON Consortium of Serbian libraries, have immensely contributed to this endeavour, making achieving the necessary degree of scholarship and precision easier and more exciting than ever. I also use this opportunity to thank the Astronomical Observatory of Belgrade, and the Future of Humanity Institute of the Oxford University, for providing an excellent and highly inspiring working atmosphere in the course of the work on this book. Occasional stays at the Petnica Science Center have helped clearer thinking about many of the issues discussed in the book.

Finally, a perennial inspiration for this kind of work comes from the many artists who enable us to experience beauty; in the magnificent words of Kahlil Gibran, 'Everything else is a form of waiting'. In particular, listening to music has always been an inseparable part of the writing experience for me. Thus, an eclectic—and perhaps oxymoronic—*sound*track for *The Great Silence* includes pieces by Johann Sebastian Bach, Franz Schubert, Richard Wagner, Sergei Rachmaninoff, Kurt Weill, Nat King Cole, John Lee Hooker, Leonard Cohen, Diana Krall, David Bowie, Jose Feliciano, Carlos Santana, Ultravox, Television, Nick Cave and The Bad Seeds, Gary Moore, Azra, EKV, Arsen Dedić, Suba, Enrico Macias, Mistake Mistake, Bebel Gilberto, Anastacia, Dream Theater, and Muse.

Many inadequacies and weaknesses certainly remain; for all of those I take full responsibility.

CHAPTER 1

# Introduction

## *The Many Faces of Fermi's Paradox*

Every journey needs a proper beginning. In this introductory chapter, we shall survey several historically and epistemologically relevant definitions of Fermi's paradox, before we concentrate on the strongest version of the problem. The hidden assumptions built in the strongest version are explicated, and a new taxonomy of solutions based on the acceptance or rejection of each of these assumptions is proposed, to be developed in detail in subsequent chapters.

## 1.1 The Famous Lunch

Fermi's paradox presents arguably the least understood of the 'Grand Questions' posed in the history of science, as well as the biggest challenge for any practical SETI activity. As is now well known (having been established by the diligent research of Eric M. Jones), the key argument follows a lunchtime remark of the great physicist, Enrico Fermi: *Where is everybody?*[1] It turns out that, when Fermi was having lunch with Edward Teller, Emil Konopinski, and Herbert York at Los Alamos on a day in the summer of 1950, at one point the conversation touched on how there had recently been a flood of UFO sightings all over the United States. Some trash cans had gone missing in New York City at the time as well, and a *New Yorker* cartoon (of 20 May, 1950) had charged the interstellar visitors for the misdeed. In the relaxed atmosphere, Fermi remarked that extraterrestrial visitations could indeed have been the single common cause of two independent empirical phenomena—in this case, UFO sightings and missing trash cans—in the best tradition of scientific methodology: searching for a small number of causes for many different phenomena. The talk then veered towards the *general* issue of existence, or otherwise, of extraterrestrial intelligence. While they did not take the flying saucers stories seriously, Fermi and his companions earnestly discussed topics such as interstellar—and even superluminal—travel. Then, after some delay—and, as one might imagine with such key moments, which are always more the domain of legend and imagination than historical fact, in the midst of some tasty dish—Fermi allegedly asked his famous question. Where, indeed, is everybody? His friends understood that the great physicist was talking about extraterrestrials.

In a nutshell, Fermi's reasoning was as follows: both time and space are large in our astronomical environment, but there is an important sense in which the temporal scale is *larger*. The Galaxy is about 100,000 light years from edge to edge,[2] which means that a star-faring species would need about 10 million years to traverse it if moving at a very modest velocity of just 1 per cent of the speed of light. Since the Galaxy is about a *thousand times* older than this, any technological civilization will have much more available time for such expansion and colonization of *all* planetary systems that exist in the Milky Way. If one species fails in this endeavour, another won't. Consequently, if intelligent species were out there in any appreciable numbers, they would have been here already. And yet, we do not see them on Earth or in the Solar System. For Fermi and many thinkers since, this constituted a paradox.

All this was happening, one should keep in mind, 7 years before Sputnik 1 and a full 11 years before the first human cosmic flight by Yuri Gagarin. Fermi, sadly, did not live to witness either of these two epoch-making events for humanity, having died in 1954, only 4 years after his Los Alamos lunch. However, his powerful intuition applied excellently to this case, since he could not see any reason interstellar travel would be impossible (we do not nowadays either, in spite of the repeated assertions[3]), and even if he underestimated the size of the Galaxy, it was at best a factor of a few orders of magnitude. In any case, he found a way to formulate, clearly and memorably, an obvious—but still not understood enough—fact: that humanity is a new phenomenon on the cosmic scene and that extraterrestrial intelligence, if it exists, is likely to be older than us on the basis of that fact alone. Adding the modern-day values for the age and size of the Galaxy and other astrophysical data just strengthens this conclusion and makes likely that the difference in ages becomes only larger.

The idea that even imperfectly known temporal and spatial scales constrain our conventional ideas about the emergence and evolution of intelligent beings and their civilizations has been so brilliant that its full ramifications have not been grasped as yet. First discussed in print by the Russian space science pioneer

Konstantin Eduardovich Tsiolkovsky in 1933,[4] although less in a scientific style and more in a philosophico-mystical style, and in the later decades elaborated in detail by David Viewing, Michael H. Hart, Frank J. Tipler, G. David Brin, Stephen Webb, and many others, the argument presents a formidable challenge for any theoretical framework assuming a naturalistic origin of life and intelligence.[5] As such, this should worry more than just the small group of SETI enthusiasts and science-fiction buffs, as it challenges some of the deepest philosophical, social, and cultural foundations of modern human civilization. It is hard to conceive a scientific problem more pregnant and richer in meaning and connection with the other 'Grand Questions' of science throughout the ages. In addition, Fermi's paradox presents a wonderful opportunity for public outreach, popularization, and promotion of astronomy, evolutionary biology, future studies, philosophy of science, and related fields.

While a consistent in-depth study of the history of Fermi's puzzle is beyond the scope of the present book, one important historical issue needs to be touched upon, since it continues to generate numerous misconceptions and confusions. There is a persistent myth that the existence of Fermi's question remained unknown for a quarter of century, from 1950 to 1975, until it was reported by Michael H. Hart in a paper published in the *Quarterly Journal of the Royal Astronomical Society* (and by David Viewing in a much-less-known contemporaneous study).[6] The myth is often used by sceptics and denigrators of the problem itself to argue that it was taken less than seriously by the originator and that it has shorter and poorer intellectual history than it seems on a first glance.[7] This is misleading and, strictly speaking, not true; while there was, indeed, little published research work on this topic in the said period, this should not be taken as proof that the questions had been forgotten, ignored, or intentionally downplayed. The 'Great Silence' problem—what I shall dub '**StrongFP**' (see Section 1.2.3)—was discussed in 1962 in Iosif Samuilovich Shklovsky's seminal monograph, which had clearly been motivated in part by Project Ozma and by Dyson's note on the eponymous spheres.[8] While an exact and detailed genealogy of ideas on this topic remains a task for historians of science, it certainly would not provide fodder for a cursory dismissal of the problem. Shklovsky was clearly influenced by Kardashev's work, codified in his 1964 paper 'Transmission of Information by Extraterrestrial Civilizations' but known informally for some years prior, and he wrote about what is effectively the 'Great Silence'/Fermi's paradox throughout the 1960s and the early 1970s.[9]

Shklovsky's discussion was elaborated and expanded by Stanislaw Lem in *Summa Technologiae*, again under the name of the 'absence of [astroengineering] miracles' or 'silence of the universe'; it is rather unfortunate that Lem's magisterial work was published in English only in 2013.[10] In the aftermath of the initial wave of debates provoked by Hart's 1975 study, David Brin, in his key 1983 review, helped spread not only the 'Great Silence' meme but also the much more important insight that the 'Great Silence' is nothing more than a generalized form of Fermi's question.

That the existence of Fermi's question was known in the scientific community, albeit 'in the underground', is shown by the fact that Sir Fred Hoyle, one of the most visible figures in science at that time, referred to Fermi's paradox quite early on, although he did not call it by that name. Hoyle, who had many close contacts in Los Alamos—and had also had personal experience with having lunch with Fermi![11]—had obviously heard the rumours about the 1950 lunch and mentioned it in the course of the Jessie and John Danz Lectures at the University of Washington in Seattle in 1964,[12] although he mistakenly ascribed the *Where is everybody?* question to Teller! He then proceeded, however, to draw some important lessons from the problem—which *he clearly perceived to be a very real problem*. We shall discuss these in Chapter 4 and elsewhere.

Since it is of paramount importance to clarify completely what we mean when we talk about Fermi's paradox, I shall try to explicate its different versions straight from the beginning. As we shall amply see, some of the confusion reigning in the literature stems exactly from conflating different versions of the problem. This does *not* mean that the whole problem—or most of it—is of a semantic nature, is a pseudo-problem, or can be resolved or avoided on the philosophical level, as some sceptics have argued.[13] In journeying towards the goldmine of research which Fermi's question has inspired and will continue to inspire even more in the future, we need to navigate carefully between the Scylla of simplistic contempt for the 'false problem', and the Charybdis of frustrated despair at the real complexity of the puzzle and its many convoluted issues. Nobody has promised a safe journey, though.

## 1.2 Different Versions of the Paradox

Since the original Fermi's lunchtime remarks are occasionally taken as the authoritative reading of the paradox,[14] it is worth formulating them here in a clear form.

### 1.2.1 ProtoFP

**ProtoFP**: The absence of extraterrestrials on Earth is incompatible with the multiplicity of extraterrestrial civilizations and our conventional assumptions about their capacities.

A serious student of SETI needs to bear two caveats in mind with respect to this version of Fermi's paradox:

1. The very practice of relying on 'the authoritative reading' of any complex statement in science is highly suspicious. The fact that a concept is called by the name 'X' need not mean that the name is more than peripherally relevant to the concept itself. In *The Astrobiological Landscape*, I used the term 'Lou Gehrig's disease' to demonstrate this fact,[15] but an even more pertinent example is the use of the term 'Copernican' to describe our current model of the Solar System—we certainly do not learn today about the Solar System from the book of Copernicus. Actually, not only would it be limiting and naive to do such a thing, it would be clearly *wrong*, since, after all, the modern, post-Newtonian view of the architecture of the universe and the Solar System is about as different from the original Copernican version as it is from the Ptolemaic geocentric universe. There is no particular importance assigned to *circular orbits* in the real world of planetary science, as there was in Copernicus's writings, and the Sun is certainly not the mystical source of all light and good, as it was described in *De Revolutionibus*. But even those errors of fact or logic pale in comparison to the fact that the introduction of the Copernican model of the Solar System was such a widely important *historical phenomenon* in the years, decades, and even centuries after the death of Nicolas Copernicus that limiting the meaning of the term 'Copernican' to what was described in *De Revolutionibus* would be a massive fallacy. Similarly, one can be an excellent relativist today without ever reading a word of Einstein's original writings on the subject, or a distinguished evolutionist without peeking into *The Origin of Species*. The same applies to Fermi's paradox, since, as some of the critics were quick to point out, Fermi did not actually do any research on the subject, and the degree to which his relaxed lunchtime remarks should be understood as giving the canonical view of the entire problem is much *less* than the degree to which the aforementioned book by Copernicus (who was doing diligent research on planetary motions for the larger part of his life) could be used as a textbook on the Copernican model or planetary science. The greatest strength of science lies in its capacity for *generalization*; if an argument could easily be made more

general—and hence stronger—it is only scientifically and intellectually honest to face the most general version, irrespective of its historical genesis and evolution.

2. Fermi's remarks should be firmly put in the context of our physical, astronomical, biological, and even sociological knowledge at the time. They have been directly motivated by the 'flying-saucer' craze, which had just only started at the time and which brought about—at least in minds of the public—the possibility of empirical verification of the long-standing pluralist speculations about extraterrestrial life and intelligence.[16] These were, at best, protoscientific speculations, far removed from the scientific practice of modern astrobiology; and we wish to put the problem in the latter context as much as possible, if we are at all serious in our attempt to resolve it. We should also take into account the very limited astronomical knowledge of Fermi's time. The concept of Hubble expansion was still relatively new and uncertain, as demonstrated in 1952 by Baade's revision of the extragalactic distance scale.[17] The age of the universe and of the Milky Way galaxy was uncertain to at least a factor of few. There were, of course, no extrasolar planetary systems known, and even early false detections such as Van de Kamp's were still some years in the future.[18]

The Cold War had already started, and the Second World War was still very fresh in everyone's memory, together with its emphasis on rough, physical power. The idea of 'spheres of influence' and their expansion, inherent in the Cold-War thinking and certainly present in the intellectual atmosphere of Los Alamos, might also be of relevance (after all, one of the participants in the lunch was Edward Teller, the 'Dr Strangelove' prototype of Stanley Kubrick's 1964 movie fame). The UFO phenomenon itself has been occasionally interpreted in terms of the Cold-War atmosphere.[19]

Even the Nazi idea about *Lebensraum* ('living space') may have exerted some indirect influence on the thinking that took place during that lunchtime in 1950. In its simplest rendering—and Nazi ideology can be confusing and often self-contradictory—it states that 'races' have an innate tendency to expand into their 'living space', even if there is no particular economic, military, or cultural incentive to do so.[20] Instead, it is always their 'destiny' to do so, and 'higher' races are, obviously, destined for much larger *Lebensraum* than their 'lower' counterparts. While these sociobiological ideas have been grounded in the naive and long-discredited social Darwinism of the Victorian era, the degree of their ominous presence in the twentieth century and especially during the Cold-War era should not be too easily downplayed. The ferocity of the hard-won battle against Nazi totalitarianism was still very much in people's minds in 1950 and, as the new Soviet totalitarian empire was being created in the heart of Europe, it was not difficult even for quite a liberal mind to find some bits of truth in such crude imperialist ideas. The long-term exposure of Fermi himself, as well as his lunchtime companion Teller, to totalitarianism and the characteristic totalitarian celebration of power, territorial expansion, and semi-mystical 'life force' might have also contributed.[21]

If anyone finds this far-fetched, we could pause here for a moment and consider a masterpiece of both literature and futuristic—and astrobiological!—thinking: H. G. Wells's *The War of the Worlds* (1897). Whatever one might think of Teller and his Cold-Warrior tendencies, nobody can deny that Wells was a sincere humanist and, for most of his illustrious career, rather far on the left of the political spectrum.[22] However, *The War* is written, from the very first paragraph, as a social-Darwinian story about parallel evolution. Since Mars was formed before Earth—according to the then-dominant Kant–Laplacian cosmogony—its biosphere and civilization must be older and, since that same difference in ages caused Martian climate to change for the worse, Martians were fully entitled to seek new *Lebensraum* on the nearest habitable planet, which happens to be our Earth. It is essentially no more or less a moral process than cutting down tropical forest for building new habitation far from flooding regions; our present-day tender ecological sensibilities should not blind us into believing that many people would consider such an act to be wrong even today—and how about Victorian times? (And we ought not to delude ourselves that cutting down tropical forests, draining swamps, or doing any other kind of melioration does not lead myriads of species to extinction, as has been realized only recently.) Thus, Wells, contrary to most trivial latter-day renditions, goes to great lengths *not* to portray the conflict between Martians and Earthlings in moral terms as an evil, unprovoked aggression. To him, it is a natural (e.g. social-Darwinian) consequence of the Martian vast intellectual superiority—coupled with the imperatives of their survival in the changing physical universe—that will always motivate the 'colonial' grab of resources and living space. The fact that such acts cause incredible suffering and destruction to humans is no substantially different from any other ecological overturns, or the fact that some beetles, wild flowers, or even pandas and polar bears are receiving the worse end of our melioration efforts right now, as you are reading these words.

While I shall discuss later how our reading of the logic of Fermi's paradox is still influenced by, admittedly much more sophisticated, sociobiological assumptions—or presumptions—of this day,[23] for the current purpose of framing **ProtoFP**, it is enough to emphasize that, while Fermi originally used to argue against the existence of extraterrestrial intelligent beings in general, there are at least two other interpretations which immediately come to mind. One is, obviously, what the great Italian and his companions rejected in the beginning of their lunch, namely that UFOs are in fact extraterrestrial visitors. They were quite correct in rejection of that interpretation, and we are more certain of that today, but there are many (maybe even majority of the public) who accept this idea—and who consequently find Fermi's paradox resolved *from the very start.* I shall discuss these ideas in more detail in Chapter 4, which deals with the solipsist explanatory hypotheses. Another is that Fermi was too firmly grounded in the materialist scientific tradition to suspect, even for a moment, that the solution to the apparent paradox might not lie in the domain of science. Consequently, his reaction to the same basic input—the difference in the degree of astronomical knowledge of the day seems not to be very important—was quite opposite to that of Tsiolkovsky who, as mentioned above, sought a solution in the half-mystical transcendence of all sentient beings. Near the mid-twentieth century, the idea that advanced intelligences would seek 'other worlds' and the transcendent Absolute instead of roaming the Galaxy and building an interstellar empire sounded hopelessly old-fashioned, even medieval. Ironically, as the wakes and tides of the history of ideas go, it is the thoughts of the Russian pioneer of astronautics which have recently acquired an ultramodern rendition in the form of hypotheses on singularity or transcension, which will be fully considered in Chapter 6.[24]

This is perhaps the best point for the clarification of a terminological issue bound to be mentioned in some circles in any discussion about either interstellar civilizations or the cosmic vision of humanity's future: the concept of interstellar *colonization.* **ProtoFP** is occasionally construed as the statement that Earth has not been colonized—rather than merely visited—by extraterrestrial intelligent beings. If it is really necessary (and I do maintain that it is not, from the point of view of an intelligent reader) to emphasize such simple things, the concept of *colonization of the universe* in general means building and maintaining of colonies of humans or other intelligent beings in otherwise uninhabited parts of the universe (or at least uninhabited by sentient beings), as well as utilization of their resources for the purpose of maintaining or developing advanced technological civilization. It is, therefore, a cosmic-scale equivalent of the colonizing activities of ancient Greek (Corinth, Phokaia, Chalcis) or Phoenician (Tyre, Sidon) city states which, when faced with overpopulation, sent some of their citizens, usually by sea, to found a new city at a convenient place. This is not an entirely fortuitous analogy; as we shall see in Chapter 7, there are several concrete physical reasons why the city states of antiquity provide the best terrestrial analogy—if *any* analogy may be useful at all—to cosmic evolution, certainly much more than the national or imperial states of other epochs of human history. Along a similar vein, visionaries of the cosmic future of humanity, from Konstantin Tsiolkovsky, to Sir Arthur Clarke, to Gerard O'Neill, to Robert Zubrin, to Ray Kurzweil, to Elon Musk envisioned the colonization of the Moon, Mars, or other planetary systems and even interstellar and interplanetary space. What scientists in SETI studies have been discussing and what I shall be discussing here under the term 'colonization' of other planets, planetary systems, or even galaxies has nothing to do with the deplorable human practice of conquering and enslaving indigenous populations or spreading narrow-minded religions by force. To generalize any of these and similar loathsome practices to the universe is a wild and entirely unfounded anthropocentrism, testifying that some people use *Star Wars* or *Starship Troopers* as sources for models of contact, instead of bothering to read serious SETI literature.[25] However, such preposterous gut reaction at the very mention of 'space colonization' is almost unavoidable in many circles.

## 1.2.2 WeakFP

Weak versions of Fermi's paradox arise when we add contemporary astronomical knowledge pertaining to the Solar System and some additional philosophical assumptions. How far is one ready to go with these varies from author to author, but a prototypical rendering of this setup could read:

> **WeakFP**: The absence of extraterrestrials or their artefacts on Earth and in the Solar System is incompatible with the multiplicity of extraterrestrial civilizations and our conventional assumptions about their capacities.

This version has been the default in most of the popular discussions of the paradox and SETI studies in general. Sceptical discussions like those of Michael Hart and Frank Tipler were based on this assumption, explicated through what

Hart called 'Fact A': the absence of extraterrestrials in the Solar System.[26] The extent to which we can be entirely certain in this premise is doubted by some—although prudence would suggest that it is accepted on the basis of existing empirical evidence; I shall discuss some (half-serious, mostly) attempts to negate it in Chapter 4. Of course, intuitive support for the premise of **WeakFP** comes from the plausible conjecture that any visitation of extraterrestrial intelligent beings to Earth would be very difficult to hide or ignore. Equally obviously, **WeakFP** enables many researchers to sidestep the real problem by claiming that it is our assumptions about interstellar travel that are in default here: if, for some currently unclear reason, interstellar travel is actually physically or socially impossible, the reasoning behind **WeakFP** dissolves. Ironically, a SETI-sceptic such as Frank Tipler has produced the most persuasive form of **WeakFP**, as we shall see.

What even many keen observers fail to perceive is that there is a simple way to immensely strengthen the substance of Fermi's paradox, by including in it the absence of extraterrestrials from Earth and the Solar System, the lack of success of SETI projects thus far, and the possibility of getting around the controversial (for some) issue of interstellar travel.

### 1.2.3 StrongFP

By taking a further bold step and generalizing **WeakFP** to our entire astronomical observations, including the negative results from SETI searches so far, we reach the strongest and the most serious form of the paradox.

> **StrongFP** (a.k.a. The Great Silence, *Silentium Universi*): The lack of any intentional activities or manifestations or traces of extraterrestrial civilizations in our past light cone is incompatible with the multiplicity of extraterrestrial civilizations and our conventional assumptions about their capacities.

This version has much stronger implications than the other ones, for the very simple reason that it assumes something about our past light cone, a large spatio-temporal region encompassing not only the Milky Way but many external galaxies as well. In addition, it considers the most general class of activities which could be ascribed to extraterrestrial intelligent beings (or *sophonts*, to use a convenient term of Poul and Karen Anderson), not just visitations or colonization, which has traditionally—and misleadingly—been the focus of Fermi's-paradox thinking. While detectable extraterrestrial presence in the Solar System implies detectable intentional activity, the implication in the other direction does not hold. The class of intentional activities is much wider and includes, for example, such important possibilities as the construction of Dyson spheres or any other macroengineering/astroengineering artefacts or auxiliary activities necessary for constructing artefacts, such as stellar uplifting or stellar rejuvenation. An example of a phenomenon possibly belonging to this wider category is the strange behaviour of objects transiting the star KIC 8462852; discovered in 2016, this remains highly controversial.[27]

This is the reason why **StrongFP** has occasionally also been dubbed 'the absence of cosmic miracles' or the 'astrosociological paradox'.[28] In the evocative words of physicist Adrian Kent, *it's too damn quiet* in the local universe.[29] From the most general case of **StrongFP**, particular cases of **WeakFP** and **ProtoFP** can be obtained by restricting our knowledge to the Solar System or Earth, and our wide class of intentional activities to visitations or leaving artefacts on our planet or in its vicinity (as symbolically shown in Figure 1.1).

For both reasons (allowing a larger spatio-temporal volume and a more inclusive class of phenomena), the premise of **StrongFP** is—in principle—easier to *falsify* than any alternatives—which is, of course, an advantage according to post-positivist epistemology, which does make it 'more scientific' in sufficiently broad terms to start with—hence the attribute 'strong'. In the rest of this book, I shall use the locution 'Fermi's paradox' in this meaning, except when explicitly stated otherwise; in addition, I shall attempt to demonstrate in several places how the research on Fermi's paradox is, to a large degree, a *search*

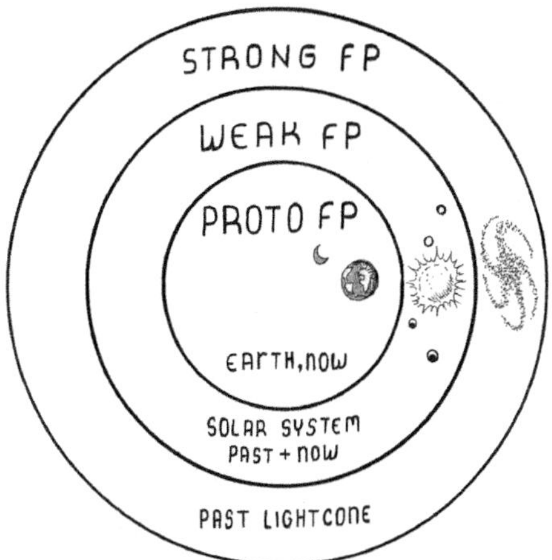


**Figure 1.1** A schematic representation of different versions of Fermi's paradox.

*Source*: Courtesy of Slobodan Popović Bagi

*for empirical falsification of explanatory hypotheses*—which should, in turn, give more support to its scientific credibility.

This way of presenting Fermi's paradox has large and obvious advantages, which go far beyond the presentational and pedagogical level. Part of the reason why the problem has not been often taken seriously lies in the fact that it has usually been presented in some weaker form. Basic intellectual honesty requires presenting any problem as strongly and forcefully as possible; otherwise, we are tearing down straw men (which will happen often enough *anyhow* along the front of research, where context is at best only partially understood and we cannot easily distinguish between viable contenders and straw men hypotheses). So with Fermi's problem—some of the conventionally suggested hypotheses do not work simply because they are tailored to weaker versions of the problem. Suppose one tries to resolve the issue by putting forward a hypothesis that there are no aliens on Earth because they are predominantly anaerobic and avoid places rich in free oxygen, like the surface of our planet. This hypothesis is worthless, since it entirely neglects the context given by formulations that are stronger than **HistoricalFP**; for instance, it fails to explain the documented absence of aliens from the Moon, where there is no free oxygen, which is part of the context set by **WeakFP** (and, a fortiori, any stronger version). While this example might sound naive and contrived, such is the case with many apparently more sophisticated hypotheses, like some of Stephen Webb's *seventy-five solutions* and many others suggested at least half-seriously in both fiction and non-fiction.[30]

### 1.2.4 KardashevFP

As with all great conundrums, there are still further alternative ways of expressing the same problematic situation. Another formulation of Fermi's paradox (one which I shall only occasionally use in further discussion), motivated by the celebrated Kardashev's classification of technological civilizations,[31] could be adequately dubbed **KardashevFP**:

> **KardashevFP**: There is no Kardashev's Type 3 civilization in the Milky Way.

Clearly, a Kardashev's Type 3 civilization would encompass the Solar System and other nearby star systems; however, this is not observed. **KardashevFP** is weaker than **StrongFP**, since it does not exclude such extremely advanced—and potentially detectable—technological civilizations in other galaxies.

Can we be certain that no Type 3 civilization exists in, for instance, M31 (the Andromeda Galaxy)? An affirmative answer is fairly safe if we suppose kinds of energetics usually imagined in the context of a Type 3 civilization. It could not be powered by stellar energy, since the lack of outgoing starlight and the anomalous mass–luminosity relationship would have been noticed. It could not be powered by a controlled gravitational collapse, since the star-formation rate in M31 seems perfectly natural for a galaxy of such size and morphological type (and no dynamical anomalies such as an excessive presence of dark matter have been noticed so far, in spite of much observation). One could rather safely claim that, insofar that any process of energy transfer and conversion is a *local* one, there are strong constraints on the amount of energy available to intelligent life in M31.[32] This, of course, could motivate further questions about alternative sources of low-entropy energy which advanced technological civilizations might use—and their detectability.

But why should *any* version of Fermi's question be paradoxical in the first place? The answer lies in the spatial and temporal scales under consideration. It is to this key issue that we must now turn.

## 1.3 Spatio-Temporal Scales and the Real Strength of StrongFP

At the time of Fermi's lunch, the size and, especially, the age of the Milky Way were known only very roughly; this even more applies to the architecture of the universe on even larger scales. In the meantime, several astronomical revolutions have come and gone and, as a consequence, we have immensely better insight into this obviously crucially relevant topic. The tremendous success achieved by astronomers such as Shapley, Baade, Zwicky, Oort, and Binney in determining the structure and evolution of the Galaxy is, unfortunately, not understood—and celebrated—enough. In any case, the spatial structure of the Galaxy may hold a key (or even *the* key, within the framework of some of the contending hypotheses) to the distribution and manifestations of the hypothetical advanced technological civilizations, *under the reasonable assumption that they arise on habitable planets around Main Sequence stars similar to our Sun.*[33] I shall discuss some of the relevant findings of Galactic astronomy in Chapter 2; for the moment, we need to assume that we understand correctly the size of our Galaxy's stellar disc, and the timescale for its traversing with a fixed mean velocity.

Tsiolkovsky, Fermi, Viewing, Hart, Tipler, and their followers argue their views on the basis of two premises:

1. The absence of extraterrestrials in the Solar System ('Fact A' of Hart[34]).
2. The fact that extraterrestrials have had, ceteris paribus, more than enough time in the history of the Galaxy to visit, either in person[35] or through their conventional or self-replicating probes.

This would correspond to **WeakFP**, as discussed in Section 1.2.2. **StrongFP** would add some further complications. The characteristic time for colonization of the Galaxy, according to these investigators, is what I shall call the Fermi–Hart timescale:

$$t_{\mathrm{FH}} = 10^5 - 10^8 \text{ years}, \tag{1.1}$$

making the fact that the Solar System is (obviously) not colonized hard to explain, if not for the total absence of spacefaring extraterrestrial cultures. It is enough for our purposes to content that this timescale is well defined, albeit not precisely known due to our ignorance on the possibilities and modes of interstellar travel. For comparison, the accepted age of the Solar System is[36]

$$t_{\odot} = (4.5681 \pm 0.004) \times 10^9 \text{ years}. \tag{1.2}$$

The age of Earth as an object of roughly present-day mass is quite similar, usually rounded to 4.5 Ga, as will be discussed in Chapter 2. The drastic difference between the timescales in (1.1) and (1.2) is one of the ways of formulating Fermi's paradox. In Section 1.7, we shall see that there is a still more serious numerical discrepancy in play when we account for the distribution of ages of terrestrial planets in the Milky Way, one of the most important recent astrobiological achievements.

As discussed in Section 1.2.3, we need not consider the direct physical contact between an extraterrestrial civilization and Earth or the Solar System (insofar as we do not perceive evidence of extraterrestrial visits in the Solar System; however, this is still an act of faith, considering the volume of space comprising our planetary system[37]). It is sufficient to consider a weaker requirement: namely, that no extraterrestrial civilizations are *detectable* by any means from Earth at present. This includes conventional radio (or optical) messages, direct travel, and presence of alien artefacts in our spatial vicinity (on Earth or in the Solar System), as well as the detectability of astroengineering projects over interstellar distances.[38] In the words of the great writer and philosopher Stanislaw Lem, who authored some of the deepest thoughts on this topic, Fermi's paradox is equivalent to the 'absence of cosmic miracles' or the *Silentium Universi* ('cosmic silence'[39]). Following the classic review by Brin,[40] we may introduce 'contact cross section' as a measure of the probability of contact—by analogy with introduction of cross sections in atomic and particle physics[41]—and reformulate the paradox as the question why this cross section in the Milky Way at present is so small in comparison to what could be naively expected.

Figure 1.2 presents a view which already has become commonplace in our age of satellites, GPS, the International Space Station, and so on: our planet as viewed from orbit at night-time. Essentially all illumination visible in the image originates with human civilization: it is an example of intentional—although *quite wasteful* (I shall return to this important point in Chapter 7, in connection with striving for computational efficiency)—behaviour. It is reasonable to assume that an alien astronomer, when faced with an analogue of Figure 1.2, would be rather quick to conclude that Earth is inhabited by a technological civilization—perhaps a very immature civilization, but this is still qualitatively very different from Earth being either an empty planet or a planet inhabited by

**Figure 1.2** Earth, as seen from space, shows unmistakable signs of being not only habitable, but actually inhabited by a primitive (i.e. wasteful) technological civilization.

*Source*: Courtesy of European Space Agency

low-complexity organisms only. The next step is to put this hypothetical alien astronomer not in Earth's orbit, but at much larger distance, first an interplanetary and then an interstellar one. The difference is, considering the explosive pace of development of observational astronomy we have experienced in recent decades, not that big. It might require further decades and perhaps centuries for humans to develop instruments capable of, for instance, observing Earth-like extraterrestrial planets with the same resolution as we can observe Earth from close orbit (provided that we do not destroy ourselves in the meantime). But the very fact that this is neither impossible nor requires ludicrously large resources—on the contrary, the advent of well-researched space-based interferometry may provide us very soon with such capacities, not to mention innovative ideas like using the Solar gravitational lens, and so on—suggests that the timescale for such advance is small even compared to human historical timescale, not to mention evolutionary or cosmological timescales. Our presumed alien astronomer could infer—from what is essentially human light pollution, an accidental *manifestation* of human civilization—the existence and some basic properties of the Earth-originating technological civilization.[42]

Here we come to the large realm of cultural aspects of Fermi's paradox. In some cases, research on this topic is obstructed by layers of prejudice towards the entire field of SETI studies, or even astrobiology as a whole. Astrobiological and SETI circles have not done enough to counter these prejudices, and this situation has indeed impeded research to this day, the glaring example of such being the perennial problems associated with obtaining observational SETI funding.[43] An additional problem—apart from the general prejudice against anybody who discusses allegedly non-serious concepts like 'little green men'—is the multidisciplinary nature of the enterprise itself, which of necessity brings into contact seemingly remote and disparate disciplines like astronomy, social sciences, future studies, and artificial intelligence. When the tough disciplinary barriers are finally overcome, there is obviously some loss of steam, which, coupled with other difficulties, causes researchers to be often too cautious and conservative. It is this sort of fatigue we need to combat in order to return SETI to the high table—to use a quintessentially conservative metaphor!—of innovative and creative science.

One could argue—as some of the philosophical critics have indeed done[44]—that what I have labelled **StrongFP** is not what we should properly call Fermi's paradox, and that, strictly speaking, one should use a new name, like Brin's 'Great Silence' paradox. I find such criticism a prototype of unproductive and disingenuous hair-splitting, which attempts to reduce serious issues to semantic (quasi)problems. While such activities have a long and venerated tradition (especially in the Middle Ages), the temptation to engage in them anywhere and everywhere should be resisted. The logical structure of all versions of Fermi's paradox, including the Great Silence, is one and the same: paradoxical conclusions remain puzzling no matter how general a definition of manifestations and traces we employ, for the simple reason that they stem from the same source—the influence of intelligent beings on their physical environment. The wide-ranging consequences for the shape of astrobiological evolution, the future of humanity, and the 'Big History' of the role of life and intelligence in the universe remain the same. Precision is one thing; hair-splitting quite another.

This type of criticism leads to a single useful conclusion: that we need to devote more attention and research to the concept of *detectability* if we wish to understand Fermi's paradox in its full strength. While **ProtoFP** and **WeakFP** hinge crucially on the concept of *interstellar travel* (and, consequently, some critics tried to resolve them on the basis of problems related to interstellar travel[45]), **StrongFP** is based on the much more inclusive concept of *detectability*. It is one of the crucial concepts of this book and, I believe, one of the crucial concepts of the entire search for life and intelligence elsewhere, and yet it had not received enough attention. While it is overambitious to expect that we can strictly define *detectability* in its most general sense, it is still rather obvious on the intuitive level—which is what most researchers in any field of science must live with *until a satisfactory theory is developed*—that we can understand detectability pretty well.

Detectability of an intelligent species is, obviously, a metric telling how easy or hard it is to detect the presence of such a species from afar. Several further specifications could be given here, while keeping it clear that each might have conceived or conceivable exception. 'Afar' here means interstellar distances, characteristic for distances between planetary systems in the Galaxy; these would range from less than a parsec to several thousand parsecs. I shall try to give further elaboration of these spatial scales when discussing the Galactic Habitable Zone in Chapter 2. Detection implies the information of species' existence at any point in the past light cone of the observer and does not necessarily mean the existence in the very present moment (which is unfeasible anyway, due to relativistic constrains). It means the information about species' existence, and no further specific information; thus, it is *the least constraining* of all the requirements discussed in connection with SETI. Finally, the notion of easiness/hardness is, of necessity, anthropocentric. To use a terrestrial example, even a very massive obstacle like a Doric column is hard for an unaided human being to detect in the darkness, while the same is quite easy for a bat, which is equipped with the ability to echolocate. In the SETI case, the criterion of easiness or difficulty is based not only on the notions inherited from our biological past but also on the complicated pathway of our cultural evolution bringing about observational astronomy as we know it. A (wo)man in the street, if at all informed about SETI searches, will tend to associate them with radio astronomy, with good, but entirely contingent, historical reasons. It is easy to imagine a counterfactual world in which optical SETI, infrared SETI, X-ray SETI, or even neutrino-based SETI (as a tip of the hat to Stanislaw Lem; see 'Stanislaw Lem: *His Master's Voice*') would be the dominant form of the enterprise.[46]

### Stanislaw Lem: *His Master's Voice*

Perhaps the best novel ever written about SETI is a dense tract indeed, and the study of the motives and ideas relevant for this field would require at least another book.[47] It is a challenging text in more than one sense; there is almost no dialogue and no manifest action beyond the recounting of a SETI project that not only failed but was *never truly comprehended* in the first place! Without revealing too much, in the novel, which is set at a time when neutrino astrophysics is advanced enough to be able to detect possible modulations, a neutrino signal repeating every 416 hours is discovered from a point in the sky within 1.5° of Alpha Canis Minoris. An eponymous top-secret project is then formed in order to decrypt the extraterrestrial signal, burdened by all the Cold-War paranoia and heavy-handed bureaucracy of the second half of the twentieth century.

An intriguing consequence of Lem's scenario is a realization that, while detectability generally increases with the progress of our detecting technology, it does so very unevenly, in jumps or bursts. Although the powerful source of the 'message' in the novel had been present for a billion years or more, it became detectable only after a sophisticated neutrino-detecting hardware was developed. And even then, the detection of the signal happened serendipitously. Thus, in a rational approach to SETI—not often followed in practice, alas—the issue of *detectability* should be entirely decoupled from the issue of *synchronization* (the extent to which other intelligent species are contemporary to us).[48]

Fermi's paradox does not figure explicitly in *His Master's Voice* (in contrast to many other Lem's works, especially his equally magnificent *Fiasco*), and for an apparently clear reason: 'the letter from the stars' has always been here. Detectability is, at least in part, a function of historical human development. And there is a very real possibility, in the context of the plot, that 'the letter' does not originate with intentional beings at all. The fulcrum of the book is reached when three radical hypotheses are presented to weary researchers, including the one attributing 'the letter' to *purely natural astrophysical processes!* But even in this revisionist case, there are other problems, especially in light of the fact that the signal manifests 'biophilic' properties: it helps complex biochemical reactions, and Lem's scientists speculate about whether it helped the emergence of life on Earth (abiogenesis). If it did so, the same occurred on many other planets in the Galaxy, so leading to an *increased severity* of Fermi's paradox.

There is another key lesson. While the discovery of even a single extraterrestrial artefact (and Lem's neutrino message can be regarded as an artefact in the sense of **StrongFP**) would be a great step forward, it would not, at least not immediately, resolve the problem. If one could conclude, as some of the protagonists of *His Master's Voice* do, that there exist just two civilizations in the Galaxy, us and the mysterious Senders, that would still require explanation. Two is, in this particular context, equal to one (see Section 1.8).

## 1.4 Structure (and Culture) of Fermi's Paradox

As discussed in Section 1.2, there is no single Fermi's paradox. Instead, we can discuss several versions of it, as they have appeared historically and as they are discussed in the contemporary literature on the subject. This property is common for all 'Big Issues' in science and philosophy, and should not surprise us; instead, it should remind us of the necessity to analyse and delineate our working concepts as carefully and precisely as possible. When the issue at hand is so pregnant with subtle concepts of nuanced meanings and usages, as well as different manners of use in scientific and vernacular jargon, or among various scientific disciplines, it is useful to try to consider as wide a picture as possible. Consequently, I shall now introduce the wider assumptions inherently present in all versions of the paradox, but only very rarely explicated and clarified.

Schematically, all of the Fermi's paradox versions discussed in Section 1.2 can be represented as follows:

**Spatio - temporal scales of the Galaxy + The apparent absence of detected extraterrestrial civilizations + Additional assumptions → A paradoxical conclusion.**

Here, under 'spatio-temporal scales', we include our understanding of the age of the Galaxy, the age of the Solar System, and the ages (incompletely known) of other planetary systems in the Milky Way. Although our level of astronomical knowledge of those varies and has a long way to go to be complete, rough values are fairly uncontroversial; the fact that they survived all revolutionary advances of modern astronomy since the time of Fermi's lunch testifies on that.

While some might wish to challenge the apparent absence of detected extraterrestrial civilizations, it seems clear that we can use it as a working hypothesis ('working fact' might indeed be better in this context), and relegate the possible weakening of this premise to additional assumptions dealing with the relationship of what is observed and what is real. Such a procedure accords well with most of the historical construals of Fermi's paradox, and by far the majority of hypotheses we shall be dealing with in subsequent chapters can be made intelligible only with this premise. The qualification 'apparent' serves a clear purpose: to underscore that we are dealing with observations, while their interpretation is still to some extent open to questioning. This is particularly relevant for the subject matter of Chapter 4.

So we need to travel further. Obviously, the content of these 'additional assumptions' must be controversial—and both the source and the resolution of the paradox must be hidden there. Clearly, those assumptions must at least in part be of philosophical nature, dealing with the relationship of our empirical knowledge and the underlying reality, dealing with the changes of the world around us in the course of astronomical time, and so on. Even here it is obvious why I have insisted from the beginning that the whole cluster of problems surrounding Fermi's question must be partially philosophical in nature.[49] Of course, knowing the number and properties of these additional assumptions is the key for judging both the seriousness of the puzzle itself and the quality of any proposed solution. The additional assumptions can be further explicated as follows:

**Additional assumptions = Scientific realism + Naturalism + Copernicanism + Gradualism + Economic assumptions.**

This is quite a heterogeneous lot. I shall consider them in some detail here, since they are vital for the central thesis of this book, that is, that there are very specific philosophical commitments necessary for any functioning solution to Fermi's paradox. Since the number of such commitments is rather small—in particular when compared to the number of alleged solutions in the literature—a corollary is that the number of working solutions is much smaller than is conventionally thought. Another corollary is that many hypotheses are much more similar than a cursory glance would suggest. This conclusion is the backbone of the proposed fourfold taxonomy in Chapters 4 through 7. And, in turn, the same conclusion indicates that the field of SETI studies and Fermi's paradox should be taken much more seriously than it usually is; after all, if *seventy-five solutions to Fermi Paradox* really exist,[50] it is easier to discard the problem either as too difficult or poorly defined.

Historically, there have been some quite important exceptions to such a discarding in modern science, especially astronomy. The review of Nemiroff published in 1994 lists no less than 118 theoretical solutions (!) to the puzzle of the origin of gamma-ray bursts (GRBs), which were entirely mysterious for about three decades, from the mid-1960s to the mid-1990s; his list is intentionally limited to models proposed before 1992. Similarly, the problem of the physical origin of the Lyman-alpha absorption lines seen in the spectra of all known quasars (and other *quasi-stellar objects*, or QSOs) has been infamous for dozens of hypotheses suggested for the state and origin of the absorbing gas, ranging from intergalactic filaments and red-shift caustics, to galactic winds from dwarfs and the gaseous haloes of normal luminous galaxies.[51] In contrast to the example of GRBs, the latter puzzle has not been entirely solved, although a great deal of progress has been made to this day. But both cases have been universally regarded as quite well-defined research questions; and, in spite of many sarcastic comments, such as the motto quoted at the beginning of Nemiroff's survey,[52] nobody has really doubted the importance of these problems, both for their respective branches and for astronomy as a whole. And, in both of these cases, the outcome is either clearly or likely to be of mixed nature, that is, including at least two competing hypothesis in a wider framework.[53] Although our astrobiological knowledge in general, and our understanding of Fermi's paradox in particular, lag very, very far behind that in gamma-ray astronomy or QSO absorption-line studies, those historical lessons should not be entirely disregarded. In their light, in fact, as I shall try to discuss in somewhat more detail in Chapter 9, it is rather surprising that, considering the immaturity of astrobiology, there are not *more* hypotheses for the explanation of Fermi's paradox!

## 1.5 Philosophical Assumptions

Now, a bit of the shape of things to come. While a more thorough discussion of the philosophical underpinnings of Fermi's paradox will be given in Chapter 3, it is useful to introduce here briefly the major players in the game.

By 'naive', 'direct', 'common-sense', or 'scientific' **REALISM**, I denote the working philosophy of most of science (as well as everyday life), implying that there is a material world out there, composed of objects that occupy space, have a duration in time, and have properties such as size, mass, shape, texture, smell, taste, and colour.[54] These properties are usually perceived correctly and obey the laws of physics as we know them at the moment (with their specific limitations taken into account, of course). While some of our perceptions might turn out to be untrue under more careful investigation—for example, it might turn out that solid bodies are, in fact, mostly empty space—we need some very weak version of realism, which is exactly what usually goes under the label of *scientific realism*.

While realism has taken a battering during the long history of philosophy, and in particular in modern epistemology in the wake of the quantum revolution in modern physics, it still is the mainstay of practically all science. Its position within the scientific method can be understood by analogy with the position of simple Newtonian (or, even better, Galilean) mechanics in the mundane life of every one of us, physicists and others. While on an intellectual level we might understand that Newtonian mechanics is not, strictly speaking, correct, and—if we are physicists—that we possess much better tools to deal with phenomena, namely relativity and quantum theory, it is still obvious that we use it when trying to avoid spilling coffee, when jumping into water, or when shooting a hoop at the basketball court. In a similar manner, while we are intellectually aware that direct realism can be misleading, especially when applied to exotic topics like non-linear dynamics, black holes, cosmology, or the nature of consciousness, we are still prone to use it and gauge every other interpretation of empirical data by comparison to it. The fact that some scientists do not, when explicitly asked, subscribe to stronger versions of realism (advocating various instrumentalist doctrines instead), does not mean that they do not use the very weak scientific realism in their daily jobs.

In the specific case of Fermi's paradox, the basic premise following from scientific realism is that there are, indeed, no traces of extraterrestrial intelligence detected either or indirectly (Hart's 'Fact A'). I shall discuss in Chapter 4 specific hypotheses which violate this realist view; an extreme example—but powerfully present in pop culture and obviously interesting from a societal point of view—of such a naively anti-realist standpoint is the view that, contrary to scientific consensus, some humans are in contact with otherwise undetectable extraterrestrial visitors and are conspiring with them.[55] As we shall easily see, all such hypotheses are what the younger among us would call 'weird'. (This in itself is not an argument against them; as Niels Bohr famously warned a young aspirant, your theory might not be *crazy enough to be true*. Fortunately, there are other criteria.)

Almost as uncontroversial is the doctrine of **NATURALISM**, namely a prohibition against invoking supernatural agents or events in scientific explanations. There is a huge literature on naturalism in science and philosophy but, for our present purpose, it is enough that naturalism implies that both the formation of life (abiogenesis) and the formation of intelligence (noogenesis) occur due to natural processes, whether on Earth or anywhere else.[56] Apart from a small cadre of scientists promoting 'intelligent design' and other concoctions, this is universally accepted; even those who, like the great French biologist and Nobel laureate Jacques Monod, believe abiogenesis to be exceedingly unlikely (and, consequently, that the probability of life elsewhere in the Galaxy is negligible), have accepted that such a 'lucky accident' occurred naturally on Earth.

Note, though, that naturalism in the sense used here does not reject the *existence* of the supernatural in some appropriate metaphysical sense. It just insists on its absence in the scientific explanation. That is why the oft-quoted 'argument' of creationists and proponents of intelligent design, that so many scientists are personally religious, is simply a non sequitur, a logical error. It would not be worthy of our attention if not for the ill-understood possibility that metaphysical hypotheses which include a 'lesser' or naturalistic Designer (understood, for instance, as a powerful computer programmer who comes from an advanced civilization and who created our universe as a simulation) might, at least in principle, help resolve Fermi's paradox. There are related debates about two kinds of naturalism, namely ontological and methodological ones. There will be more on that in Chapter 4.

Scientific realism and naturalism are methodological assumptions, usually used in any scientific research. As such, they are almost never explicated or discussed in research publications. Arguably, in a strange-enough context, they need to be discussed as fully or else as any other assumptions. And I wish to argue that Fermi's paradox is indeed one of the strangest contexts ever. Consequently, the role that such philosophical assumptions could play is tremendously enhanced. There is nothing new under the Sun here: while cosmology was a young field perceived mainly as a branch of abstract mathematics, the role of philosophical considerations was immense, as documented in excellent historical studies conducted by Helge Kragh.[57] There are some reasons for making an analogy between the state of cosmology at the time of Friedmann, Eddington, or Lemaître, and the state of astrobiology at present. Thus, it is to be expected that philosophical concerns will play some role in the unfolding of events in astrobiology and SETI studies.[58]

In contrast to the pair of philosophical assumptions introduced so far, **Copernicanism** and **gradualism** are somewhat more specific tenets, stemming more from our experience in the history of physical science than from the general epistemology. Copernicanism (often called 'the Principle of Mediocrity', the 'Principle of Typicality', etc.) in the narrow sense tells us that there is nothing special about Earth, the Solar System, or our Galaxy within large sets of similar objects throughout the universe. In a somewhat broader sense, it indicates that there is nothing special about us as observers: our temporal or spatial location, or our location in other abstract spaces of physical parameters, chemical parameters, biological parameters, and so on, are typical or close to typical. Hence our *observations* (with the help of scientific realism), including all the results of our sciences, are typical as well.

Two important points to be made here and to be elaborated in more detail in Chapter 3:

1. These tenets do not imply that our locations in these spaces are *random*. This is clearly not the case, since a random location in configuration space is practically certain to be in the intergalactic space, which fills 99.99...per cent of the volume of the universe. This is a source of a long-standing confusion and the reason why Copernicanism is most fruitfully used in conjunction with some expression of the observational selection effects, usually misleadingly known as the anthropic principle.
2. What is typical, and how it is operationally defined? Usually, this does not trouble scientists too much, since there is some theoretical insight into the problematic situation under consideration suggesting at least vaguely what kind of distribution we expect. For simplified cases, the answers are clear: if something is distributed along the normal distribution (a Gaussian distribution, or 'bell curve'), typical values are close to the maximum or centroid of the distribution; if the distribution is uniform, *any value* in the relevant range of variable is typical. Time and again, in strange problem situations—such as Fermi's paradox—our insight may easily fail.

**Gradualism**, on the other hand, is often expressed through the motto that 'the present is key to the past' (with corollary that 'the past is key to the future').

This paradigm, emerging from geological science in the nineteenth century with the work of Charles Lyell—and expanding, through Lyell's most famous pupil, Darwin, into the life sciences—has been the subject of fierce criticism in the last quarter of century or so. While it has not certainly been replaced by its old rival, catastrophism, there is more and more talk about a new, *neocatastrophic* paradigm, which unites some of the historically successful features of gradualism with the new and dramatic data on mass extinction episodes, great catastrophes of both terrestrial and extraterrestrial origin, faunal overturns, large-scale changes of evolutionary regimes, and punctuated equilibria.[59] We shall return to this new paradigm, with its enormous importance for SETI, in Chapters 3 and 6.

The **ECONOMIC ASSUMPTIONS** are the first thing which comes to mind when we expose a proverbial 'man (or woman) in the street' to Fermi's paradox. Is it feasible to undertake interstellar travel at all? Maybe the colonization of space is impractical for any intelligent species, at all epochs. Or are the preferred habitats of extraterrestrials dramatically different from those in the focus of SETI studies and surveys? How about astroengineering artefacts—maybe they can be ruled out by cost–benefit analysis? Above all (and quite relevant from the point of view of future studies here on Earth), is it possible that some of the actions most likely to be detectable over interstellar distances are carriers of large, even existential, risks for those intelligent beings who undertake them? All those questions cause us to doubt our intuitions both about the universe around us and about the capacities we ascribe to advanced technological civilizations. We shall consider hypotheses exploiting some of these doubts in Chapter 7.

Finally, the role of **NON-EXCLUSIVITY** (or 'hardness', in some of the literature[60]) needs to be elucidated. Non-exclusivity was introduced and discussed in some detail by Brin in his seminal 1983 review article. According to this construal, it is simply a principle of causal parsimony applied to the set of hypotheses for resolving Fermi's paradox: we should prefer those hypotheses which involve a smaller number of local causes. The paradox is eminently *not* resolved by postulating that a single old civilization self-destructs in a nuclear holocaust. The paradox *is* resolved by hypothesizing that *all* civilizations self-destruct soon after developing nuclear weapons, but the major weakness of such a solution is obvious: it requires many local causes acting independently in uniform way to achieve the desired explanatory end. In other words, such solution is exclusive (or 'soft'). As long as we have any choice, we should prefer non-exclusive (or 'hard') solutions, that is, those which rely on a small number of independent causes. For instance, the hypothesis, as we shall discuss in more detail in Section 1.9, that a GRB can cause mass extinction over a large portion of the

Galaxy and thus arrest evolution towards advanced technological society is quite non-exclusive. Obviously, non-exclusivity is highly desirable property of a hypothesis.

## Non-Exclusivity and the Value of *N*

Even from this simplistic sketch, it is clear that non-exclusivity cannot be separated from the overall astrobiological background. Notably, when we state that we prefer a small number of independent causes, the meaning of 'small' crucially depends on the relevant astrobiological dynamics, which determines whether, for instance, we expect 10, 1000, or 1,000,000 technological civilizations to arise in the Milky Way in a fixed temporal window prior to applying the solution. Suppose that a hypothesis H1 explains the absence of manifestations (or resolves any other strong-enough version of Fermi's problem) in 80 per cent of cases; clearly, its usefulness is dramatically different in $N = 10$ versus $N = 1{,}000{,}000$ situations. In the first case, it seems entirely acceptable, given the limits of our astronomical knowledge, to conclude that we have in fact explained the paradox, since the remaining two civilizations could easily go unnoticed in the vastness of the Galaxy even if they were of Kardashev's Type 2; no *further* explanatory hypothesis seems necessary. In the second scenario, overlooking the remaining 200,000 technological civilizations is hardly feasible, and further explanatory hypotheses H2, H3, and so on, are required to account for Fermi's paradox.

This is necessary to keep in mind, in particular when we allow for the possibility of a 'patchwork-quilt' solution, where different explanatory hypotheses fractionally contribute to the overall outcome. Such is, for instance, the solution (#50!) favoured by Webb in his 2002 book.[61] While it may sound more realistic and commonsensical to invoke multiple hypotheses to account for what is obviously a high-complexity problem, considerations of non-exclusivity highlight the fact that *a chain is only as strong as its weakest link*. The non-exclusivity of each individual hypothesis comprising a patchwork-quilt solution needs to be assessed separately. We shall encounter some examples of this in Chapter 8.

There is another sense in which non-exclusivity can be understood. It is a finer filter through which the set of hypotheses which are logically coherent and empirically non-falsified needs to pass in order to sort them by likelihood. We shall encounter particular hypotheses with different degrees of non-exclusivity in the further course of this book, and we shall see that there is a significant amount of clustering depending on which kind of scientific concepts are involved. Highly non-exclusive hypotheses are obviously the best targets for future research programmes, so the heuristic role of non-exclusivity should not be neglected.

## 1.6 The Null Hypothesis

As is often the case in science, we might need a null hypothesis, upon which the results of our research can be gauged. In contrast to what may first come to mind, I shall not use Fermi's conclusion—that there are, in reality, no extraterrestrial intelligent beings—as the null hypothesis. There are several reasons for that. First of all, while it is undoubtedly a resolution of the paradox, as a bland statement it does not do much to advance our understanding; fortunately, we nowadays have a whole family of 'rare-Earth' hypotheses to try to account for the obvious question: *why*, pray tell, is nobody else around, when we (realistically and naturalistically) are obviously around? I shall discuss that family of related hypotheses in Chapter 5. Secondly, in accordance with the Levy Strossian motto from the beginning of this book, in order to have any hope of building a real scientific theory of that part of astrobiology dealing with high-complexity phenomena (which is SETI, for good or bad), we must have taxonomy. The relevant taxonomy here is one of *ideas*.

Suppose we find a well-educated person in the street and ask her or him about Fermi's paradox. There will, most probably, occur a moment in our explication where s/he is likely to protest our framing of it as a paradox. Therefore, I'll explicate the 'null hypothesis' in this particular domain and give it a specific label:

> **Hermit Hypothesis:** Intelligent beings never expand beyond their home planetary system, communicate, or in any other way become detectable over interstellar distances.

There are ways of making the **Hermit Hypothesis** more precise. Since we do not see—now, as well as in Fermi's time—any physical reason why expansion (or just messaging!) over interstellar distances should be impossible, the **Hermit Hypothesis** actually means that they truly *do not wish* to expand or make their presence known. Under 'physical reason', I here subsume everything which would be universally valid for all places in the Galaxy and in all epochs. Counterfactually, if the Milky Way disc were full of marble-sized interstellar rocks in a density of about 1 per cubic kilometre, any interstellar travel (unless it were *very* slow) would be impossible; similarly, if interstellar space were filled with too much hot plasma, any long-range radio messaging would be impossible. We *know* such things are not true. And, while one should never be entirely dogmatic about the issue that some exotic physical effect may not lurk in the depths of interstellar space, the string of successes of galactic astronomy and astrophysics in the last century or so clearly makes it quite improbable, at least.

Another aspect of the problem is whether there is something universal in the cultural evolution itself and that it is this that makes the required tasks of interstellar travel and interstellar messaging impossible. That we do not know of and have not envisioned such a forbidding factor so far does not mean that such a factor does not exist. On the other hand, a mere glance at Figure 1.2 tells us that, to some extent, the complaint is vacuous: we have already, in the visible part of the electromagnetic spectrum as well as in radio waves, announced our presence to the universe. The wavefront containing information about our existence expands at the speed of light throughout the Galaxy and there is nothing we can do about it.[62] The same, by very basic Copernicanism, is true for other technological civilizations, at least in their immature phases analogous to our stage in cultural evolution. So, why don't we perceive them—or is it just the size of our 'eyes and ears' which is insufficient for the moment? The latter would mean that we may expect the great discovery (and the resolution of Fermi's paradox) any time a new and more sensitive astronomical detector comes online. This in itself has some important ramifications, to be discussed in more detail in Chapter 9.

By itself, the **Hermit Hypothesis** is nothing new, and not only has been mentioned in reviews,[63] but also is a sort of the 'default' reaction to Fermi's paradox, especially if it's given in its SETI version:[64] *Why, they just don't want to communicate (with us)!* It has been present in musings of the SETI 'founding fathers' on the Galactic Club of advanced civilizations.[65] Like Sparta in Hellenistic times, like China under the Ming (after the Zheng He exploratory voyages in the fifteenth century, which were regarded as aberrations) and the Qing dynasties, like Japan during the Tokugawa Shogunate, or like the totalitarian 'hermit kingdoms' of Albania (1945–90) and North Korea (1953–today), extraterrestrial societies do not wish to have any contact with the outside universe.[66] An arguably extreme version of the **Hermit Hypothesis** has been described in science-fiction prose by Greg Egan in his novel *Incandescence* and his novella *Riding the Crocodile*:[67]

> The disk of the Milky Way belonged to the Amalgam, whose various ancestral species had effectively merged into a single civilization, but the central bulge was inhabited by beings who declined to do so much as communicate with those around them. All attempts to send probes into the bulge—let alone the kind of engineering spores needed to create the infrastructure for travel—had been gently but firmly rebuffed, with the intruders swatted straight back out again. The Aloof had maintained their silence and isolation since before the Amalgam itself had even existed.

In contrast to human historical experience, in which isolated societies usually lag behind their neighbours, Egan's Aloof represent an example of technologically superior hermit sophonts.[68]

Note two important details about the **Hermit Hypothesis**, which are conveniently ignored in most discussions:

1. The emphasis on *will*. It implies that advanced technological civilizations, in spite of their wildly heterogeneous presumed evolutionary origins, have a unique will which could be enforced over all their individuals and groups. This is a rather strong constraint, for the following simple reason. By analogy with human history, the history of technological progress has been the history of empowerment of individuals and groups, in both a societal sense and in the sense of control over the natural environment. The average individual in Europe in 2018 has at his or her disposal resources comparable to or greater than members of aristocracy or even royalty did in, say, 1218. And in any case, the modern-day individual has much more time at his or her disposal to use such resources, taking the increase in life expectancy into account.
2. The **Hermit Hypothesis** is not just the passive refraining from activities, such as colonization or communication. As discussed, the existence of a technological civilization, either a present-day human-level or an advanced civilization, causes detectable changes in its physical environment. Those changes are located at many levels but, for the sake of example, let us consider just Earthshine (Figure 1.2). It is a generic product of urbanization, ultimately resulting from the increase in human population and industrial activity. All this means that the **Hermit Hypothesis** implies *active suppression* of at least some generic aspects of advanced technology. It requires expenditure of effort and resources to maintain the hermit state through long

> periods of time. Unless the singleton state or any other way of total societal control of Point (1) is achieved, there seems to be no way how such a situation might evolve.[69]

I have devoted much attention to the **Hermit Hypothesis** since it is a very important case study, not for research on Fermi's paradox as such, but for its *reception* in both specialist and lay audiences. What is usually shrugged off as unanswerable (*Why would they want to communicate/manifest themselves in the first place?*) is, in fact, a quite specific scenario with a lot of additional assumptions, not satisfying most of the desiderata discussed in Section 1.5. It is quite reasonable to state that, taking all into account, the **Hermit Hypothesis** is already very unlikely, relative to many other proposed solutions. This shows how biased is the usual mundane perception of Fermi's question.

Two sources of this particular bias exist, which again analysis of the **Hermit Hypothesis** clearly reveals. One is oversimplification. When people, upon hearing for the first time about Fermi's paradox, invoke the **Hermit Hypothesis** in some form, it is because they think that the problem is much simpler than it really is. The simplicity of the idea that extraterrestrials do not want to communicate is quite appealing. The other is the anthropocentric concept of will and willingness. Since we all have experienced that some people do not wish to communicate with us, for reasons we often find mysterious, irrational, or inexplicable, we are likely to transfer this mode of behaviour to extraterrestrial intelligent beings as well. This is going to mislead us systematically more often than not.

The problem with 'xeno-sociological' hypotheses (e.g. the **Hermit Hypothesis**, or the **Zoo Hypothesis** or the **Interdict Hypothesis** to be described in Chapter 4, among others) is that they depend on the unknowable details of social organization of advanced technological civilizations. We know so little about the generic social organization of human societies, that any claim in the more general astrobiological context sounds preposterous and seriously decreases the credibility of any such hypothesis.

That said, the hypotheses are not all on the same footing. The degree of 'xeno-sociological' speculation obviously varies. It is one thing to postulate uniformity of behaviour over thousands of parsecs, millions of years, and the unimaginable diversity of evolutionary parameters, as in the **Hermit Hypothesis**, and quite another to postulate agreement between presumably a small number of independent agents required to 'maintain the zoo' in the **Zoo Hypothesis**.

## 1.7 Why Now?

It is one of the basic contentions of this book that Fermi's paradox has become significantly more serious, even disturbing, of late. Of course, this is not due to any real change in the astrobiological picture of the Milky Way, but to our improved insight into the astrophysical and biological groundings of life and intelligence, roughly since the mid-1990s and the onset of the astrobiological revolution.

The key example of fascinating new astrobiological result is the work of Charles Lineweaver on the age distribution of terrestrial planets in the Milky Way.[70]

In particular, his calculations show that Earth-like planets began forming more than 9 Ga ago, and that their median age is:

$$t_{\text{med}} = (6.4 \pm 0.9) \times 10^9 \text{ years,} \tag{1.3}$$

which is significantly greater than the Earth/Solar System age given by (1.2). Now we can formulate Fermi's paradox in at least roughly quantitative terms: its main insight is that the age difference $t_{\text{med}} - t_{\oplus} = (1.9 \pm 0.9) \times 10^9$ years is large in comparison with the Fermi–Hart timescale in (1.1):

$$t_{\text{med}} - t_{\oplus} >> t_{\text{FH}} \tag{1.4}$$

This means that not only the oldest ones, but also a large majority of habitable planets are much older than Earth (as will be discussed in more detail in Chapter 2). Naive Copernicanism would then immediately suggest that the stage of the biospheres and even the stage of evolution of advanced technological civilizations must be, on the average, older than the stage we see on Earth by the amount in (1.4).

**This constitutes a paradox.**

And the paradox has been slowly growing over recent decades, due to several independent lines of scientific and technological advances:

- The discovery of more than 3700 extrasolar planets so far (15 December 2017), with new ones being discovered on an almost daily basis (for regular updates, see <http://exoplanet.eu/>). Although some of them are 'hot Jupiters' and not suitable for life as we know it (some of their satellites could still be habitable, however[71]), many other exoworlds are reported to be parts of systems with stable circumstellar habitable zones.[72] This tally includes spectacular discoveries such as the TRAPPIST-1 system with six planets, of which three are considered to be within the circumstellar habitable zone.[73] It seems that only the selection effects and capacity of present-day instruments stand between us and the discovery of thousands of true Earth-like extrasolar planets, envisioned by the new generation of orbital observatories. In addition, this relative wealth of planets decisively disproves old cosmogonic hypotheses regarding the formation of the Solar System as a rare and essentially non-repeatable occurrence. These have been occasionally used to support scepticism on issues of extraterrestrial life and intelligence.

- Improved understanding of the details of chemical and dynamical structure of the Milky Way and its Galactic Habitable Zone.[74] The significance of this development cannot be overstated, since it clearly shows that the naive naturalist, gradualist, and Copernican view *must be wrong*, since it implies that millions of planets in the Milky Way are inhabited by giga-anni-old supercivilizations, in clear contrast with observations.
- Confirmation of the *rapid* origination of life on early Earth;[75] this rapidity, in turn, offers a strong probabilistic support to the idea of many planets in the Milky Way being inhabited by at least the simplest life forms.[76]
- Discovery of extremophiles and the general resistance of simple life forms to much more severe environmental stresses than it had been thought possible earlier.[77] These include representatives of all three great domains of terrestrial life (*Bacteria*, *Archaea*, and *Eukarya*), showing that the number and variety of cosmic habitats for life are probably much larger than conventionally imagined.
- Our improved understanding in molecular biology and biochemistry, leading to heightened confidence in the theories of naturalistic origin of life or abiogenesis.[78] The same can be said, to a lesser degree, for our understanding of the origin of intelligence and technological civilization—*noogenesis*.[79]
- Exponential growth of the technological civilization on Earth, especially manifested through Moore's Law and other advances in information technologies.[80] This is closely related to the issue of astroengineering: the energy limitations will soon cease to constrain human activities,

just as memory limitations constrain our computations less than they once did. We have no reason to expect the development of technological civilization elsewhere to avoid this basic trend.

- Improved understanding of the feasibility of interstellar travel in both the classical sense,[81] and in the more efficient form of sending inscribed matter packages over interstellar distances.[82] The latter result is particularly important since it shows that, contrary to the conventional sceptical wisdom shared by some of the SETI pioneers, it makes good sense to send (presumably extremely miniaturized) interstellar probes, even if only for the sake of communication.
- Theoretical grounding for various astroengineering/macroengineering projects potentially detectable over interstellar distances.[83] Especially important in this respect is the possible combination of astroengineering and computation projects of advanced civilizations, like those envisaged by Anders Sandberg and collaborators.[84]
- Our improved understanding of the extragalactic universe. This has brought a wealth of information about other galaxies, many of them similar to the Milky Way, while not a single civilization of Kardashev's Type 3 has been found, in spite of the huge volume of space surveyed, albeit only in a cursory fashion.[85]

Although admittedly uneven and partially conjectural, this list of developments (entirely unknown at the time of Tsiolkovsky's and Fermi's original remarks and even Viewing's, Hart's and Tipler's later reissues) testifies that Fermi's paradox is not only still with us more than 80 years after Tsiolkovsky and more than 60 years after Fermi—it is more puzzling and disturbing than ever.

One is tempted to add another item of a completely different sort to the list above. The empirical fact that humans have survived for almost three-quarters of a century after the invention of the first true weapon of mass destruction gives us at least a vague Bayesian argument countering the ideas—prevailing at the time of Fermi's original lunch—that technological civilizations tend to destroy themselves as soon as they discover nuclear power or any other technology leading to the possibilities for global mass destruction. This is closely related to the crucial issue of the likelihood of various evolutionary trajectories of advanced civilizations, to be considered in some detail in Chapter 8. For now, I emphasize that the fact of our survival thus far in spite of the persistent threat of nuclear annihilation should not be construed to deny that the longer part of the road towards safety for humankind is still in front of us.[86]

The list also shows that quite widespread (especially in popular press) notion that there is nothing new or interesting happening in SETI studies in recent years is deeply and decisively wrong. Part of the problem, however, is that, due to widespread prejudice, the relevance for SETI of many extremely interesting developments is difficult to perceive. Obviously, apart from 'the plague of journalism' (N. N. Taleb[87]), a large chunk of the responsibility for such sad state of affairs lies on the mainstream SETI community, which has been rather insular in the last couple of decades, a state of affairs noted by many observers.[88]

In Figures 1.3 and 1.4, we schematically present a version of Fermi's paradox based upon the scenario of Frank Tipler, using self-replicating, von Neumann probes which, once launched, use local resources in visited planetary systems to

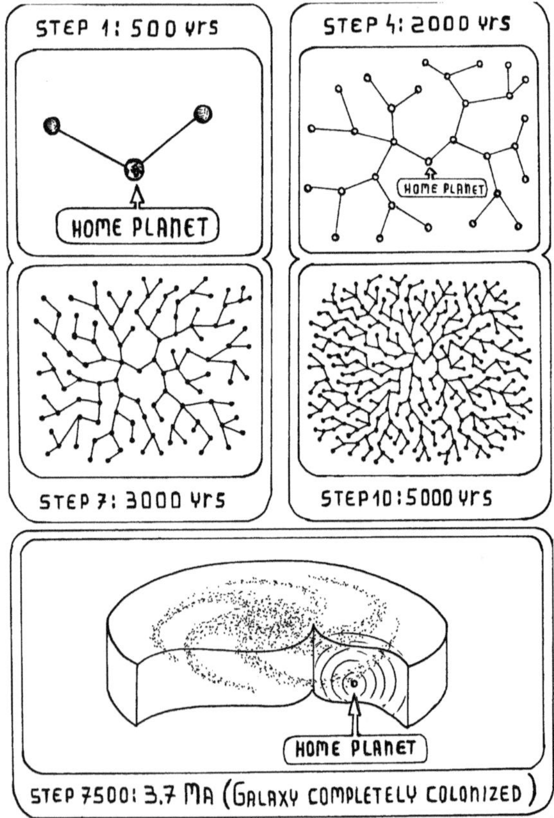


**Figure 1.3** A scheme of Fermi's paradox in a model with slow von Neumann probes, giving a typically low Fermi–Hart timescale for the colonization of the Milky Way. While first steps have been shown in upper panels, we quickly reach late steps in this pan-galactic expansion, still corresponding to a minuscule Fermi–Hart timescale. Some of the relevant timescales are also shown.

*Source*: Courtesy of Slobodan Popović Bagi

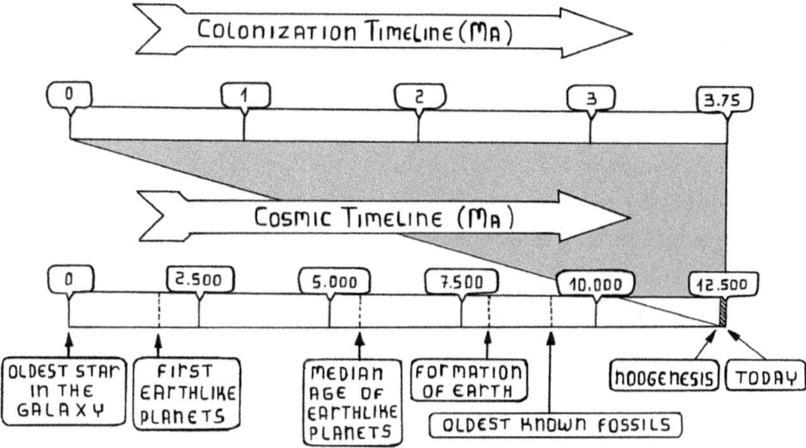


**Figure 1.4** The elaboration of temporal scales relevant for the toy model of Galaxy colonization from Fig. 1.3.

*Source:* Courtesy of Slobodan Popović Bagi

create copies of themselves.[89] It is clear that the exponential expansion characteristic for this mode of colonization leads to the lowest values for the Fermi–Hart timescale (1.1). It is important to understand, however, that Fermi's paradox is *aggravated* with von Neumann probes, but it is not really dependent on them. The paradox would still present a formidable challenge if at some stage it could be shown that interstellar von Neumann probes are unfeasible, impractical, or unacceptable for other reasons (possibly due to the danger they will pose to their creators, as speculated by some authors; see the **Deadly probes** hypothesis in Chapter 6). I shall discuss details of individual scenarios in specific chapters, as a function of specific philosophical assumptions.

There is another important concept I shall be using throughout this book, and that is *advanced technological civilization*.[90] While it certainly is premature to give any formal definition, which would be in the domain of social sciences, by advanced technological civilization I shall assume a hypothetical society of intelligent beings, either biological or non-biological in nature, which possesses at least such internal stability to last for the Fermi–Hart timescale and maintain the degree of technological sophistication needed to engage in SETI programmes or to be a viable SETI target. An advanced technological civilization can be understood as a generalization of a Kardashev's Type 2 civilization, the one using the energy output of its parent star for industrial or cultural purposes.[91] There is an incredible variety of conceivable forms (and perhaps even more at present inconceivable ones) an advanced technological civilization might take: it could be biological or post-biological in origin, inhabiting one or many planetary systems, having a large or a small number of individuals, and so on. But, in assuming a long-term continuous stability, we have assumed not only the technological capacity which facing physical and ecological changes at such timescales requires but also a vast amount of knowledge which could be gathered in the course of time, including the one on the astrobiological state of various other locales in the Galaxy. This will play an important role in determining whether some of the hypotheses for resolving Fermi's paradox, as discussed in Chapters 4 through 7, satisfy our philosophical and methodological desiderata.

Advanced technological civilizations would be the primary targets for any practical SETI enterprise, if they exist, for the obvious reason that their longevity would make temporal synchronization with them trivially easy. However, while the energy used by such civilizations, similar to that used by the conventionally imagined Type 2.x civilizations, should also help in their detection, the detection itself might be subject to severe selection effects; some of these civilizations could act to decrease their actual detectability and could thus present us with hypotheses for resolving Fermi's puzzle. Properties of detectability will be discussed further in Chapters 7 and 8.

## 1.8 Two Is Equal to One: Fermi's Paradox and the Success of SETI

Among dozens of prejudices, misconceptions, and confusing statements usually made about Fermi's paradox, there is one very common blind spot of high both theoretical and practical import. Many authors tend to confuse some form of success of SETI endeavour with the solution of the paradox. This stems from the usual, narrow perception of the problem itself—confusing the wider issue with what I have dubbed **ProtoFP** in Section 1.2.1. While it would give a sort of unique answer to Fermi's original question, a success of SETI in detecting *an* alien signal would not resolve the paradox in its stronger formulations. A naive hope to the contrary is expressed, for instance, by the protagonist of Adam Roberts's intriguing novel *The Thing Itself*:[92]

> 'The simplest solution to the Fermi thing,' I said once, 'would be simply to pick up alien chatter on our clever machines. Where are the aliens? *Here* they are.'
>
> 'Don't hold your breath,' he said.

On the other hand, philosopher David Lamb perceives the subtlety of this point, writing in the introduction to his important book on philosophical aspects of SETI:[93]

> Arguments in this vein frequently invoke what is known as 'Fermi's Paradox', named after the Italian physicist, Enrico Fermi, who responded to arguments in favour of a galaxy populated with ETI [extraterrestrial intelligence] with the retort 'Where are they?' While one authenticated contact would resolve this problem there is, nevertheless, an intellectual requirement that SETI researchers respond to the assertion that so far the universe has been silent.

So, this has to be clear from the outset: Fermi's paradox, when posed in its strongest, starkest form, *represents a problem far, far beyond the simplistic hopes and fears of SETI success.* Contrary to most perceptions, the discovery of, for instance, a single artificial signal *would not resolve the paradox*; it would dramatically change the background and our way of thinking about the issues involved and it would certainly falsify *some* of the hypotheses proposed for the solution of the paradox thus far—but it would not resolve it clearly, cleanly, and definitely in the way that, for instance, the discovery of cosmic microwave background (CMB) resolved the 'great cosmological controversy' between the Big Bang and the steady-state theories in the mid-twentieth century.[94] This is not to downplay the importance of such a discovery: the success of SETI would undoubtedly represent one of the greatest triumphs in the history of humankind, on a par perhaps only with emergence of language, the discoveries of agriculture and of industry, the first spaceflights, and a handful of similar moments of tremendous and incalculable import. But even such an epochal victory of science and reason will not of itself or necessarily answer all our questions about the relationship of life and mind to the universe at large, and notably will not necessarily throw light on the distribution of life and mind on large spatial and temporal scales.

It is easy enough to imagine a scenario in which a detected SETI signal does not tell us *much* concerning those distributions (and hence does not contribute *much* to the resolution of Fermi's paradox). If a signal originates from an Earth-like planet several kiloparsecs away orbiting an otherwise inconspicuous Sun-like star, for example, and if the signal's properties are such-and-such as to reflect a technological civilization roughly similar to the present-day human civilization, we would essentially add another data point, very similar to the one we already have, to any diagram describing the distribution of intelligent living beings in the Galaxy (or to the overall astrobiological *landscape*[95]). Now, while two data points are, in any case, much better than a single data point, it is obvious that we are left with a huge number of curves ('astrobiological theories') which still pass through and accommodate or explain these two data points. Let us give this scenario a specific name:

> **Bare Contact:** Humanity detects a signal of undoubtedly artificial origin, from a planet similar to Earth and a planetary system similar to the Solar System, and located sufficiently far away in the Galaxy. The signal is not deciphered, but its strength, spectral properties, and so on, are best explained by positing a technological civilization similar to the present-day human one.[96]

Clearly, **Bare Contact** would immediately lead us to reject all those hypotheses positing the uniqueness of Earth and Earth's biosphere in the Galaxy or in the known universe (cf. the hypotheses discussed in Chapter 5). As such, it would boost our confidence in the Copernican assumption that Earth is indeed rather typical in astrobiological, as well as in planetological and cosmographic senses. Possibly (depending on the timing of its realization), the same scenario would falsify some other hypotheses pretending to resolve Fermi's paradox as well, such as the idea that all technological civilizations inevitably destroy themselves before being able to properly communicate over interstellar distances (cf. Chapter 6). However, the core of the paradox would be unaffected even after the **Bare Contact** comes to pass. The 'Great Silence' would not be so completely and bafflingly *silent* as previously, but it would still need an explanation. Where are much older Galactic civilizations, those close to the Lineweaver limit in (1.3)? Where are the traces of their astroengineering projects? Why haven't they visited and colonized the Solar System for resources? Those Fermi-related questions are left unanswered by **Bare Contact**. In the same manner, hypotheses posed in the literature to explain Fermi's problem, other than those explicitly falsified by this scenario, would remain largely unaffected. In that sense, since the paradox is essentially a statistical (or a 'large-number') problem, *two is sufficiently close to one!* Two civilizations in an otherwise empty galaxy are slightly less puzzling than a single civilization in an otherwise empty galaxy—but only *very slightly* so.[97]

Of course, information received in the case of successfully deciphering an interstellar *message* might indeed resolve the problem for us, especially if it comes from an older and wiser Galactic civilization with much better empirical insight into the true astrobiological landscape. But this goes way beyond the **Bare Contact** scenario and is correspondingly less likely to happen actually in our future. Besides, it might be considered cheating—or a clear admission that our cognitive capacities are insufficient for reaching the right answer. In essence, this would not be very different from praying for a sudden supernatural enlightenment.[98] We can always return to this council of despair if all other attempts of finding an adequate solution are eventually exhausted.

That prospect is remote at present, as will be amply demonstrated in the rest of this book. A true lesson here is that we need advances in *both* empirical and theoretical SETI to be able to face the great puzzle with proper tools of scientific methodology. An empirical success, in spite of its tremendous importance, will not really resolve even today's 'Big Puzzles', not to mention

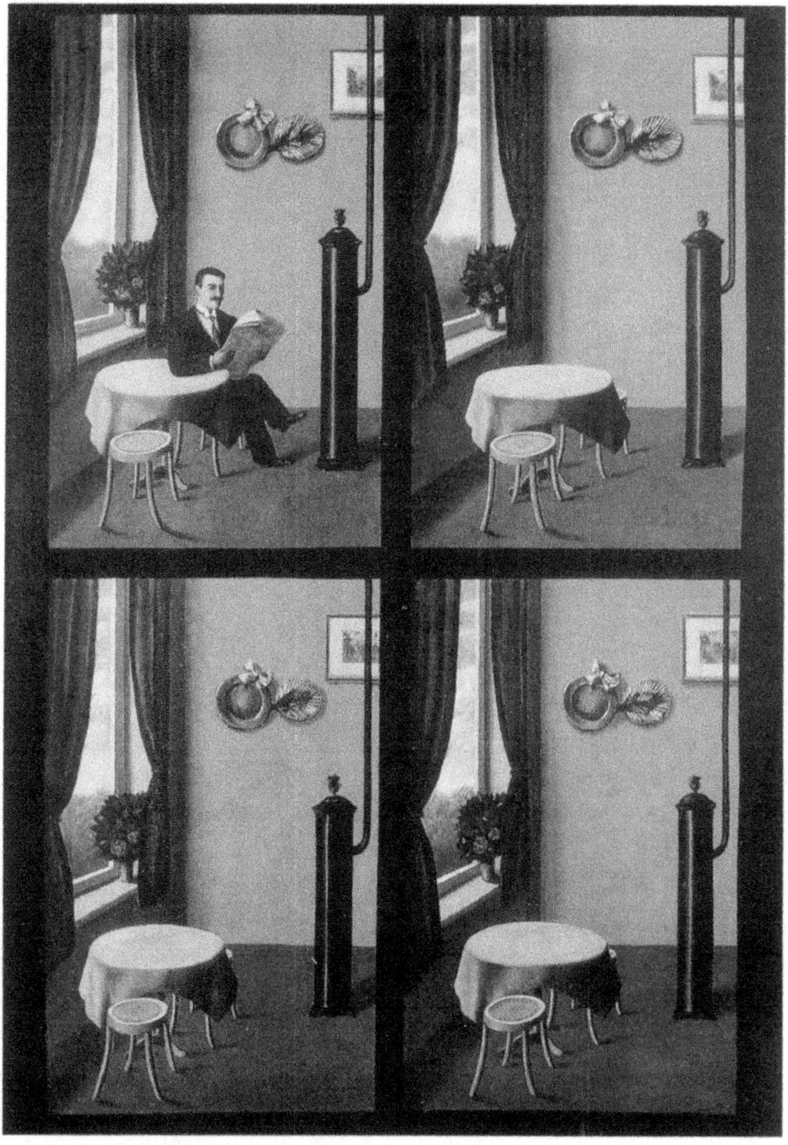

**Figure 1.5** René Magritte's *Man with a Newspaper* (1928).

*Source*: Courtesy of Tate Modern

the ones of tomorrow. It is exactly for this kind of reason that Fermi's paradox is such an encompassing mystery, a riddle of truly cosmological scope and transgenerational importance, joining in its very formulation the central concepts of *any* serious thinking about the world: *universe*, *life*, *mind*, and *technology*.

Fermi's paradox is a paradox of *absence*. As in the famous painting of René Magritte, *Man with a Newspaper*, it queries our conventional perceptions of words, intuitions, and reality (Figure 1.5). Like 'inhabited universe', the concept 'man with a newspaper' might be absent from three of the four panels, but is present in the title of the work. Who would *not* wonder, at even the slightest glance at the artwork, why is he absent from perfectly 'habitable' other panels? Or why is he present in one in the first place? The fact that the panels are not truly identical, but only very similar, adds to the disquieting nature of the picture. Each of the panels is a perfectly legitimate image of something mundane, usual, conventional, and uncontroversial; together, they create a subversive impression of the 'Big Mystery'. So it is with Fermi's paradox as well: each of its component assumptions might be mundane and conventional, but together they present a deep puzzle, a mystery undermining the apparently smooth weaving of the scientific narrative. In it, astrobiology has posed its last and possibly biggest challenge to Copernicanism—after planetary and galactic astronomy, evolutionary biology, biochemistry, and cognitive and computer sciences, has the steam run out of it with Fermi's puzzle?

## 1.9 X-Factors and Navigating Spaceship Earth

People who dismiss Fermi's paradox and the rest of SETI studies as 'just science fiction' (as if that would have changed the inherent value of ideas!) or 'nebulous theorizing' (as if there is anything more practical than a good theory!) are simpletons, in for a nasty surprise. If we stick to at least a shred of Copernicanism—and I shall consider some minor concessions people are willing to make in the opposite direction in Chapter 5—then, obviously, we should reason as if humanity is a typical member of the set of all intelligent species evolved in a naturalistic manner in all epochs. Therefore, if there is a particular reason, let us call it X, which prevents other, much older intelligent species from being detectable, that same reason *X must operate on humanity as well*. For instance, if X is construed, as many Cold-War pessimists argued, as

- **X**: technological civilizations are very likely to destroy themselves through nuclear warfare,

then the conclusion that humanity is also very likely to be destroyed through nuclear warfare is inescapable. While 'very likely' still does not mean 'certainly', the consolation is quite small, since (as we shall see in Chapter 6), the fraction of those civilizations escaping the nuclear holocaust must be small indeed to account for the observed Great Silence. The same applies to many other construals of **X**, such as the following:

- $\mathbf{X_1}$: technological civilizations tend to become extinct due to the global greenhouse effect their industrial activities necessarily cause
- $\mathbf{X_2}$: technological civilizations are likely to succumb to external natural hazards, like impacts of comets and asteroids, or nearby supernova/GRBs
- $\mathbf{X_3}$: technological civilizations are likely to become locked in a permanent totalitarian state, possibly as a response to some other existential (e.g. ecological) risk, which stops scientific progress and bans any space exploration
- $\mathbf{X_4}$: technological civilizations are likely to attempt expanding through space, but that endeavour consumes too many resources, so such civilizations are likely to fail and return to a pre-industrial/primitive state

Such a list would include almost all items commonly suggested as solutions to any version of Fermi's paradox, and we shall consider them in detail in Chapters 4–7. All of them are, on the simple Copernican construal, applicable to the future of humanity as well, since human technological civilization is—obviously and necessarily—among the youngest members of the set. Therefore, many possible reasons for non-detection/decreased contact cross section are not operative yet, but will become operative with high probability in the future.[99]

This simple, almost trivial property of Fermi's problem—that it presents a sort of 'magic mirror' in which we can see the future of our own species—has long been ignored by both researchers and wider public. One could speculate that wider awareness of this crucial link would lead to an increased interest in foundational work in astrobiology and SETI, as well as to much quicker advance when modelling and SETI theory in general are concerned. If the future of humanity is something we care about (anybody can frame it in any number of plausible ethical ways, 'the value of future generations', 'caring about our grandchildren', 'responsibility towards posterity', etc.[100])—then we should care about

Fermi's paradox as well! This conclusion is unavoidable even if we fully understand that any argument about the future of humanity derived from the analysis of Fermi's paradox is a probabilistic one: it is not *our destiny* for any $\mathbf{X}_k$ to ensue, it is just *very likely* that some of them will (namely, one which is responsible for the true solution, or at least some true combination of solutions). Probabilistic arguments are, barring metaphysical determinism, all we can do in future studies anyway. And no amount of rhetoric or disparaging of science fiction will make them go away.

As the great visionary Richard Buckminster Fuller emphasized, 'There is one outstandingly important fact regarding Spaceship Earth, and that is that no instruction book came with it.'[101] A similar predicament is likely to be faced by wardens of any other Galactic biosphere. Lacking the manual, we might just try to steer according to rational rules of probabilistic navigation; and if we truly wish to steer between the devil and the deep blue sea facing all cosmic civilizations, we better draw all possible lessons from the Great Silence.[102]

CHAPTER 2

# 'What's Past Is Prologue'

## *Cosmological and Astrophysical Background*

When Antonio delivers the following famous phrase in Act II of *The Tempest*,

> (And by that destiny) to perform an act
> Whereof what's past is prologue; what to come,
> In yours and my discharge

in trying to persuade Sebastian to murder his father, the King of Naples, and assume the crown, what he meant was quite a bit different from what is suggested by the modern renditions. In arguing for such a heinous act, he tries to convey the message that what has already happened merely sets the scene for the *really important* stuff, which is the stuff 'our own greatness' will be made on. Therefore, in contrast to the past, the future is not fixed or fated, but entirely open to the free will of the actors on the scene. This is a very apt metaphor for the elucidation of the relationship of intelligent life in the universe to the physical processes which brought about the conditions for its existence during the course of cosmological history. In contrast to other physical phenomena, life and intelligence can—in a sufficiently relaxed but quite relevant sense—choose their own future destiny. In a still wider sense, the phrase perfectly captures the dilemma inherent in studying Fermi's paradox: we can only learn so much from studying the already-fixed historical background, in contrast to a vast number of possible evolutionary trajectories an intelligent community can take in the future. We have only recently begun to even conceive of this huge space of possibilities in anything remotely resembling a scientific manner, although, as it is often the case, some artists and philosophers were there first. The present and the future are so much larger than the past.[1]

To a lesser degree, a similar conclusion holds for space, as well as time. The heterogeneity of local conditions throughout the universe (or even when we restrict the analysis to the Galaxy) dwarfs anything evolution as we know it has encountered thus far on Earth. This implies that thinking about the astrobiological landscape as a set of a larger or smaller number of Earth-like planets is necessarily limited in both the quantity and the quality of its outcomes. In turn, this insight could be liberating as well: wider-ranging speculations, while in accordance with laws of physics and some general considerations such as Copernicanism, need not be burdened by past history and local specificities.

This does not mean that we should not take into account all cosmological and astrophysical evidence in our search for the proper assessment of high-complexity parts of the astrobiological landscape; we just need to be cautious in taking extrapolations from the past and our local astrophysical and astrobiological evidence. Such evidence, as presented in this chapter, represents a powerful tool brought into play by the dramatic increase in our astronomical knowledge since Fermi's time. Not only has the astrobiological revolution since 1995 dramatically changed our general outlook towards the search for life and intelligence elsewhere; it achieved such status by showing that hard science can proceed along the road envisioned by what a couple of decades ago was considered wild speculation and science fiction. There is no reason to assume that such sort of partnership between research and speculation, including purely science-fictional discourse, cannot be continued in the twenty-first century at an even more intense pace.

A major difference in comparison with the epoch of Fermi's lunch is that now we have much better insight into the *cosmological and astrophysical background* necessary for the emergence of intelligent beings. While the general issue of the importance of cosmology for astrobiological research has been discussed in some detail in *The Astrobiological Landscape*,[2] I shall attempt here to present just a crude snapshot of some of the relevant data and concepts from the general SETI point of view, with an emphasis on those topics relevant for Fermi's paradox. In particular, I shall first present the new standard cosmological paradigm as a general background for evolution of any kind of complexity, especially those peaks of complexity which present SETI targets, that is, communities of intelligent observers. One of the key concepts of relevance for Fermi's paradox is the existence of cosmological horizons, delineating the spatio-temporal volume in which it makes sense to search for traces and manifestations of such intelligent communities. Finally, on the local astronomical level in the Milky Way, we shall encounter a particular subregion, the Galactic Habitable Zone, where abiogenesis and noogenesis are most likely to occur. Although this does not limit the extent of the region which contains SETI targets, it certainly does influence its shape and evolution. The concepts introduced in this chapter will be extensively used in formulation and assessment of various particular hypotheses for resolving Fermi's paradox in Chapters 4–7.[3]

## 2.1 The New Standard Cosmological Model

Post-1998 cosmology has been marked by the emergence of a new paradigm, often called the new standard cosmological model, a revision of the standard cosmology which appeared after 1965 and the discovery of CMB radiation. Observations of cosmological supernovae, published in 1998 (and for which the researchers involved were awarded the Nobel Prize in Physics in 2011), supported by data from the Wilkinson Microwave Anisotropy Probe (WMAP) and the Planck mission, have resolved the 'three-numbers' puzzle of classical cosmology and opened up the present epoch of modern research into the origin of structure in the universe.[4] While it is exactly this new standard cosmology (also called the Lambda Cold Dark Matter -or λCDM- model) which provides the background for our investigation into Fermi's paradox, some historical notes are in order here for several reasons, including highlighting the oft-mentioned analogy of our problem and that old cosmological chestnut, Olbers's ('dark night sky') paradox.

The first cosmological revolution actually occurred in the 1920s and 1930s, when it was first possible to account for an astonishing new fact—the global expansion of the universe—through the application of Einstein's local theory of gravity, general relativity. Friedmann models, the dynamical solutions to Einstein's field equations in the cosmological case, showed that the universe undergoes evolution from a primordial state of extreme density and that the past has been finite. The initial singularity, first deemed a mathematical artefact without any link with physical reality, gave way to Lemaitre's 'primordial atom' and Gamow's 'cosmic egg', which have subsequently evolved into Big Bang cosmologies as we know them today. After a period of 'great controversy' (1948–65), this finite past duration of our cosmological domain—and a fortiori the finite age of any structure within it—was completely vindicated. This period witnessed the emergence of cosmology as firmly grounded empirical physical science, through a series of experimental tests of various paradigms, culminating in the discovery of CMB by Penzias and Wilson in 1965.[5] However, numerical precision was only reached in the post-1998 period, with the use of advanced instruments such as the COBE, WMAP, and Planck observatories, observations of supernovae at cosmological distances, and the compilation of huge galaxy catalogues like the Sloan Digital Sky Survey.[6]

In addition, secondary theories—with first-rate results!—grew around the emerging Big Bang paradigm, such as the theory of primordial nucleosynthesis (very successful), the λCDM theory of structure formation (partially successful), and various inflationary scenarios (still quite speculative, but with great explanatory potential). The emergence of the new field of quantum cosmology, as well as the ever-closer synergy between particle physics and the early universe cosmology, contributed to the veritable explosion of cosmological results we have witnessed in the last three decades or so.

The most important *process* in the history of our universe is the formation of structure, which can be interpreted as a *violation* of the cosmological principle of universal homogeneity and isotropy on large spatial scales. This is a specific instance of the key fundamental process of *spontaneous symmetry breaking*, which accounts for the entire—or almost entire—information content of the universe. We can imagine a very early state of the universe as homogeneous 'soup' of elementary particles and fields, entirely describable by a small amount of information (e.g. particle spectrum, density, and temperature), measured in dozens or hundreds of bits at most. Today, the universe is mind-bogglingly complex, the part within our cosmological horizon requiring perhaps as much as $10^{120}$ bits to be adequately described.[7] A large part of this complexity is located in living beings, both on Earth and wherever else they might exist; an even greater potential 'reservoir' of complexity is *intelligent* beings and their creations—some of them being exactly those which could be detectable by present-day and future SETI endeavours.

The age of the universe, according to the most recent high-precision cosmological data, is[8]

$$\tau = \left(13.798 \pm 0.037\right) \times 10^{9}\,\text{years}. \tag{2.1}$$

This result has been obtained from the concordance of various measurements, most notably from the Planck and WMAP satellites, with the observations of early Type Ia supernovae and the amount of clustering of normal, luminous galaxies at different red shifts (so-called baryon acoustic oscillations). At the very least, this is an upper bound to the age of any object or any kind of structure within our cosmological domain (which is a larger set than just the *visible universe*; the latter means the universe within our cosmological horizon; see Section 2.4). Of course, more relevant from the astrobiological point of view is the epoch of the beginning of habitability, for which the requirements, notably those related to Earth-like planet formation, must be satisfied. In a recent paper, Abraham Loeb has argued that there was a brief period of about 7 Ma in duration in the very early universe, about 10 Ma after the Big Bang, when the energy required for life could have been provided by the CMB radiation (which, of course, was not really *microwave* at that epoch, having its maximum emissivity in the infrared part of the spectrum, equivalent to the temperature range 273–373 K, in which liquid water exists).[9] This would have enabled planets—if any existed that early on, which is unclear—at the distant periphery of their planetary systems to be habitable, independently of their parent stars. The fact that the metallicity necessary to form not only life forms but the terrestrial planets or moons themselves would have been extremely low that early on, as well as the brevity of this epoch, argues strongly against the possibility of such early life taking root. However, even if just on a conceptual level, Loeb's hypothesis contributes to the overall spatio-temporal extension of the conventionally understood habitable zones; there will be more on this in Section 2.3. For the moment, we should note that the existence of galaxies and other levels of the cosmological structure seems to be a *prerequisite* for any kind of life and intelligence. However, it need not be a *requisite* for the latter, since we can imagine not only advanced technological civilizations living and travelling outside of conventionally understood structures (e.g. engaging in *intergalactic* travel), but also, following in the steps of Olaf Stapledon's *Star Maker*, even remaking/refurbishing such structures to their advantage.[10] This is still another instance of our need to resist the temptation to judge the future (and possibly the present) on the basis of the mainly lifeless and intelligence-less past.

While our universe as a whole is obviously habitable, it is very relevant to ask which parts of the cosmological structure are capable of supporting life—and how probable is the emergence of life in any given epoch. Since most of the spatial volume of the universe is intergalactic space, which certainly does not support life, we need to seek those matter density peaks which correspond to structures such as galaxies and groups and clusters of galaxies. The predominant paradigm of structure formation in the new standard cosmology is still the cold-dark-matter (CDM)-dominated 'bottom-up' approach, which suggests that smaller structures formed first and were subsequently assembled into larger structures. Notably, individual galaxies formed before small groups (like the Local Group in which we live) and clusters (like the Virgo cluster, the closest rich galaxy cluster, containing about 2000 galaxies) of galaxies did; galaxies themselves had been seemingly assembled from pre-existing protogalactic fragments and subsequently grew by swallowing further such fragments, as well as many small galaxies in the course of their history. Our Milky Way is rich in the remnants of such episodes of galactic cannibalism.[11]

## 2.2 The Size and Age of the Galaxy

Since Fermi's paradox is usually—although, as we have seen, not always or necessarily—formulated in the context of the Milky Way, it is worthwhile to recapitulate important astrophysical facts about our Galactic habitat. Simultaneously, such discourse presents a good introduction to the important concept of the Galactic Habitable Zone (henceforth, GHZ), one of the most important research subjects in the contemporary astrobiology, as well as a concept of obvious relevance for SETI and Fermi's puzzle.

The Milky Way is a barred spiral galaxy, usually classified as 'SBc' in the Hubble 'tuning-fork' morphological classification of galaxies.[12] Due to the obvious fact that we cannot observe the Milky Way from the outside, this property—as well as some other descriptive parameters of our stellar system—remains somewhat controversial. What we can see from our vantage point deep within the disc is shown in Figure 2.1: there is a strong planar concentration of both stellar and interstellar matter. While the size of the visible disc of the Galaxy (roughly corresponding to what astronomers call the *Holmberg radius*[13]) is thought to be about 20 kpc, we know today that the really dynamically important matter associated with any galaxy extends much farther. In particular, for galaxies similar to the Milky Way (giant spirals), the dark matter halo extends to the galactocentric distance of at least 50 kpc (direct and uncontroversial evidence), and probably as much as 300 kpc (via a plethora of indirect evidence, growing by the year). While most of the mass in this dark halo is composed of non-baryonic CDM, there is a minor amount of baryons, probably in form of extremely rarefied and extremely hot hydrogen–helium plasma.[14] There are very few stars or any other 'things' or 'objects'—as conventionally understood—in the vast volume that is the Galactic halo; any astrobiologically and SETI-interesting events are certain to unfold within the Holmberg radius of the Galaxy.[15]

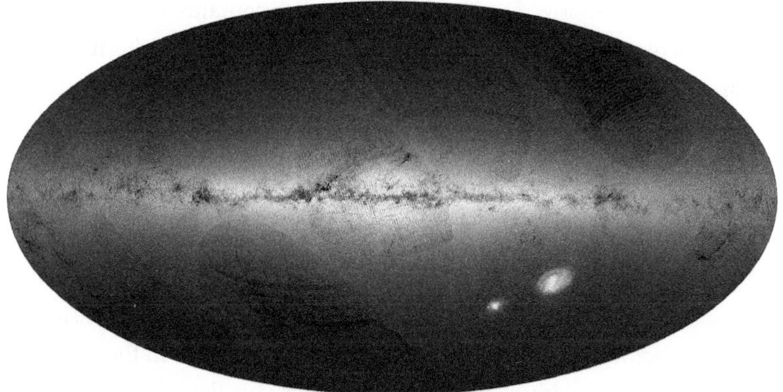

**Figure 2.1** The Milky Way, as seen by the *Gaia* observatory of the European Space Agency (ESA) during its first year of observations, from July 2014 to September 2015. Other Local Group galaxies are visible in the image, notably the Large Magellanic Cloud and the Small Magellanic Cloud, as well as M31 and M33. Some *artefacts* of the observing procedure are also visible—curved features and darker stripes are not of astronomical origin but rather reflect the scanning procedure. As this map is based on observations performed during the mission's first year, the survey is not yet uniform across the sky; this reflects both the incremental nature of our astronomical knowledge, as well as the all-pervading nature of selection effects, biases, and artefacts—very useful lessons for astrobiology and SETI studies.

*Source*: Courtesy of ESA/Gaia/DPAC

The conventional estimate is that Galaxy contains about $3 \times 10^{11}$ stars. This includes visible stellar objects in the disc of the Milky Way—in the halo, there is a poorly known number of sparse Massive Compact Halo Objects (MACHOs), which may or may not be the remnants of previous generations of stars, like white or black dwarfs or old neutron stars, as well as substellar objects (brown dwarfs/giant planets). Stars of different spectral types are differently distributed throughout the Milky Way, and those of spectral classes most interesting to astrobiologists—roughly, the F, G, and K stars—are concentrated towards the plane of the Galaxy.[16] While the planetary census is, obviously, empirically much poorer, preliminary estimates suggest that the total number of planets is huge, probably significantly higher than the number of stars and not much smaller, even in very conservative models.[17] This is particularly important in the context of 'extended habitability', which encompasses habitats similar to Europa or Titan as at least potentially interesting. There can easily be more than $10^{12}$ such habitats in the Milky Way, not to mention more speculative possibilities, like free-floating planets, which are a few orders of magnitude more numerous still.[18]

The Galaxy consists of several components: a halo, a bulge, a bar, a thin disc, and a thick disc. The component of highest interest for astrobiologists is the thin disc. The thin disc contains all the metal-rich stars, which are similar to our Sun and are categorized as Population I. (It also contains all giant molecular clouds, comets, and other entities sometimes speculated to be alternative, non-planetary habitats for life; and, insofar as chemistry is in any way important for those alternative forms of life, interesting chemistry is to be found *only in the thin disc.*) The thin disc is markedly younger than the Galaxy itself; while the oldest objects in the Milky Way, evolved globular clusters, are about 13 Ga old, the age of the thin-disc population is in general lower than 10 Ga.[19] As the name suggests, the thin disc is a very flattened system; in fact, it is so flattened that it makes sense in many numerical models to take only two dimensions into account—galactocentric distance and an angular coordinate in the plane of the Galaxy. If a process is isotropic in the sense that, *on the average*, there is no dependence on the angular position, only a single coordinate, galactocentric distance, is required for the specification of spatial coordinates. In our present, quite poor, understanding of the prerequisites for astrobiological evolution, it makes sense to use this one-dimensional model for most of the relevant processes. While we should acknowledge possible exceptions (e.g. vertical oscillations about the Galactic plane might lead to gravitational perturbations on the Oort cloud, thus causing cometary impact showers on habitable planets; or the spiral arms of the Galaxy might have an increased risk of having supernovae or encounters with giant molecular clouds, with consequences for life forms[20]), for now, even models with only radial dependence give rich and unexpected insights, most of which have been revealed only since the beginning of this century.

The Sun is located near the inner rim of the Galaxy's Orion Arm, at a distance of 8.33 ± 0.35 kpc from the Galactic centre. This distance has slowly 'shrunk' in the course of twentieth-century astronomical progress, from an initial rough estimate of 10 kpc, to the IAU-accepted value of 8.5 kpc, to the present-day value. The age of the Solar System is usually quoted as[21]

$$\tau_\odot = \left(4.5681 \pm 0.0004\right) \times 10^9 \,\text{years}. \tag{2.2}$$

Of course, one needs to take into account the fact that there might not be a well-defined 'age of the Solar System' at all, since various parts of our planetary system originated at different times; the clearest example of this is our Moon, which, according to the dominant theory, originated in a slow impact on Earth 20–100 Ma after the formation of our planet.

The Sun, with its retinue of planets and other Solar System bodies, orbits around the Galactic centre with a period of about 250 Ma, and it oscillates both in the Galactic plane (radial oscillations) and perpendicularly to it (vertical oscillations). The small radial oscillations mean that, from the point of view of an observer in the inertial system of the Galactic Centre, the orbit of the Sun does not close on itself but forms a rosette figure. As far as the vertical oscillations are concerned, our planetary system is *currently* very close to the Galactic plane, and in general moves vertically about it with an amplitude of about 100 pc and with a period of about 87 Ma.[22] Whether these astrophysical timescales are in any way causally related to the history of the terrestrial biosphere has been a perplexing and intriguing question since the mid-1980s, unresolved to this day.[23] In any case, these timescales play some role in the concept of the Galactic Habitable Zone to be considered in Section 2.3.

As far as the temporal size of the Milky Way is concerned, it spans most of the history of the universe given by Eq. (2.1). In 2013, it was reported that a well-known subgiant star HD 140283, belonging to an extremely old and metal-poor halo population of stars, has the best-fit age of 14.5 ± 0.8 Ga, equal within errors to the age of the universe in Eq. (2.1)![24] Although it is, of course, preposterous to think that a star can be older than the universe (the uncertainties are underestimated here, if anything), it testifies that the process of structure formation was relatively quick and that first stars formed only several hundred mega-anni after the Big Bang. Our Sun, as we know from Eq. (2.2), is a relatively young disc star. And, on the other end of the scale, we observe many regions of active star formation in the Milky Way (Orion, Taurus, etc.), where we can find stars at about zero age. Star formation will continue in the future of the Galaxy for a long time to come—the estimates in the domain of physical eschatology range up to $10^{12}$–$10^{14}$ years in the future.[25] In contrast to the alleged absence of intentional influences in the Galactic *past*, such absence is by no means warranted—and probably not a good assumption anyway—for the *future*, in which intelligent beings, terrestrial and extraterrestrial (if any), are likely to exercise a large influence on their astrophysical environment. For reasons that may be either straightforward or inscrutable, advanced technological civilizations might choose to extend or shorten star-formation timescales. We shall discuss some of the more straightforward reasons for doing this in subsequent chapters.

Planetary habitability, as defined by the terrestrial kind of life (and its moderate extensions), follows from the properties of specific stellar and planetary populations of the Galaxy. In particular, it is the chemical composition and dynamical properties of the Population I stars in the Milky Way thin disc which make their planetary systems *potentially* habitable. One important distinction has to be made here between *necessary* and *sufficient* conditions for habitability. Obviously, we are in a position to establish some of the necessary conditions for habitability, even without having a deep theoretical understanding of the phenomenon of life. However, it is highly doubtful, at best, whether the entire set of sufficient conditions can be established without a deep understanding. This is sometimes confusing, especially in popular presentations; conflating necessary and sufficient conditions is one of the major sources of mistakes, fallacies, obfuscations, and prejudices of our age. The correct way to separate the two lies in one of the most interesting new concepts introduced by the astrobiological revolution.

## 2.3 The Galactic Habitable Zone

One of the milestone developments in astrobiology is related to the concept of the *Galactic Habitable Zone*. The term originated with the seminal paper of Gonzalez, Brownlee, and Ward in 2001, although the idea that not all places in the universe are equally hospitable to life is an old one.[26] There have been some notable predecessors, including Alfred Russel Wallace, who, in his famous 1903 book *Man's Place in the Universe*, argued that only a restricted number of positions that Earth and the Sun could occupy within our stellar system would have enabled abiogenesis and the evolution of life on our planet. Interestingly enough, Wallace came to the dramatically erroneous conclusion that the habitability of our Solar System must be due to its location near the centre of the Milky Way! But the reasoning behind it, namely that there must be—as soon as we accept that Earth and our planetary system are not closed boxes—variation in the probability of evolving life and intelligence over the set of all possible spatial coordinates in the Galaxy is perfectly sound. The twofold problem of (i) correctly identifying all spatially varying causal factors influencing the habitability of a planet, and (ii) correctly quantifying the impact of those factors on the distribution of potentially habitable planets, has proved a tough nut to crack, even with all the advances of contemporary astrophysics.

Since the failure of Kapteyn's (and Wallace's) model of our stellar system, and the pioneering work of Shapley, Hubble, and others on the size and shape of the Milky Way and other galaxies, people have been facing the implicit question of whether the position of the Sun about halfway across the radius of the stellar disc has to do something with our origin and evolution. In 1983, a distinguished Soviet/Russian astronomer Leonid Marochnik suggested that the formation of planetary systems in general, and Earth-like planets in particular, occurs mainly near the co-rotation radius of our Galaxy.[27] The causal mechanism in Marochnik's picture includes the impact of Galactic density waves and spiral-arm crossings, which occur least frequently near the co-rotation circle.

Marochnik's work explicitly recognizes the importance of observation-selection effects, something we shall encounter often in the rest of this book: we should regard what we observe as typical *only after taking into account all preconditions for our emergence as intelligent observers at this cosmic epoch*. The last sentence of Marochnik's paper emphasizes that 'the extrapolation of results obtained for the solar vicinity for the whole Galaxy can prove to be incorrect'.[28] About the same time, similar ideas have been suggested by the Hungarian astronomer Béla Balázs, who coined the term 'Galactic belt of life' for what would eventually be called the GHZ.[29] While the detailed investigation of this early history is obviously a task for future historians of astronomy, it is important to keep in mind that a cluster of related ideas has been around for quite some time.

In 2001, two important papers were published in back-to-back volumes of *Icarus*, which is the world's premier journal in planetary science and includes many astrobiological studies (although less now than back in the late 1960s and 1970s, when Carl Sagan served as the editor-in-chief). The papers contained, by the irony of time and editorial processes, two key components of any conceivable attempt to make habitability a more precise notion than it had hitherto been: temporal and spatial constraints. The paper by Charles Lineweaver, 'An Estimate of the Age Distribution of Terrestrial Planets in the Universe: Quantifying Metallicity as a Selection Effect', focuses on the temporal distribution of potential habitats for life.[30] It joined vast improvements in our understanding of chemical evolution with the nascent statistics of extrasolar planets and new insights into observation-selection effects. The main conclusions of this seminal paper were that (i) Earth-like planets start forming after a delay of ~1.5 Ga after the onset of thin-disc star formation, which can be treated as the birth epoch of GHZ, and (ii) the *median* age of Earth-like planets is the one given by Eq. (1.3). As previously mentioned, this gives rise to the crucial timescale in the whole Fermi's paradox quandary, which I shall call the Lineweaver timescale:

$$\tau_L \equiv \tau_{med} - \tau_\odot = (1.8 \pm 0.9) \times 10^9 \text{ years.} \tag{2.3}$$

(since uncertainties in Eq. (1.3) obviously dominate the total error). The strong discrepancy between (2.3) and the Fermi–Hart timescale in (1.1) is a generator of most of the difficulties in trying to answer *Where is everybody?* Overall, Lineweaver's results—and subsequent elaborations I shall deal with—delineate the temporal boundary of the GHZ and make Fermi's paradox more serious still.

The paper published in the subsequent volume of *Icarus*, by Guillermo Gonzalez, Peter Ward, and Donald Brownlee, is usually taken as the beginning of the modern studies of the GHZ. The GHZ in its modern rendition has the form of an annular ring, with its inner and outer boundaries determined by various astrobiological processes at any given epoch. There is a consensus that the outer radius, $R_{out}$, is determined by the gradient of metallicity in the Milky Way disc; this gradient is currently measured as roughly $\nabla Z = 0.07$ dex kpc$^{-1}$ (the decade logarithmic unit, dex, denotes that the metallicity falls off by such-and-such fraction of an order of magnitude per kiloparsec traversed outwardly from the Galactic centre).[31] Very low metallicity will prevent the formation of terrestrial planets or will cause only very small, Mars-like terrestrial planets to form. Since Mars, at 0.107 $M_\oplus$, is at the very lower limit of planetary mass which allows habitability in the conventional sense, we expect no habitable planets beyond some galactocentric distance in the disc.[32] As metallicity increases with the passage of time, the GHZ slowly expands outward, as clearly shown in Figures 3 and 4 of the Lineweaver et al. 2004 paper and schematically represented in Figure 2.2.

While the outer boundary of the GHZ is rather uncontroversial, there is significant uncertainty over the radius and the causal determinant of the inner boundary. There are at least three causal mechanisms precluding habitability in the innermost regions of the Galaxy: (i) frequent stellar collisions and close encounters, (ii) a higher frequency of supernovae/gamma-ray burst explosions, and (iii) suppression of Earth-like planet formation by *excessive* metallicity (leading to the excessive growth of protoplanets and the formation of, say,

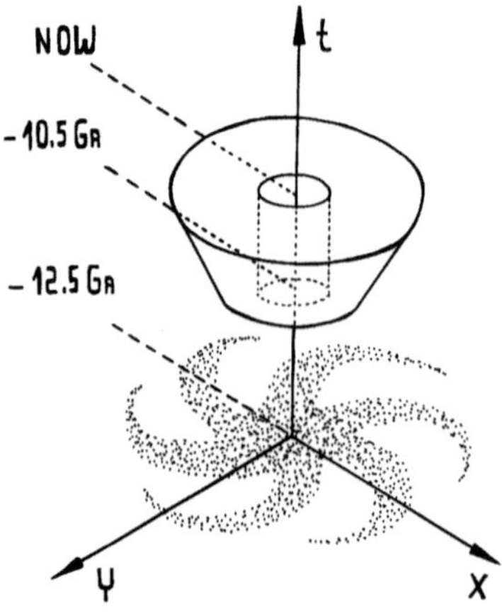


**Figure 2.2** Gradual widening of the Galactic Habitable Zone in a simplified model. In this model, the major cause of widening is gradual chemical enrichment of the Galactic gas, enabling formation of Earth-like planets at progressively larger distances from the centre of the Milky Way. Additional contributing effects are the decrease in supernova—and other cosmic explosion—rates, and slow orbital diffusion of systems possessing Earth-like planets, from the dense inner regions to sparsely populated outskirts.

*Source*: Courtesy of Slobodan Popović Bagi

ocean planets instead).[33] In addition, another rather vague mechanism pertains to high levels of ambient UV ionizing flux, which might photoevaporate protoplanetary discs as well as suppress the dust-grain growth and adhesion necessary for the formation of Earth-like planets[34]—and since the ambient UV flux is higher in the inner regions of the Milky Way, this might be a contributing factor as well. Obviously, the *real* inner GHZ boundary is determined by the most restrictive process of all involved. It is not yet entirely clear which one is that, and the situation is worsened by the fact that these processes are secularly evolving in a way which is more complex than the simple build-up of metallicity characterizing the evolution of the outer boundary. There are no comprehensive numerical studies so far incorporating all the relevant physical processes within a single family of models—and this remains a significant challenge for theoretical astrobiology.

Since 2001, our handle on the astrobiological history of the Milky Way has vastly improved.[35] Two studies which elaborate upon Lineweaver's results are of particular interest for our present purpose. Peter Behroozi and Molly S. Peeples of the Space Telescope Science Institute in Baltimore, MD, calculated that the Solar System formed after 80 per cent of existing Earth-like planets had formed (in both the Milky Way and the universe in general), and that we should expect there to be $\sim 10^{20}$ Earth-like planets in the entire visible universe (the Hubble volume). In 2016, Erik Zackrisson and his co-workers repeated Lineweaver's study of the inventory of Earth-like planets, with improved models of galaxy formation and evolution. The results are striking, since they *increase* the Lineweaver scale to ~2.7 Ga for parent stars belonging to the standard F, G, and K spectral types; if we include red dwarfs of the M spectral class—like the parent star of the TRAPPIST-1 system—this jumps even further, to ~3.8 Ga![36] While we can—and should!—question the precision of these newer studies, it is important to notice the general lesson: a large chunk of the total spatio-temporal volume of the Milky Way is habitable, even under our rather parochial construal of habitability.[37]

Overall, the best evidence that astrobiological research on habitability is on the right track is offered by cosmological simulations which start as close to the 'first principle' as possible. Two such works appeared recently, by Branislav Vukotić with coworkers in 2016 and by Duncan Forgan and coworkers in 2017; these opened a new line of research by pressing cosmological *N*-body simulations into service for establishing the origin, extent, and evolution of GHZs. That the results obtained are rather similar, in spite of different software, initial conditions, and methodology, and that both are broadly consistent with the semi-analytic work of those such as Lineweaver and his collaborators, Zackrisson and his collaborators, and Behroozi and Peeples, clearly shows that astrobiological *theory* has reached maturity. Prototype results of these large-scale numerical simulations are shown in Figures 2.3 and 2.4.

In any case, it seems certain that the innermost regions of the Galaxy are not conductive to the emergence of complex biospheres and intelligent lifeforms. However, while the GHZ is crucial for the *emergence* of intelligent beings, it does not really constrain their subsequent *expansion*. In fact, there are many

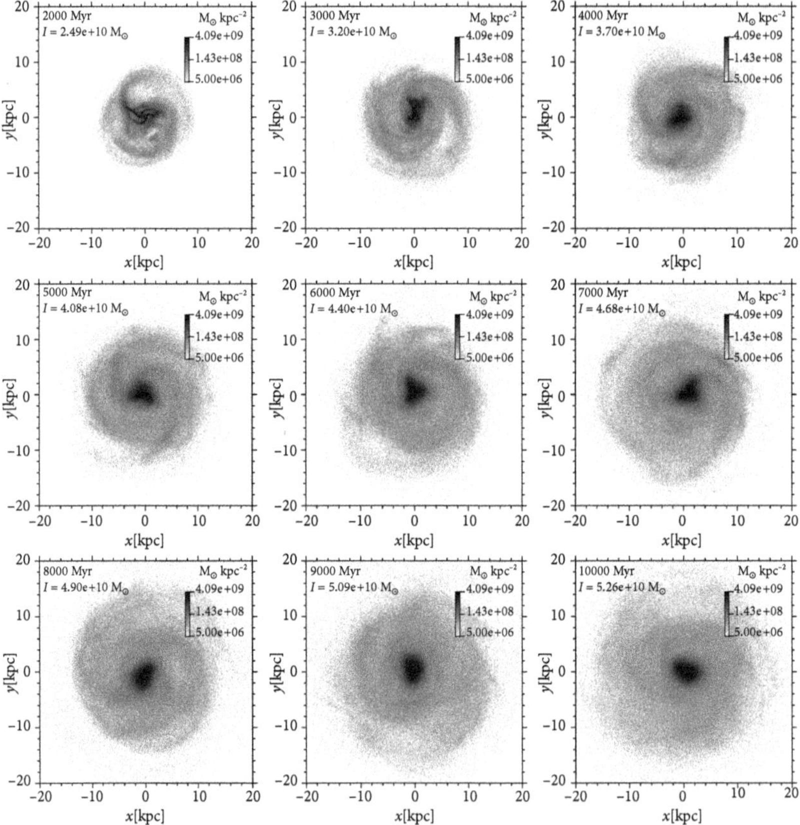


**Figure 2.3** Stellar mass surface density in a simulated Milky-Way-like galaxy. This *N*-body smoothed particle hydrodynamical simulation was created with the explicit purpose of studying habitability. Results presented here—in $10^9$ year increments, starting with the disc-formation epoch—reflect the necessary astrophysical background of any search for life. It is exactly this kind of numerical astrobiology which can improve our understanding of the Galactic Habitable Zone and its observable properties.

*Source*: Vukotić, B., Steinhauser, D., Martinez-Aviles, G., Ćirković, M. M., Micic, M., and Schindler, S. 2016, '"Grandeur in this view of life": N-body simulation models of the Galactic habitable zone', *Mon. Notices Royal Astron. Soc.* **459**, 3512–24. Courtesy of Branislav Vukotić

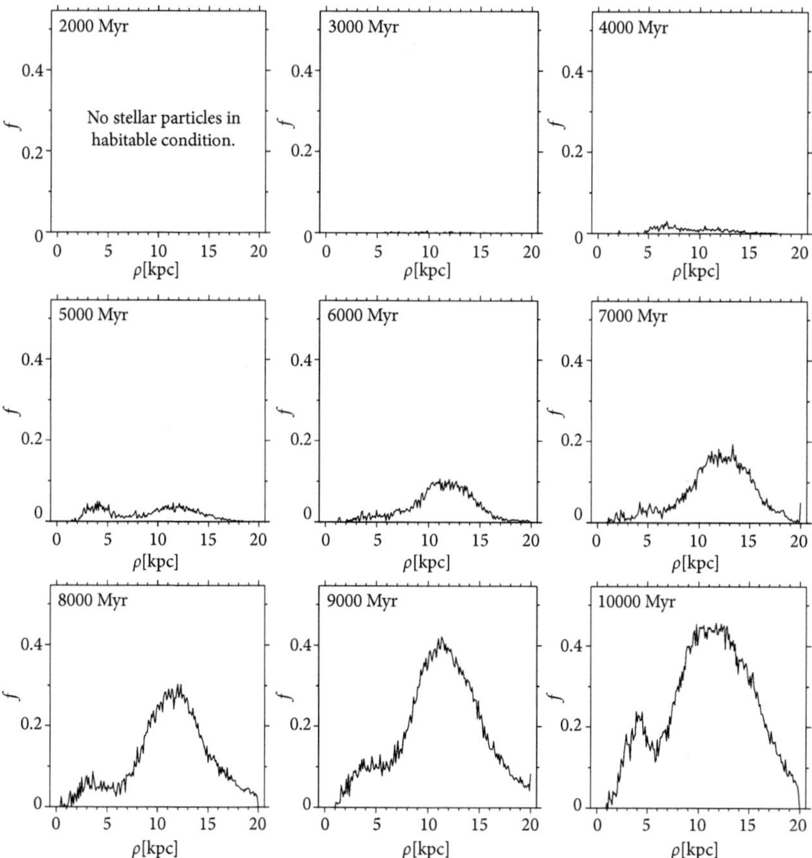


**Figure 2.4** The fraction of habitable grid cells in a simulated Milky-Way-like galaxy, as a function of the galactocentric distance $\rho$ for various epochs of galactic history. This is *not* the number of habitable planetary systems, since the grid cell is still significantly larger than the amount of space properly assigned to a single planetary system; in this respect, this numerical model still resembles a 'coarse-grained' grid. One can think about these cells as those containing *at least one* habitable planet among its stars, as opposed to the 'dead' volumes of space. We notice that, at a later time, the Galactic Habitable Zone becomes pronounced and obtains structure that is more complex than that of a naive 'annular ring'.

*Source*: Vukotić, B., Steinhauser, D., Martinez-Aviles, G., Ćirković, M. M., Micic, M., and Schindler, S. 2016, '"Grandeur in this view of life": N-body simulation models of the Galactic habitable zone', *Mon. Notices Royal Astron. Soc.* **459**, 3512–24. Courtesy of Branislav Vukotić

conceivable reasons why intelligent beings, upon reaching the status of an advanced technological civilization, would expand beyond the GHZ bounds. Some of the reasons are due to pure economics: after all, the amount of Galactic resources located outside of GHZ is significantly larger than the amount inside it. While the exact difference depends on the definition of the term 'resources' (e.g. whether advanced civilizations will have access to cheap nuclear transmutation of chemical elements[38]) and the poorly understood GHZ inner boundary, it is clear that the inequality would be a strong motivation for expanding. There might also be safety and ethical reasons: after all, if there are other civilizations emerging in the GHZ, the probability of conflict increases if the resource utilization is carried out solely in their common astronomical vicinity.

In the context of SETI studies, one could go a step further and postulate the *Galactic Technological Zone* (GTZ), which could, but need not, partially overlap with the GHZ.[39] That would be a set of spatial locales where advanced technological civilizations exist and/or keep the bulk of their hardware (in a sufficiently generalized sense). The very fact that any technology is limited by the laws of physics—even if it could, in a sufficiently advanced form, offer a different *impression*, as noted by Sir Arthur C. Clarke—means that the GTZ is, obviously, the target area for practical SETI projects. Rather trivially, SETI cannot really probe the distribution of intelligent life if, for any reason imaginable, it does not develop technology (again, sufficiently generalized, since it could be vastly different from what we conventionally understand as technology[40]). If there is some truth in the arguments flaunted in the early days of SETI that marine species could evolve intelligence but not technology, due to hydrodynamic constraints and a lack of oxidants—such marine meditative intelligences would not be detectable through SETI projects. Whether the GTZ diverges much from the GHZ is dependent on the overall astrobiological landscape of the Galaxy—if we knew that, we would have already established the correct solution to Fermi's paradox, and this book would be rather superfluous. Some conclusions can be drawn even in our state of ignorance, though. Considering the size of the GHZ, it seems clear that quite advanced interstellar travel would be required to leave the GHZ from an average point within. This would imply very advanced technology, comparable to at least Kardashev Type 2 civilizations. Searches for signs of extraterrestrial technological activity in external galaxies (like the Ĝ infrared survey, or the work on possible outliers on the mass–luminosity diagram) are actually attempts to detect GTZs; although such unorthodox SETI projects are still in their infancy, the negative results so far have already tremendously limited the parameter space describing very advanced—Kardashev Type 3 and possibly some 2.x—civilizations.

There are some criticisms of the concept of GHZ, encountered in recent literature or in the less formal context of conferences or blog debates. This comes on the top of occasional 'loud silence' about it: for instance, the term 'Galactic Habitable Zone' is conspicuously absent from the indices to *Complete Course in Astrobiology*, a recent and technical reader entitled *Fitness of the Cosmos for Life*, and the popular and jovial *Confessions of an Alien Hunter*.[41] The criticism can be divided into two categories: general critiques based on the alleged conservatism with respect to alien life, and technical criticism of the usefulness of the concept itself. The arguments in the first category are analogous to the criticism that focusing only on circumstellar habitable zones is too constraining, since it does not take into account the possibility of subglacial life on Europa and other gas giant satellites which are heated by tidal dissipation instead of sunlight. When we go to spatial scales larger than those of a planetary system, other wildly diverse possibilities open up. Possible habitats based on radioactive decay or even gravitational collapse have been speculated upon by both scientists and science-fiction authors. Thus, according to this form of criticism, the GHZ concept is based upon a mistaken reification of habitability for *terrestrial forms of life*.[42]

In the second category, we can encounter a very sound criticism to the effect that the term 'GHZ' implies a gradualist mode of change, which is an assumption not really supported by evidence. On the contrary, large local fluctuations could, in principle, obviate the correlations upon which the concept is based. Consider, for example, the chemical abundance gradient in the Galaxy. It is obviously an outcome of an averaging process, in which metallicity over various objects and lines of sight are measured and collated into a general pattern exhibiting the gradient. The picture is incomplete for two different reasons: (i) we obviously cannot ever hope to sample all locations in the Galaxy and have to rely on finite and usually small samples, and (ii) the nature of chemical evolution of matter in the Galaxy is such that the chemical enrichment occurs inhomogeneously and the mixing of the enriched interstellar medium never occurs perfectly. Both these reasons leave room for fluctuations beyond the measured average gradient, meaning that, for instance, a local high-metallicity region could occur beyond the outer boundary of the GHZ and could give rise to a habitable, and indeed inhabited, planet.

(Similarly, it *may* happen that an already formed and inhabited planetary system—or a system in formation—is ejected by a close encounter with another passing star. Such a slingshot mechanism has been well studied and, in the fullness of astronomical time, must occur with reasonable frequency.[43] Thus, we could reasonably expect to find *some* habitable or, indeed, inhabited planetary system outside the GHZ, possibly even outside the Galactic disc. Also, only an Earth-like planet could be ejected; calculations performed by Gregory Laughlin and Fred C. Adams show that there is a very small, but finite, probability of Earth having such fate before our planet becomes uninhabitable for complex life, due to the runaway greenhouse effect in ~1 Ga, or before the Sun enters its asymptotic giant branch evolutionary phase and destroys our planet in ~6 Ga.[44])

But that is just the beginning of the story, since, if we take into account effects taking place after the formation of a habitable planet and abiogenesis on it, we may face a prototypical 'butterfly-effect' situation in which small local causes could have large global consequences. This is disturbing from the point of view of the consistency of GHZ, since the small probability of an initial occurrence could be compensated for by its large subsequent impact. If the ancient panspermia doctrine is valid in any of its modern reissues,[45] an inhabited planet could become an 'emitter' of life soon after abiogenesis—and, if that had happened quickly enough, large areas outside the GHZ could have been 'infected' before the present time. In addition, those 'infected' sites could become secondary 'emitters'—again, possibly beyond the original GHZ boundaries.

Therefore, the term 'Galactic Habitable Zone' should always be used with caution. The criticisms raised against it are certainly justified to some extent—but each can be answered in terms of theoretical and practical usefulness. Reification of terrestrial habitability is inevitable to some extent, since we have no empirical evidence and no serious theoretical model of a radically different form of life and hence no radically different concept of habitability. If and when the situation changes in this respect, we shall have reasons to generalize or even reject the GHZ as a concept. In the meantime, as with most other astrobiological research, we ought to bear in mind the limitation of the GHZ to concepts related to terrestrial life, namely the (bio)chemical basis required for such life, the necessity of terrestrial planets, the stability of conditions, and so on. While judging whether such a framework is too narrow or too wide is in the eye of the beholder, it is indisputable that this is a necessary foundation and grounding for *any* kind of astrobiological thinking. Even if we are the most radical proponents of alternative forms of life—silicon-based life or Sagan–Salpeter floating Jovian fauna—the only way in which we could speak of its specific properties is in contrast with the terrestrial life forms.[46] Therefore, the terrestrial habitability is still the *norm*, in its adequately weakened methodological sense; therefore, the GHZ is justified as a working concept.

The same applies to other lines of critical attack. While the area encompassed by the GHZ can never pretend to contain *all* possible habitats with conventional habitability, it still will contain *most* of such habitats. Essentially, it is a statistical concept, with exceptions and outliers acknowledged, as in any other statistical concept whatsoever. Thus, any hope of drawing *sharp* boundaries for the GHZ is illusory, but this does not demean the idea itself; after all, the fact that there is no sharp boundary for Earth's atmosphere does not demean the concept of 'atmosphere', nor does it impede research in atmospheric science or on the interaction of the atmosphere with interplanetary space. If we go one step further, and allow for the possibility of intelligent beings evolving into an advanced technological civilization and then colonizing other planetary systems, we are obviously in the realm of highly non-linear and unpredictable processes. (After all, if they were predictable, we would not be facing such uncertainty about Fermi's paradox and you would probably not be reading this book!)

An important review by Gonzalez in 2005 generalized various concepts of habitable zones into somewhat unified approach.[47] In accordance with the general physical desiderata, habitable zones have full four-dimensional spatio-temporal volume; alternatively, we can speak about *habitable eras* and reserve the term 'habitable zone' only for the 3-D region of space (as I shall do in the rest of this book). This tendency to generalize is an important signpost of the things to come, since further development of theoretical astrobiology could hardly proceed without gaining more quantitative and numerical precision, achievable only within a sufficiently broad space of model parameters.

We can conclude that the idea of the GHZ is a very useful and interesting theoretical concept—but one in the very early, almost embryonic stage of development. It requires an enormous amount of further work and elaboration. Clearly, it sets an important stage for more quantitative discussions of the emergence and evolution of life in a wider, Galactic context. In the same time, it should not be overrated or taken too literally. As a vehicle for intriguing research, it has served astrobiology excellently in more than a decade since its introduction; examples are novel and provocative studies on the habitable zones in early type galaxies and in M31.[48]

## 2.4 Horizons and Temporal Scales

Part of the confusion surrounding not only lay discussions but also some technical analyses of Fermi's paradox concerns the limitations imposed on biological and cultural evolution by cosmological temporal and spatial scales. One of the examples is the often quoted, allegedly sceptical, view that we are indeed alone *in the universe*. Unless we abandon naturalism with respect to the origin of life and intelligence (an unscientific and implausible move; see Section 3.2), it is simple to show that such a view *must* be false—irrespective of any further astrobiological knowledge. Modern cosmology postulates a spatially infinite universe which is uniform and homogeneous on large scales.[49] In such a universe, there are an infinite number of galaxies and an infinite number of stars of any given category, including an infinite number of Sun-like stars. Within this particular infinity of Sun-like stars, there is a fraction of them containing Earth-like planets, and again an infinity of them. Among infinite number of Earth-like planets, there must be some whose history and all other particulars are arbitrarily close to those of Earth. Even if this fraction is as small as $10^{-100}$, their number will still be infinite, so there must be infinity of biospheres in the universe. Again, even if the fraction of biospheres evolving human-level intelligence is extremely small, this would still mean infinite number of extraterrestrial intelligent species. This is just an application of Borel's ('infinite monkeys') theorem, the reasoning of which in a vague form has been known since the time of Epicurus and Lucretius, if not earlier.[50] This argument was explicitly endorsed by the great Russian physicist and human rights activist Andrei Dmitrievich Sakharov in the moving last paragraph of his Nobel Prize lecture:[51]

> In infinite space many civilizations are bound to exist, among them civilizations that are also wiser and more 'successful' than ours. I support the cosmological hypothesis which states that the development of the universe is repeated in its basic features an infinite number of times. In accordance with this, other civilizations, including more 'successful' ones, should exist an infinite number of times on the 'preceding' and the 'following' pages of the Book of the Universe. Yet this should not minimize our sacred endeavors in this world of ours, where, like faint glimmers of light in the dark, we have emerged for a moment from the nothingness of dark unconsciousness of material existence. We must make good the demands of reason and create a life worthy of ourselves and of the goals we only dimly perceive.

Note that Sakharov does not prejudice whether the infinity is temporal or spatial or both. He also appeals to a broad Copernicanism—to which I shall return in Chapter 3—in assuming diversity of civilizations. Although the great dissident physicist did not address Fermi's paradox specifically, his reasoning tightly resonates with many of the strands of hypotheses suggested for the explanation of the 'Big Puzzle'. This argument would be valid even in the case there were only one possible (in either physical or logical sense) type of intelligence—namely the human one. Of course, this would lead to a drastically decreased frequency of intelligent species, in comparison to the optimistic view that there are many possible forms of intelligence and many different evolutionary pathways leading to it. But again, no matter how small, a finite fraction of infinity gives again—infinity.

So, we are *certainly* not alone *in the entirety of the universe*. But a sceptical view can still be justified, at least in principle. What truly is of interest to us in astrobiology and SETI studies is the number of species and communities we can be in *causal contact* with. Therefore, we need inquiry about the causal structure of the universe on large scales in order to delineate the part of space–time containing possibly detectable intelligent species. This is the question of cosmological horizons. Horizons are, roughly speaking, surfaces separating those space–time points with which we can be in communication from those which the expansion of the universe makes inaccessible. Depending whether this inaccessibility pertains to a particular epoch or to all times, we can talk about different kinds of horizons. For our present use, it is relevant that our realistic universe, described reasonably well by the λCDM paradigm, possesses both the particle horizon (which is applicable to a particular epoch) and the event horizon (which is fixed for all times by the dark energy magnitude λ).[52]

The Hungarian astronomer and SETI expert Ivan Almár has explicitly discussed the analogy between Fermi's paradox and Olbers's paradox.[53] The latter is also linked to a simple, dramatic question which challenges some of our deepest assumptions about the world. In Fermi's case, it was *Where is everybody?* while in Olbers's case it was *Why is the night sky dark?* Why, indeed, is the night sky dark when, on the basis of naive Newtonian assumptions about a static and infinite universe, it should be as bright as stellar surfaces? In other words, why is there such a huge contrast between bright stars (and other sources of light) and the darkness of most of the universe, in the face of the strong thermodynamical tendency of all systems to equilibrate in a finite—and usually short—relaxation time?[54] In modern cosmology, this is resolved through the finite age of sources; after all, the entire universe is 'only' 13.7 Ga old, and most sources are considerably younger. There simply has been no time for the hot sources to warm up the cold depths of (intergalactic) space. A minor contribution comes from the expansion of the universe and the corresponding cosmological red shift; but the finite temporal duration is the main explanation.[55] The analogy with Fermi's paradox is rather clear: why do we perceive the 'Great Silence' when life—and especially intelligent life—tends to spread, communicate, build artefacts, and fill the universe with traces of its existence?

One possible sceptical complaint can be rejected straightforwardly: the fact that there are many known stars and just one species of intelligent life makes no difference as long as we subscribe to naturalism about life and intelligence. After all, the number of stars known—or indeed *observable*—in the time of Olbers was many orders of magnitude smaller than the number of stars necessary for the counterfactual paradoxical conclusion to arise. And, on the other hand, as mentioned in Chapter 1 and as will be argued in more detail in subsequent chapters, a *single* extraterrestrial civilization high enough on the Kardashev scale is a better SETI target than a million or more present-day human-level civilizations. A simple (counterfactual) Kardashev's Type 3 civilization in the Milky Way would be no less obvious than the (counterfactual) *brightness* of the night sky.[56] So, the numbers do not make the difference.

How about responses analogous to those addressing Olbers's paradox? The expansion of the universe is unlikely to play a major role, unless the density of life- and intelligence-bearing sites is excessively low, something which would require an explanation in the first place! There is a version of this idea, to be considered under the 'rare Earth' heading in Chapter 5. Remarkably, in this case, as well as for Olbers's paradox, Hubble expansion plays just a subsidiary role to the *main* explanatory mechanism. The main role must be reserved for something that suppresses the emergence of detectable intelligent beings in the Milky Way as well. Is it the finite age of the sources (understood, in a generalized manner, as *sources of detectable emissions*)? Inhabited sites must be younger than the structure in the universe, so there is clearly such a limit. One might still argue that so many requirements for the emergence of intelligent species further bring the 'starting point' of detectable sources closer to the present, thus relaxing the problem somewhat.

In the aftermath of Lineweaver's results in Eq. (1.3), however, this proviso has become largely irrelevant. The role of the light-travel time from the sources in Olbers's paradox is now played by the Fermi–Hart timescale in Eq. (1.1), which is so much smaller than the Lineweaver timescale (1.4) that it is justified to treat the signal as eventually reaching 'anywhere'; of course, this is even more pronounced if we, following the idea of **StrongFP**, treat the manifestations and traces of extraterrestrial astroengineering as *signals* propagating with the speed of light. So, there is indeed an analogy and, as much as Olbers's paradox has played an important role in the development of our cosmological views, Fermi's paradox has the potential to play a similar role for astrobiology and SETI studies.[57] I shall continuously follow this strand of thought when exposing different explanatory hypotheses in Chapters 4–7. Before that, we need to consider another important concept bearing on the relationship of observers and the universe at large: extragalactic SETI.[58]

The cosmological horizon represents the largest spatial scale on which any two observers could, in principle, communicate. Thus, it represents the absolute outermost limit of any SETI survey. The cosmological horizon conceptually and absolutely limits the *extragalactic SETI*. A devil's advocate may bring into question the very point of extragalactic searches, when searches much closer to home, within the Milky Way, are fraught with so many problems and difficulties. Extragalactic distances are orders of magnitude larger than the interstellar distances within our Galaxy, so that the flux received by our detectors is twice as many orders of magnitude smaller. Depending on the method used (mainly the choice of waveband in the electromagnetic spectrum), the signal-to-noise ratio is, of necessity, minuscule, and especially if we expect a signal varying in time on the scales of human observations—and, indeed, human lifetime!—positive detection might be plainly impossible. If one deems the prospects of success of conventional (= *intraGalactic*) SETI faint, the prospects of finding something at megaparsec-scale distances must be considered negligible indeed.

This certainly is not the whole story, however, because there are circumstances *favourable* to extragalactic SETI. The most important of them is the fact that in the extragalactic observations we could sample a much larger volume of space. By looking at many external galaxies instead of the single Milky Way, we are sampling orders of magnitude more potential sites of advanced civilizations. In principle, the gain in sampled volume more than compensates for the loss in signal strength (i.e. received flux), since the sampled volume increases with increasing distance $R$ as $R^3$, while the received flux decreases as $R^{-2}$. (At least these Euclidean approximations would be valid at distances up to a few times 100 Mpc, before effects such as dimming of sources due to the accelerated expansion of the universe become important.) Therefore, if SETI targets were, on average, homogeneously distributed in space, time, and intrinsic power, it would obviously be advantageous to go as far out in the extragalactic space as possible.

In practice, however, things are different because there are clear instrumental detection thresholds, and for most conceivable methods of search these are too high for our present and near-future equipment. The *price* of a detecting instrument in a realistic human economy rises much more steeply with the decrease in threshold sensitivity necessary for registering extremely faint signals. In astronomical practice, that would mean larger telescopes (radio, optical, or infrared ones), preferably located in space, and even more preferably at large distances from all terrestrial sources of noise, for example, on the dark side of the Moon—all of which means tremendous costs and requires constant funding at levels unlikely to be available to practical SETI in the near future.

One particular class of solutions to Fermi's paradox, the one dealing with 'rare Earths' (on whose structural details more in Chapter 5), relies in part on the existence of cosmological horizons. This is manifested in two different ways. Firstly, as emphasized by Canadian astronomer Paul Wesson, whose

## A Scenario for Extragalactic SETI

The increased volume of sampled space, remains an important motivation in a particular class of SETI searches, in which one seeks for a particularly energetic instance of astroengineering, detectable over large distances.[59] For example, suppose that we search for civilizations capable of building Dyson shells, but those are in fact (unknown to us) quite rare: there is only 1 in a $10^9$ chance of building one in an $L^*$ galaxy like the Milky Way *per million years*. Let us also suppose—not implausibly—that each Dyson shell built has a very long lifetime, comparable to the Hubble time. Essentially, once it is built, it just stays there forever. It is straightforward to show that, in this simplistic scenario, we would have a probability of only about 1 per cent of having a *single* possible target for SETI searches in the entire Milky Way. On the other hand, if a similar scaling relation holds for all galaxies with luminosities larger than 0.1 $L^*$, irrespectively of morphological type and other particularities, the expected number of targets in the cube with a side of 50 Mpc centred on our Galaxy is about 16. This is obtained by integrating the rate of Dyson shell emergence over the galaxy luminosity function.[60] I have restricted luminosities to $L > 0.1\ L^*$ in order to avoid dwarf galaxies which are mostly chemically unevolved and hence lack the prerequisites for life and intelligence; see the discussion of GHZ above. The emergence of Dyson shells in this example is a Poisson process, starting with the beginning of GHZ about 10 Ga ago and neglecting the look-back time effects.

Obviously, in such an admittedly contrived scenario, *intragalactic* SETI does not make much sense, in contrast to the *extragalactic* one! The realistic situation is certainly somewhere between this extreme and the total rejection of extragalactic SETI which has been the dominant view until very recently.

work has incidentally been instrumental in our modern view of Olbers's paradox as well, the average density of observers might be so low that the number of galaxies containing even a single technological civilization within our cosmological horizon is unity or at least some very small number.[61] Such rare civilizations would clearly be undetectable by means of present or close future, even if some of them have already achieved Kardashev's Type 3 status. Note that this is somewhat philosophically appealing, since the total number of civilizations might still be *infinite* in any particular moment of time (starting from some early time), provided that the universe is spatially infinite, as in open or flat Friedmann models. However, this infinity would not produce any observable and a fortiori no paradoxical properties.

In addition, by looking towards distant galaxies, we are looking in the *past*. The particle horizon is, *very roughly*, the set of points we are seeing at the time of the initial Big Bang singularity (or whatever masquerades for singularity in modern particle/cosmological theories). Thus, we are seeing galaxies in their earlier stages of evolutionary development—and this applies to morphological, chemical, or *astrobiological* evolution. For instance, we observe M31 as it was roughly 2 Ma ago, which is a minuscule interval of time in the cosmological context. Still, it is enough that observers in M31, even if they were in possession of instrumentation with *infinite sensitivity*, would not have been able to detect *Homo sapiens*, which is only 200–300 Ka old. A hypothetical Andromedan would have been able to detect only those intelligent species in the Milky Way which are more than 2 Ma old *right now*, from our present perspective. Thus, our exact analogues (and anything else younger than the margin of 2 Ma) would be missed in the survey.[62] The simplest—and naive—view of astrobiological evolution would suggest that, the deeper we go into the past, the more we are likely to have a smaller number of complex biospheres and, consequently, a smaller number of technological civilizations (hence, SETI targets) in the overall astrobiological landscape.

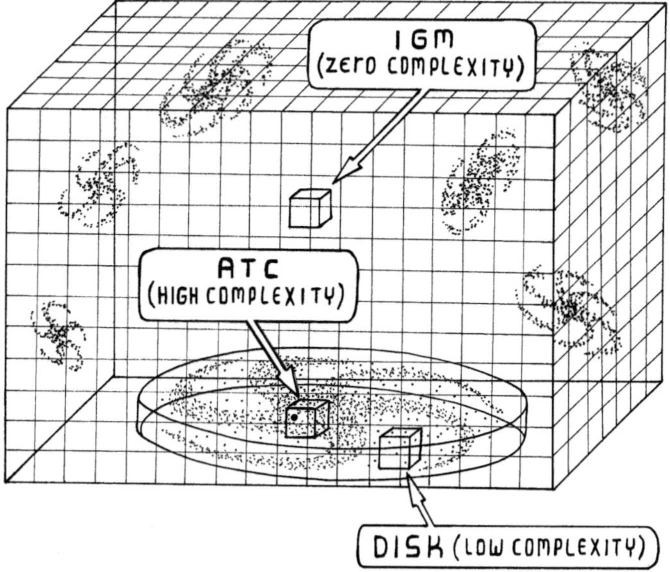


**Figure 2.5** A simple presentation of the three 'regimes' of astrobiological complexity we can expect to encounter, in principle, on the background of existing cosmological history and structures; ATC, advanced technological civilization; IGM, intergalactic matter.

*Source*: Courtesy of Slobodan Popović Bagi

While such a conclusion might not hold in general, since the Lineweaver timescale in (2.4) is sufficient for many 'cycles' of rises and falls of civilizations in any single galaxy, the loss in temporal 'depth' becomes more and more pronounced as we go further out to more and more distant galaxies. For targets whose distance is much smaller than $\tau_L c$, we still have ample theoretical reasons to engage in extragalactic SETI—even if we might not possess satisfactory observational resources, sensitivity-wise. There are many millions of such potential targets out there.

As an appropriate conclusion to this background chapter, in Figure 2.5 I schematically present the discrete distribution of expected astrobiological complexity. While the exact definition of such complexity might elude us for some time to come, an intuitive understanding is quite clear: the 'dead space' outside of cosmological structures (mainly galaxies) is filled by extremely rarefied intergalactic matter (IGM), is obviously uninhabitable and has zero astrobiological complexity. While IGM can demonstrate some interesting physics, it is essentially a very simple system, describable through a small number of physical parameters. On the exact opposite on the complexity scale are hypothetical advanced technological civilizations, which might be spatially quite compact (although it is of course not necessarily) and minuscule in cosmological terms, but have very high density of information processing and require a huge list of parameters to be described in any detail. In between these two extremes are those regions of galactic thin discs which contain habitable planets and, perhaps, biospheres like the terrestrial one. These are still rather low on the complexity scale, since the intentional design space accessible to advanced technological actors is likely to dwarf the morphological space of conventional, naturalistic, Darwinian evolution. The figure is drawn on purpose to suggest that conventional discrete analysis in terms of lattices and cells—including methods such as cellular automata or multigrids—should be applicable to astrobiological complexity as well.

CHAPTER 3

# Speaking Prose

*Realism, Naturalism, Copernicanism, Non-Exclusivity*

## 3.1 Why Philosophy?

Molière's famous satire *Le Bourgeois Gentilhomme* (*The Bourgeois Gentleman*) was first performed on 14 October 1670 for the court of Louis XIV at the equally iconic Château de Chambord, whose architectural and *artifactual* complexity, allegedly designed by Leonardo, could serve as a fitting metaphor of the astrobiological complexity we are seeking for in the SETI context. The comedy contains the following charming dialogue between a highly aspiring, but not very sophisticated, protagonist and his philosophy teacher:[1]

> MONSIEUR JOURDAIN: There is nothing but prose or verse?
>
> PHILOSOPHY MASTER: No, Sir, everything that is not prose is verse, and everything that is not verse is prose.
>
> MONSIEUR JOURDAIN: And when one speaks, what is that then?
>
> PHILOSOPHY MASTER: Prose.
>
> MONSIEUR JOURDAIN: What! When I say, 'Nicole, bring me my slippers, and give me my nightcap', that's prose?
>
> PHILOSOPHY MASTER: Yes, Sir.
>
> MONSIEUR JOURDAIN: By my faith! For more than forty years I have been speaking prose without knowing anything about it, and I am much obliged to you for having taught me that.

This small gem should be read and reread each time any scientific (or religious or political, etc.) vigilante tries to put down or sneers at philosophical analysis of any problem under discussion. In contrast to the conventional—and wrong—understanding of this scene, while speaking in prose is in itself trivial, the insight itself is highly non-trivial, especially in the context of education and outreach.[2] Neither is the insight that intelligent beings are doing many things without being aware of their place in sophisticated classification schemes. As it happens, vigilantes are in the right *sometimes*, and some philosophical analyses are tantamount to inventing fancy words for well-known concepts and clearly justified procedures. More often, though, especially when 'big issues' are concerned (problems like the origin of life and intelligence; universe or multiverse; Fermi's paradox), vigilantes manage to sound ridiculous like Monsieur Jourdain by failing to position their thoughts and actions in a wider scheme of things. After all, M. Jourdain needs help (the wider context in which the cited passage is embedded) with writing a love letter—something clearly of high importance and complexity in his life and social aspirations.

A philosophical perspective is not optional when young fields like astrobiology and SETI are concerned; it is unavoidable.[3] Some of the recent alleged quarrels between science and philosophy have been singularly unproductive, as always throughout the history of ideas. It is quite obvious that, even if one were to be prejudiced against philosophy—and here and elsewhere in this book—it is serious, analytic philosophy which is considered, not some amorphous mass of 'perspectives' and 'narratives' which is peddled in postmodern circles under the same name—one does it at the peril of being a 'philosopher in spite himself', in the wise words of the Nobel laureate Frank Wilczek, one of the greatest living physicists.[4] Additional impetus for philosophical discussion comes from the simple fact that, in *any* scientific discipline, as the history of science teaches us, philosophical considerations (often disguised as aesthetical or ethical judgements, bold metaphors, or 'leaps of imagination') have always played a key role when a discipline is young. This was the case with chemistry in the time of Lavoisier, geosciences in the time of Hutton, psychology in the time of Wundt, and cosmology in the time of Friedmann and Lemaître. Subsequently, those things might fade in the background—as they did in cosmology, for instance, after the 'great controversy' between the standard relativistic cosmology and the steady-state alternative was over in the mid-1960s—and become remote from everyday work of practising researchers, but a crucial role they did play, as even a cursory non-myopic insight into the basic tenets and concepts of the field will tell.

This applies with obvious force to astrobiology and SETI studies. One might go further and claim that the whole idea of unifying various strands of research and discovery into something which is clearly recognizable as the 'astrobiological revolution' since the mid-1990s is essentially and profoundly philosophical. There is absolutely nothing wrong in that: on the contrary, the fruitfulness of such a synthetic endeavour clearly testifies that the time for picking up the Voltairean torch has not entirely passed. Since the very beginning, the search for life elsewhere has had an important epistemological content: the problem of recognizing life which evolved in a sufficiently different environment in comparison to the terrestrial one is obviously something which cannot be resolved on purely empirical level, with our present and near-future insight into fundamental biological processes. The same applies to more complex questions about general astrobiological complexity and its evolution. To an even larger degree, it applies to concerns which are encountered specifically in SETI studies and which I shall discuss in the rest of this chapter.[5]

Although it is clear that philosophical issues are necessary in any discussion of the question of life and intelligence elsewhere in the universe, there is a well-delineated part of philosophical baggage which we shall leave at the entrance. Part of it is the misleading insistence on the definitional issues. To precisely define either life or intelligence is impossible at present, as readily admitted by almost all biologists and cognitive scientists. This, however, hardly prevents any of them from doing so in their daily research activities or even in synthetic thinking about the cutting-edge problems in their fields. An evolutionary biologist making models of predator–prey co-evolution in the history of life will not stop her work—nor, perhaps, lose much sleep—over the fact that she could not, if hard pressed, tell *exactly* what those entities she is studying are. Neither will a cognitive psychologist doing research on visual perception be often tempted to switch to banking because, ultimately, we do not have a strict definition of perception or any number of other mental phenomena.[6] They intuitively understand what the history of science has amply demonstrated so far, namely that formalizations and definitions come *after* most of the research in a field is done and not before. The deeply misguided idea of 'definitions first!' is a relic of the epoch of logical positivism and its explicit or implicit faith in inductions and formalizations; such an approach characterizes the pre-Gödelian and pre-Popperian thinking about science and truth. The more universal and ubiquitous a concept is, the tougher is the definitional challenge. Most prominently, the intuitive concept of *a number* has enabled fruitful research in mathematics for millennia before the advent of modern set theory as the axiomatic foundation for modern mathematics finally enabled general formal definition of number (by personalities such as Frege, Russell, Whitehead, Gödel, Turing, Church,

Kleene, and Post).[7] There are many other examples of the formalization of paradigms (including precise definitions) occurring only at later stages of mature disciplines hidden in the treasure troves of the history of science.[8]

There is no reason to doubt that astrobiology and SETI studies will conform to the same general picture. If anything, the 'exotic' (in the vernacular sense) nature of the field will make caution in formalization more rather than less pronounced. The real issue is to what extent we can achieve an *intuitive grasp of problems in the field*, which is, of course, a prerequisite for successful research, something very different from having a formal definition of any kind. By the same token, the lack of formal definitions will not impede or obstruct research in these fields, as the lack of formal definition of life did not impede or obstruct research in evolutionary biology, nor has the definitional ambiguity about the definition of the universe impeded or obstructed research in cosmology. The powerful and widely present desire to castigate SETI as 'definition challenged', therefore, tells us more about deeper and irrational animosities towards SETI than about the enterprise itself.

The fact that SETI has recently celebrated its half-centennial is a red herring in this respect, similar to the oft-repeated and flattering[9]—but, strictly speaking, *wrong*!—claim that ancient druids, the builders of Stonehenge and other megalithic monuments, were engaged in *astronomy*. They did possess important astronomical knowledge, and were possibly even able to predict particular astronomical phenomena like the rising and setting of celestial bodies of cultural importance for them. Astronomical knowledge is just a necessary, not a sufficient condition for engaging in astronomy as a research discipline, however. Systematics, theoretical framework, the concept of causes, and the desire to understand them, all that is the source of the enterprise we call science—including astronomy—and can be dated back only to the time of the late Renaissance, notably Copernicus, Tycho, Vesalius, Kepler, and, above all, Galileo. The same applies to SETI: *something* has been around since the late 1950s, but it was not sufficiently well grounded until much later.

One can go further and claim that the real SETI has hardly started (which explains, in more than a formal sense, and no pun intended, why it hasn't been successful thus far).[10] It is exactly this boundary state of flux which makes it more interesting and challenging in these days of the astrobiological revolution. So, on the definitional issue, there is no discernible reason why we should take a different approach in astrobiology and SETI studies than what has been clearly demonstrated in the history of other scientific disciplines. *Intuitive* concepts of life and intelligence are developed enough to enable fruitful research in these fields, in the same manner as the intuitive concept of life enables research in the terrestrial biology and other life sciences. Even more pertinently, the intuitive recognition of cultural properties such as advanced technology, willingness to communicate, xenophobia/xenophilia, and so on, is necessary for formulating SETI activities, although, like the scaffolding, it can be discarded at a later stage and in the face of better understanding.

It is clear, for instance, that the Darwinian evolution on Earth brought about, at best, a few intelligent species[11] and only one with technological capacities for engaging in SETI and similar cosmic activities. In this context, a precise definition of intelligent species (much less a conscious one; see the disturbing comments of people like Julian Jaynes, David Raup, Tor Nørretranders, and Christof Koch, showing that consciousness is, in any case, much less than what is colloquially presumed[12]) is unnecessary; while the awareness that this might be radically different in the SETI context is desirable, we need to proceed along the same, broadly operationalist lines. For this reason, we shall use the terms 'extraterrestrial intelligence', 'intelligent beings', 'sophonts', and so on, in their non-technical or vernacular meanings, roughly as placeholders for beings we are interested in meaningfully interacting with. David Lamb condensed this pragmatic approach as follows:[13]

> What do we normally mean when we speak of 'intelligence'? It is difficult to define, but should include several of the following: faculties for reasoning, creativity, inventiveness, imagination, foresight, reflection, an aptitude for learning and problem-solving and some communicative abilities. These should be employed in a manner which indicates a degree of coordination. In this respect it is a mistake to start from a definition of intelligence; rather, it is best to indicate what it actually does, and what it actually requires in order to function.

This intuitive understanding is valid even if a specific interaction is deemed undesirable on safety, moral, or some other grounds; it is also valid if the interaction is purely one-sided or 'archaeological', that is, if we discover traces and remnants of an ancient technological civilization which has gone extinct by now. In all these cases, there seems to be no doubt about the *intelligent* aspect of beings under consideration, either existent or non-existent at present, either friendly, indifferent, or hostile, and so on.

In the rest of this chapter, I shall consider four all-important philosophical assumptions built (usually tacitly or even unconsciously) into all formulations of Fermi's paradox. After that, I shall consider the key methodological principle—non-exclusivity, as outlined in Section 1.5, which is essential for evaluating solutions. Finally, several auxiliary issues surrounding any relevant discussion of the problem are briefly explored, notably the impact of postbiological evolution and the relevance (or else) of the notorious Drake equation.

## 3.2 Philosophical Naturalism

It is not necessary to enter into voluminous debates about metaphysical naturalism here.[14] It is enough to use a greatly reduced version of the philosophical doctrine of *methodological naturalism*, which tells us that we should not invoke supernatural agencies and capacities in searching for an explanation for observed phenomena. This clearly does not presume any attitude towards the existence of such supernatural agencies, only that our explanatory mechanisms do not invoke them either explicitly or implicitly. Emerging from Enlightenment secular thought, methodological naturalism is best encapsulated in Laplace's famous sentence directed to the First Consul of the French Republic, Napoleon Bonaparte: *Je n'avais pas besoin de cette hypothèse-là*.[15]

Obviously, methodological naturalism is the basis of all science, the very bread and butter of both everyday research and our grand theoretical syntheses. It has been immensely successful and has not encountered any significant obstacle so far. (Here, I neglect the ideological parroting of creationists and ID-ers, as it is devoid of any interesting content.) While this in itself does not imply that it will continue to be successful in the future—that would be a bad form of inductivism—we certainly do not have any reason to doubt it as a *working hypothesis* in our considerations of Fermi's paradox.

The successes of science since the so-called Scientific Revolution of the seventeenth century have led to a world view which could be called *naturalistic*, since it assumes the absence of supernatural forces and influences on the phenomena science is dealing with.[16] Here, as in the case of intelligence or detectability, I am using a rough, non-technical definition which is entirely sufficient for meaningful discussion.

There have been some historically relevant exceptions to this. For instance, Alfred Russell Wallace, co-discoverer (with Darwin) of natural selection, which is, in several regards, a precursor to contemporary astrobiology, famously argued that, while natural selection can produce the huge diversity of microbial, plant, and animal life we perceive in the terrestrial biosphere, it is insufficient for producing the human mind and its more complex characters.[17] He could see no selective advantages in stuff such as a capability for abstract mathematics, for aesthetic appreciation of music, or for inquiring into the origin of the world (if anything, some of those could be construed, Monty Python-like, into *dis*advantages!). What Wallace wrote about humans could, in principle, be generalized for discussing other intelligent species in the universe; the fact that he himself argued against the existence of extraterrestrial life is irrelevant in this respect.

While does not commit us to *metaphysical* (or ontological) naturalism which rejects the existence of supernatural agencies and capacities, there is one point of importance to keep in mind, especially when we discuss intentional or cultural artefacts in the context of Fermi's paradox. Metaphysical naturalism is, from a modern perspective, closely associated with physicalism, the doctrine which rejects non-physical causes of physical effects (especially in the philosophy of mind, but it is applicable more generally). Our discussion of the possible effects of extraterrestrial intelligence on the cosmological environment will be entirely physicalist in letter and spirit, but this still does not commit us to metaphysical naturalism. Any reader surviving into the next chapter will encounter an eminent *supernatural* (in metaphysical terms) hypothesis in Section 4.3, together with some even more interesting counterexamples in Sections 4.5 and 4.7.

Even modest methodological naturalism can lead us into subtleties and difficulties concerning, for example, the role of philosophical inquiry in formulating synthetic hypotheses about the world.[18] In the context of Fermi's paradox, this might be interesting, since many of the proposed solutions are, indeed, located at the crossroads of science and philosophy, and some of them could be regarded primarily as philosophical hypotheses. In this matter, I propose to 'bite the bullet' and accept the idea that philosophy can generate fruitful synthetic hypotheses about the world, to be tested by subsequent expansion of our epistemic capacities and novel empirical data.

A possibly relevant subtlety lies in assuming that the dichotomy between natural and supernatural applies to the 'internal' perspective of observers located within a cosmological domain, or 'a universe'. The distinction is important for those discussions which do not question naturalism per se, but accept different construals of 'natural' depending on an 'internal' versus an 'external' perspective. For example, Max Tegmark argues in several papers for a radical ontology in which all possible mathematical structures are realized in some part of the (obviously huge) multiverse. Some parts, like ours, based on structures such as the Hilbert spaces of standard quantum mechanics, support life and observers, which Tegmark calls 'self-aware substructures'. He emphasizes that there is a key difference between internal and external perspectives, the former being what we as substructures are condemned to and the latter being a 'Platonic' view from outside of the realm of any particular low-energy, local physics.[19]

Obviously, from an internal perspective, the agencies and capacities from the realm beyond one's own—possibly other evolved self-aware substructures, but now substructures of the *larger structure*—will not be limited by the local laws of physics, and thus, technically, could qualify as supernatural. Of course, this does not entail any *real* violation of the laws of physics at the highest level, that of Tegmark's 'ultimate ensemble theory'. However, it might *look* to any number of internal observers as if the laws of their local physics were violated. Although Tegmark's scenario is extreme, it does convey a general and important lesson that what counts as naturalist explanation—and looks like 'pure methodology' at first glance—comes with usually tacit ontological commitments, especially where our best current physical theories (such as string theory and inflationary cosmology) are concerned. This is an important lesson for considering some of the weirder candidates for the resolution of Fermi's paradox, especially in Chapter 4.

Of course, we could easily have both supernaturalism and Fermi's paradox. Old pluralist views of the Creator endowing many stars with inhabited planets—or even, as William Herschel believed, inhabited *stars*—and not having any particular preferences in the multitude of His/Her creations, face perhaps even stronger forms of the paradox than do any naturalistic views discussed in this book.[20] But that would violate an even more fundamental guideline: usefulness. Any solution to Fermi's question found in a completely naturalistic picture would still be valid if we allow for such benign (relatively to the problem at hand) form of supernaturalism. The only reason one would posit supernaturalism in the context of the present debate would be the belief that the supernatural Creator ordered things in such way that the paradoxical conclusion does not follow. For instance, She/He could have created only one intelligent species, since that was the purpose of the whole Creation, as Western monotheistic traditions have held for most of their histories (cf. Section 4.3). Perhaps distant stars and their planets we discover nowadays do not 'really' exist, but are just optical illusions, as in a giant planetarium; perhaps they exist and are uninhabited, since the Creator did not provide them with the necessary 'spark of life'; perhaps they exist and are inhabited by different living creatures, none of whom possess the 'spark of intelligence'; and so on. Ironically—and that does show the intellectual bankruptcy of such views—nowadays we *don't even need supernaturalism to conceive such a scenario*! We shall encounter some versions of *naturalistic design* in Chapter 4 as the **PLANETARIUM/SIMULATION** hypotheses. Therefore, supernaturalism is effectively a non-starter in the Fermi's paradox-related discussions. The very best one can do is simply to reject it.[21]

## 3.3 Scientific Realism

Realism has, as a term, been so widely used in the history of philosophy, denoting so many different things, that many openly refrain from its further (ab)use.

Luckily enough, what we need here is not any deep meaning or general epistemological construal, but the rather mundane observation that we achieve best results in most cases by following empirical evidence and interpreting it in the most direct and simple manner, not just in the scientific enterprise of explaining and predicting, but in other fields like sports, banking, or sex. Of course that there are many examples of 'our eyes failing us' in everyday life, in science, and in the arts, but their measure in the overall set of observations is quite small, at least until we reach very complex domains of theories such as quantum mechanics or general relativity (or works of artists at least partially inspired by those theories, such as M. C. Escher). In other words, in what follows, I shall follow the Quinean recipe of using scientific knowledge as a yardstick for realism: what is real is that which has been established by using empirical science and, in particular, by using methods such as observation and experiment. Thus, Uranus, inosilicates, *Pseudoceros dimidiatus*, and the Osaka/Kyoto airport are undoubtedly real, while unicorns, ghosts, black magic, and round squares are not. Many other things might be real, but we still do not know for sure; among such things are, of course, extraterrestrial intelligent beings. Note that this is an extremely weak understanding of realism, not referring at all to purely theoretical or abstract entities, making no commitment to the usual-suspect chestnuts such as mathematical objects or universals, and making no claims at all about theories and their models. Rather, it is related to the aspirational attempts to define science: science aims to produce true descriptions of things in the world.[22]

Realism in this weak sense is so central to the entire post-Renaissance scientific endeavour that there are very few cases in which it has been seriously questioned. And, even in such cases, as in some more extreme subjective versions of the Copenhagen interpretation of quantum mechanics, it was not that the existence or validity of the empirical results obtained in the course of research was denied, but only our construal of the underlying physical reality.

In contrast, non-realism in the context of Fermi's paradox requires that we reject the validity of a large amount of empirical data, amassed through decades or even centuries of careful work in observational astronomy.

Some recent work surrounding the debates about scientific realism might, in an indirect way, be relevant for astrobiology and SETI studies. For example, Michael Friedman and Philip Kitcher have argued that the project of unification of various phenomena under a few relatively simple explanatory theories, successful as it has been thus far, justifies scientific realism.[23] If the reality of electricity and magnetism were much different from our scientific accounts, the fact that a single classical theory, namely Maxwell's electrodynamics, successfully explains thunderstorms, the workings of electric trams, the behaviour of the compass needle, and magnetic reconnection in the Solar corona would be quite surprising and unexpected. In so far as astrobiology is another 'unification' project, and there are indeed many reasons to believe this is so,[24] it might give another reason for accepting those more fundamental doctrines of realism. It is an optional benefit, however. For practical purposes, and especially for the purpose of analysing and evaluating possible solutions to Fermi's paradox, what we need is much weaker form of the same underlying idea. Still, it is too strong for some, and in Chapter 4 we shall see that some are willing to part with it in order to obtain—at steep price, to prejudice things a bit—a workable solution.[25]

## 3.4 Copernicanism

Copernicanism is usually the only philosophical element explicitly present in discussions of Fermi's paradox; that this is a dangerous oversimplification should not derail us from perceiving its central role in the paradox. Copernicanism is sometimes referred to as the 'Principle of Mediocrity' or something similar. This is misleading, and not only for the semantic reason that 'mediocre' is usually defined and understood to mean 'of moderate or low quality',[26] which has, obviously, nothing to do with the astrobiological context. An even bigger problem is that it presupposes the ordering of elements in some well-defined way, which, even if possible, is certainly not *obvious*. The 'Principle of Typicality' would, perhaps, be a better designation, and it, indeed, got some currency in the recent quantum and string cosmologies.[27]

If we accept Copernicanism, then, within reasonable temporal and physical constraints, we expect the status of biological evolution on Earth to reflect the Galactic average for the given age of our habitat. We do not expect—unless we obtain specific reasons to the contrary!—that we have evolved exceptionally early or exceptionally late *in the interval within which evolution of intelligence is physically possible.*[28] We can also formulate this in an equivalent manner as follows: the timescale for the evolution of intelligence on Earth is close to the median of the distribution of physically possible timescales for the evolution of intelligence anywhere in the universe. This is clearly controversial, as the debate surrounding so-called Carter's anthropic argument against SETI testifies.[29] In my view, the constraint about *physical possibility* is actually sufficient to reject the general pessimistic conclusion of Carter's argument, but there is no need to enter into that complex debate here. Instead, we just note that Copernicanism offers not only potentially testable hypotheses, but whole research programmes whose time is clearly yet to come. So much is admitted even by researchers who are moderately sceptical towards the (over)use of Copernicanism, such as Caleb Scharf in his recent, popular book *The Copernicus Complex.*[30]

In the same manner, we have no reason to believe that the current astrobiological status of the Milky Way—whatever it may be—does not reflect accurately the average astrobiological status of the current universe, or at least its habitable subset.[31] This is exactly the rationale for the assumption (widely used in the orthodox SETI since the pioneering ideas of Iosif Shklovsky and Carl Sagan) that most of the members of the hypothetical 'Galactic Club' of communicating civilizations are significantly older than ours.[32] This distribution reflects the underlying distribution of the ages of the terrestrial planets, coupled with a simple observation-selection effect: even slightly younger civilizations will not be detectable from interstellar distances, and the older ones—if they do not go extinct—will be. Since our own civilization has essentially just appeared on the cosmic scene, it would make no sense whatsoever to search for a younger civilization (at present, though; things may become very different if humanity survives as a technological civilization for any astronomically relevant amount of time, like a few million years).

Now, the intuitive impression may not accurately represent the quantitative reality. *Exactly how much* older than humanity do we expect an average Galactic civilization to be? Just Copernicanism will not be able to tell us that, and neither will any other '-ism'—we need hard empirical data at this point. This question, together with similar ones (e.g. 'Is the Sun a typical star?'), clearly defines an important research program.[33] The magnitude of the age difference has been, however, constantly underestimated even before Lineweaver's results became available in 2001. The orthodox SETI literature does not discuss age differences on the order of billion years, which is indicative of the optimistic bias on part of the authors. There have been honourable exceptions, of course. Most notably, Nikolay Kardashev wrote more than 30 years ago what can almost verbatim be repeated today:[34]

> Most experimental searches for extraterrestrial civilizations proceed from a position of 'Terrestrial Chauvinism'. Thus, in spite of criticism that the probability of finding a civilization at our level of development and—moreover—among the nearest stars is in fact close to zero, the search for Earth-type civilizations is continuing. The solution of the problem has not and will not be advanced until the initial concepts and therefore the search strategies are changed...Extraterrestrial civilizations have not yet been found, because in effect they have not yet been searched for.

(Note that what Kardashev calls 'initial concepts' are exactly *philosophical* items like those we are discussing in this chapter!) This testifies, indirectly, how poorly Copernicanism has been analysed in the astrobiological context so far—or even not taken seriously enough. Leaving aside the issue of (im)practicality of any specific SETI activity in the case of giga-anni-old targets for the moment, the very fact that the first naive SETI 'models' of the founding-fathers epoch did not take this option into consideration (bluntly assuming that our civilization is close to typical instead), testifies that, under a wide range of conditions, Copernicanism will be unthinkingly oversimplified. On the opposite extreme, contemporary 'rare-Earth' theorists often unreflexively ignore or downplay Copernicanism, even when the level of our empirical understanding of the issues involved is obviously low and there is no reason whatsoever to doubt the Copernican assumption in advance. So we need to navigate between the Scylla of naiveté and the Charybdis of ignoring the issue.

It is important to separate the considerations of preference or even expediency from our analysis of what could we reasonably expect to be typical. Obviously, not all things we could, in principle, see are typical. As discussed in Chapter 1, the fact that we observe no supercivilizations (e.g. of Kardashev's Type 3) in the Milky Way in spite of ample time for their emergence is prima facie easiest to explain by postulating the vanishing probability of their existence in general. On the other hand, if, for some hitherto unknown reason, the process of emergence of such supercivilizations lasts much longer, comparable to the Hubble time, than it is entirely Copernican to conclude that they are commonplace *averaged over all times*. Some of the small civilizations existing now—humanity, for instance—could reasonably hope to become a supercivilization in the future and to stay in that state for a very long time.

An obvious pitfall for applications of Copernicanism in science in general, and astrobiology in particular, is to assume—explicitly or, almost always, tacitly—that the underlying distribution in parameter space is normal or Gaussian. This is usually wrong for some physical reason; for instance, the distribution of ages of terrestrial planets cannot be normal, since they are obviously bounded from above by the age of the thin disc, as discussed in Chapter 2.[35] How big a mistake one makes in *assuming* a normal distribution around some mean value is, however, unclear in the general case. Unfortunately, in this unfamiliar context, the outliers are likely to be 'black swans': rare occurrences with tremendous and unforeseeable impact.[36]

Of course, Copernicanism is just a principle: it cannot on its own ('in isolation') do the explanatory work for us. It needs to be coupled to correct empirical knowledge and theoretical ideas about the world. If it is coupled to incorrect ideas—or false empirical results—it will not help us and may just additionally confuse things. A good example of such wrong-headed Copernicanism is the assumption, widely held in Victorian times and encapsulated in the internal logic of Wells's *War of the Worlds*, that, since Kant–Laplacean cosmogony implied that Mars is older than Earth, a Copernican consequence is that Martian civilization must be technologically more advanced that the terrestrial one. Martian 'channels' were, thus, interpreted as global macro-engineering projects of this Martian civilization, giving proof to this Copernican hypothesis.[37] Given the premise, the conclusion is quite reasonable: even more than a century later, we still know too little about the dynamics of cultural evolution to be able to put forward a cogent alternative model of technological development versus time. But the premise (the validity of the Kant–Laplace theory) was wrong, so the Copernican conclusions, which would nowadays be classified as belonging to SETI studies, were irrelevant. We need to be wary of encountering similar fallacies in the much more complex, and intuitively more difficult to grasp, context of contemporary astrobiology.

## 3.5 Gradualism (and Red Herrings)

While we shall consider gradualism in the sense intuitively—and brilliantly—captured by Lyell's slogan *the present is the key to the past!*, one should keep in mind that various other construals are possible and, indeed, have been proposed in the long history of this doctrine. Gradualism essentially suggests that we should take into account only processes observable by our systematic inquiry (at present, since all *observation* happens at epochs close enough to be, for all practical purposes, simultaneous[38]) in explaining the features of the past record. In other words, we need not postulate any epochs substantially different from the present one in terms of acting forces and processes. If we seem to observe such an epoch, this must be due either to our incomplete understanding of present processes—which is always in the cards—or our lack of theoretical sophistication in building an explanation based only on present processes.

This was, of course, formulated in the context of geosciences and this has some important limitations which are often overlooked. Firstly, Earth is indeed quite small in cosmic terms, and everything on it happens roughly at the same place. If we wish to generalize, we must allow for the possibility of our literal looking into the past, that is, cosmology. Gradualism has motivated the strongest historical rival to the standard cosmological model—the steady-state theory of Bondi, Gold, and Hoyle.[39] And it has not turned out very well there, although the challenge itself was a milestone in history of the discipline. It would not be an overstatement to claim that the defeat of gradualism in cosmology brought about the emergence of the latter as modern observational science.[40] Our cosmological explanations are nowadays firmly grounded in reality of past epochs dramatically different from everything we observe today.

In geosciences, gradualism was more firmly entrenched after Lyell's victory over catastrophists in the mid-nineteenth century.[41] After 1980, however, it has been undermined from many sides, notably by the understanding that the history of life on Earth has *really* (and not just apparently, as was clear long ago) been punctuated by large catastrophes, leading to mass extinction episodes. Some of those catastrophes were caused, as has been accepted after a fierce and prolonged battle, by extraterrestrial causes, specifically collisions of small Solar System bodies with Earth. This has been most clear in the case of the end-Cretaceous mass extinction, but other mass extinction episodes have been discussed in this context as well.[42] Further speculative hypotheses suggested as causes of catastrophic events in Earth's past include close supernova/GRB explosions or encounters of the Solar System with interstellar clouds, resulting in dramatic climate change on Earth. Even that newly acquired liberty to discuss catastrophic explanations has quickly become too constrained. In recent times, catastrophic events have been invoked as the best theoretical explanations for extremely diverse historical phenomena, including the origin of the Moon, the origin of various families of asteroids (as well as the *missing* Ceres family), the systematic altitude difference between the northern and southern hemisphere of Mars, the 98° tilt of the axis of Uranus, the retrograde orbit of Triton, and the formation of observed large-scale features on Mercury, Iapetus, or Dione.[43] In a sense, in the last quarter of a century, catastrophic explanations have almost become the rule rather than the exception. The radical nature of this sea change has been downplayed for many reasons, and obviously has not (yet) resulted in the formulation of (neo)catastrophism *as a unified doctrine*—but that is perhaps unnecessary. The very context of modern research in planetary and geosciences is such that general philosophical principles gradually fade and play a lesser role as more empirical instances of general regularities and laws are discovered on a routine basis.[44]

In fact, the present-day position is such that one could ask an almost inverse question: after so many *empirical* developments, in Earth sciences and around them, and after so many debates in learned scientific journals, starting with Hutton's *Theory of the Earth*, how has it come to pass that gradualism is treated here as a *philosophical* assumption (or principle), and not as a straightforward question to be empirically resolved, one way or another? Is not that a step backward in regard to our desire to have Fermi's paradox resolved in a satisfactory scientific manner? Do we really think that the Voltairean 'torch of science' has fallen and need to be picked up at this point?

Actually, the situation is more complicated. Gradualism is regarded as a philosophical principle here, rather than a straightforward empirical matter, since there have been many historical ambiguities surrounding its definition, resolution, and domain of applicability. It is obvious that it means different things for different people in different epochs. That the part of the analysis clearly necessary for disentangling various nuances of meaning and doctrine must be epistemological and methodological goes without saying. It is an empirical claim that a particular instance of the general phenomenon X occurred in small increments over long period of time. If empirical research indeed tell us so, we might conclude that *this particular instance of X has been gradual*, and it might give us some confirming evidence about the gradual nature of X *in general* (although how weighty such a piece of evidence really is depends on further assumptions about probability, inference, and justification). To prove the general claim for X in all instances is a completely different and perhaps unfeasible task; in the astrobiological context, the problem is compounded by the fact that one reasonably expects the diversity of possible contexts in which instances of X arise to be much larger than what has traditionally been considered in the geosciences. For example, the frequency and strength distribution of earthquakes on Earth are, at least in principle, determined by the details of plate tectonics, which in turn are set by global parameters fixed at the time of the accretion of Earth, 4.556 Ga ago; such global parameters are the total mass and density of the planet, and its chemical composition, especially the amount of long-lived *r*-nuclides such as $^{238}U$ and $^{232}Th$. On a different terrestrial planet in the Milky Way, where these planetary parameters are different, the frequency and the range of intensities of quakes will be different; that much seems uncontroversial. Hypothetical inhabitants of such a planet could have vastly different conclusions about the nature of quakes in comparison to the historical development of geosciences on Earth; although their consequences are often *catastrophic*, the existence of quakes have not been regarded as vindication of catastrophism versus gradualism, since they both are local in terms of consequences and happen often enough to be subsumed under the Lyellian 'present' (which is the key to the past). We can imagine an Earth-like planet with quakes happening only once per century or even rarer, but always with high intensity. Would they be considered a gradual phenomenon?

Therefore, an Achilles' heel of gradualism is its dependence on non-uniquely defined timescales; such timescales are invented by humans and are cultural artefacts, rather than something built into nature itself. Planetary and Galactic histories do not impose any particular preferred timescales. Thus, whether we estimate the probability of a decisive catastrophic event per year, per million years, or per billion years, this should not influence our conclusions and their interpretation—but, obviously, gradualism on longer timescales looks pretty much 'punctuated', and its explanatory power is greatly diminished.

Indeed, exactly how long must be period of time for a protracted phenomenon for us to conclude that it is gradual? A year? An average human life span? A thousand years? A length of time comparable to a geological period, like the Permian or the Jurassic? A length of time comparable to the Hubble time? In contrast to some simpler eras, we do not indulge in God-given timescales of reference. As the distinguished palaeontologist David M. Raup writes:[45]

> An interesting aspect of the history of meteorite study is that the uniformitarian doctrines of Lyell and his followers have been able to absorb the new facts and concepts without seriously changing the basic catechism. Meteorites fall and they make craters, often big ones. Once accepted, this pattern became part of the uniformitarian doctrine and was no longer considered catastrophic... One can argue, therefore, that catastrophism really means something that is unfamiliar. As soon as it becomes familiar, the awful label need no longer burden it. Scientists are adaptable people.

Think: if this is so complex an issue when we are dealing with geological phenomena on Earth, which has been the playground for debates surrounding gradualism since Hutton's time,[46] how much more complex must it be when applied to the emergence and evolution of intelligence and technology throughout the universe! That very complexity justifies our temporary treatment of gradualism as a philosophical principle, in substantially the same manner as the completeness of formal systems was treated before Gödel's epoch-making discovery in 1931. (The comparison does not presume that gradualism will be eventually proven wrong—it is just to show how complex an issue it is for our present, modest level of understanding of the problem situation.)

## 3.6 The Non-Exclusivity Principle

Now we come to one of the focal points in this book; here, to paraphrase Stephen Jay Gould, one desires to write in all capitals or in screaming red ink, or using some fancy font.[47] A source of almost infinite confusion in countless debates on Fermi's paradox and SETI in general, the stuff of this section is perhaps the simplest and the least technical major issue in this book—but only when we remove the particular type of spectacles we are accustomed to wear when dealing with less strange problems.

By far the most successful athlete on Earth, as of the time of writing (August 2016), is the Jamaican sprinter Usain Bolt, holder of world records in both the 100-metre and the 200-metre races. Sprint running is an individual sport[48] and a rather clear-cut case in the sense of ascribing both honour and responsibility. If Bolt runs well, it is mostly due to his actual prowess, allowing some rather minor impact of weather conditions or other extraneous factors, as well as of what might be called the historical record: his trainers, previous work, and so on. Ditto for those rather rare cases when his time was worse than expected (or he missed a track meet). When a very distinguished soccer team like FC Barcelona, Arsenal, or Bayern Munich plays badly and loses, on the other hand, the spectrum of possible reasons for this is very wide indeed, and the responsibility for the loss can be distributed in many different ways. Maybe all of the team members were in bad form; maybe a goalkeeper was particularly weak and made a beginner's mistake in a goal kick; maybe the coach made some huge tactical error; or one or more of many other different hypotheses might turn out to be true. Forget for the moment that it is very difficult in practice—in contrast to the views of casual fans—to really determine the causal structure of any team event. Suppose, in contrast, that we have a kind of 'Laplace-demon' insight into the causes of each particular match outcome in terms of any individual player's performance. For example, in the Arsenal–Aston Villa 5–0 result

(1 February 2015, Emirates Stadium), our insight tells us that the performance of the goalkeeper Ospina was 75 per cent, the striker Giroud 90 per cent, and so on. Would this knowledge be enough to give a causal account of the outcome of the entire championship (say the Premiership or the EuroLeague)? And, in particular, to what extent will the same causes underlying the outcome of each individual game determine the final outcome of a series of matches comprising such a 'higher-order' event? If Team A wins the championship in spite of mediocre performances in most matches, shall we claim that there are *additional* causes of this unexpected success, going over and above the usual causal properties? Can occasional 'flashes of brilliance' confer the cup on an average team in a prolonged championship? Such questions become particularly interesting and even dramatic if we introduce the situation in which we could not, for some objective reasons, say family or job troubles, study *all* matches in the season, studying instead—with perfect insight in each individual case—just a more or less representative *sample* of the season's matches.

The sporting analogy is apt for an additional reason, whose importance will be clear soon: elements of regularity (the baseline quality of the players and the coach, the enthusiasm of the fans or the lack of it, etc.) usually *persist* throughout a season, on timescales much longer than the timescale of any single match. On the other hand, a season *consists of* individual matches—any positive or negative fluctuation does count into the average. So, is constancy really more important for winning cups than what we might call a potential for surprises? In spite of the pundits, this is very hard to tell in a complex environment such as a football competition. Ditto for astrobiology, where it is even harder to pinpoint where exactly the surprises may come from.[49] But it is *reasonable* to conclude that, lacking deeper insight, constancy of form against diverse challenges in different matches is pre-eminent.

Hence, we need another general principle. Non-exclusivity has been introduced and elaborated upon by David Brin in his seminal 1983 study of Fermi's paradox. He writes:[50]

> We are tempted to add one more, rather tentative, 'principle'...a 'Principle of Non-exclusiveness', which states that diversity will tend to prevail unless there exists a mechanism to enforce conformity.

Subsequently, Brin demonstrates how this works on many examples of particular hypotheses posed to explain Fermi's paradox:[51]

> Non-exclusiveness would seem to apply to Resource Exhaustion. Even if some [extra-terrestrial intelligent species] were wastrels, at least a few others would see the crunch coming and plan for it...

> To overcome non-exclusivity we need a mechanism which might affect [the fraction of intelligent species which develop detectable technologies] systematically, so that there are few if any exceptions to slip away and fill the Galaxy with the commerce we do not observe.

As another oft-encountered example, consider the classical hypothesis (let us call it Hypothesis A) that all cosmic civilizations self-destruct upon developing nuclear or biological weapons. It is quite exclusive: it assumes that the same global outcome arises independently from a wide variety of local properties. The same disastrous history, the same global holocaust, gets repeated over and over again, on millions of planets separated by kiloparsecs and millions of years. This is intuitively improbable, and Brin's principle of non-exclusivity gives a formal statement of such intuitions. If, however, there is a single causal mechanism acting on *all* civilizations over *all* history of the Galaxy, we would have a properly non-exclusive explanation of the observed Great Silence. Similarly (Hypothesis B), if the very first Galactic civilization actually quarantines all the subsequent ones, including humanity, into local 'zoos', taking care to isolate them from the others, then we have a single causal explanation for the non-observation of many different Galactic civilizations and their activities, as well as the fact that none have visited Earth and the Solar System thus far. It does not matter, from our present standpoint, that we might find Hypothesis A more probable and Hypothesis B more improbable *on other grounds*—it is just the illustration of an *additional desideratum* which the principle of non-exclusivity offers us. (Indeed, Hypothesis B was conceived specifically for this purpose out of several related far-fetched explanatory hypotheses for Fermi's paradox, while Hypothesis A has a long and venerated tradition in SETI studies.[52]) As with other philosophical criteria considered in this chapter, it makes little sense to use non-exclusivity alone for judging the solutions to the puzzle; we need all—or most—of them to arrive at the really strong contenders.

Now, isn't non-exclusivity just a newfangled name for good old Occam's razor? While some similarities are obvious, like preferring a small over a large number of causal factors, the answer to this question is no, for the following two reasons. The first one is that Occam's razor, as it is usually construed, talks about the simplicity of explanatory hypotheses (or assumptions built into such hypotheses) and not about the *scope* of such hypotheses. In the usual and mundane circumstances of our terrestrial lives, it is easy to conflate the two, since complexity and scope are often correlated: larger things tend to have more complicated composition and function—and therefore require more complex explanations if everything else stays the same. It is certainly easier to understand

and explain the economic functions of a bakery on your corner than those of the London Stock Exchange. However, there is no such correlation for the astrobiological scene. If a simple probabilistic cause acts uniformly on $10^9$ terrestrial planets in the Galaxy in the course of various epochs (as in Hypothesis A), we do not have rational reasons to prefer it over a complex cause which acts once for all time and achieves the same explanatory purpose (as given by Hypothesis B).

The other reason to resist unthinking identification of the non-exclusivity principle with Occam's razor is that the latter properly applies only to situations in which the relevant assumption is what philosophers dub 'ceteris paribus': all other things are kept equal. Occam's razor suggests that, after we all agree on the empirical facts of the matter of how combustion proceeds in the real world, it is better to explain the phenomenon of combustion by oxidation than by release of the hitherto unknown mysterious substance called 'phlogiston'. In other words, the hypothesis of oxidation is better than the hypothesis of phlogiston, since it is simpler, ceteris paribus. No ceteris paribus = no Occam's razor. This is likely to cause tremendous difficulties for the application of Occam's razor in studying Fermi's paradox, since it involves comparisons of the evolutionary status of hypothetical wildly varying biospheres from all over the Galaxy and over billions of years of cosmic time. The prospects of having common conditions in such a situation are non-existent. Moreover, it is doubtful that ceteris paribus is applicable to hypotheses for explaining particular features of biological (macro)evolution even on a single planet, Earth! The whole controversy surrounding contingency versus convergence in the course of the terrestrial biological evolution testifies that unambiguous definition of causes of and influences on the particular traits of any species is, at best, extremely hard to achieve, at least outside of the laboratory. In such cases, rivalling hypotheses for the explanation of the origin of almost any particular biological trait *openly violate* the ceteris paribus clause, making Occam's razor applicable only exceptionally if at all.[53] Since (as emphasized in Chapter 1) Fermi's paradox is essentially an evolutionary problem, this limitation applies to the study of its solutions as well—a circumstance which very few authors writing about the paradox, apart from Brin (and the ubiquitous Stanislaw Lem), seem to have understood.

For these reasons, I maintain that we need to consider non-exclusivity as a specific criterion for the evaluation of hypotheses explaining Fermi's paradox, independently of Occam's razor. If pressed, we might vaguely construe the non-exclusivity principle as a *modified* Occam's razor applied specifically to the evolutionary context of astrobiology and SETI studies.

Non-exclusivity is a particular application of a wider principle of rationality, rejecting solutions which are worse than the problem itself. The Australian astronomer Luke Barnes calls such solutions *cane toad solutions*, from an ecological example too beautiful not to be quoted:[54]

> In 1935, the Bureau of Sugar Experiment Stations was worried by the effect of the native cane beetle on Australian sugar cane crops. They introduced 102 cane toads, imported from Hawaii, into parts of Northern Queensland in the hope that they would eat the beetles. And thus the problem was solved forever, except for the 200 million cane toads that now call eastern Australia home, eating smaller native animals, and secreting a poison that kills any larger animal that preys on them. A *cane toad solution*, then, is one that doesn't consider whether the end result is worse than the problem itself.

In the context discussed in this book, such cane toad solutions would be all requiring an improbable conspiracy of causes, acting over long intervals of time and huge volume of cosmic space, to produce the observed Great Silence. The **Hermit hypothesis** described in Section 1.6 is one such cane toad solution, which violates the non-exclusivity principle with gusto. Therefore, the **Hermit hypothesis**, as presented, is not really a serious candidate solution to Fermi's paradox: at best, it requires an improbable conjunction of causes. As usual, one can return to it—and other hypotheses violating non-exclusivity—*if all else fails*; but we better not hurry there. In addition, we can regard those solutions requiring that we give up solid scientific evidence or even the scientific method (e.g. those presented in Sections 4.2 and 4.3) itself as particularly egregious cane toad solutions.

A crucial point for understanding the rest of this book is the following: non-exclusivity might not be *true* (in the same sense that the vast majority of scientists hold that naturalism or scientific realism are true, as they are reliable guides to the scientific truth in any particular matter of investigation), but we still have solid reasons for using it as a tool for sorting out the jungle of hypotheses and performing the taxonomical task set out in Chapter 1. In conjunction with other philosophical criteria—scientific realism, naturalism, Copernicanism, and gradualism—it forms a set of desiderata for the true solution of the 'Big Puzzle'. This does not imply that its status is necessarily the same; quite the contrary.

As tirelessly (and vacuously) preached by creationists and their ideological allies, no amount of empirical research could ever prove the truth of naturalism. In contrast, only further research and expanding the front of knowledge will enable us to establish the truth value of the non-exclusivity principle—after all, as emphasized in the introductory parts of this book, Fermi's paradox is a specific *scientific* problem, one which is solvable, in the final analysis, through empirical means. If humanity does not destroy itself, but expands throughout the Galaxy, either in person or via robotic probes, it is inconceivable that we will *still* be puzzled by this problem at that future epoch, in, say, AD 22018. So, at that time, our descendants will be able to correctly assess whether the correct solution or solutions of the problem are, indeed, non-exclusive or not. Whichever is the case, it is a *contingent fact of the (astrobiological) history of the Milky Way*. Therefore, non-exclusivity is not on a par with metaphysical (such as realism) or epistemological (such as methodological naturalism) assumptions; nor it is dependent on the spatial and temporal framework we *choose* to work in (such as gradualism). As is the case with Copernicanism, we can reasonably hope to establish the truth value of non-exclusivity by future research—and to explain why it has one truth value or the other. However, since that prospect is not immediate, and perhaps lies in the quite distant future, I feel justified to treat it here as a philosophical assumption.[55]

## Non-Exclusivity versus Hardness

One final point needs to be made in connection with non-exclusivity principle. A most general idea expressed by various authors, under different names, is that we should strive for a general (or 'hard') explanation. But the history of science teaches us that this is sometimes impossible. For example, the temperatures of the surfaces of planets and other bodies in the Solar System are generally explained by their distances from the Sun, and their albedos. Objects closer to the Sun are warmer than those farther away for a fixed value of albedo; those with a lower albedo are warmer for a fixed distance. This seems quite a neat explanatory scheme, with ample empirical support. The temperature on the surface of Venus cannot be explained, however, in the same manner—it is simply too hot given its distance from the Sun and its (very high!) albedo. The reason for its high surface temperature remained mysterious until Carl Sagan and others realized that it is a particular, extreme form of the greenhouse effect which keeps the Venusian temperature so high at all times.[56] In this admittedly sketchy example, while we would philosophically prefer that all planetary temperatures be non-exclusively explained by the distance–albedo explanatory hypothesis, nature itself provided us with an exception where the initial conditions (or conditions in a sufficiently distant, unobservable past) led to a different explanatory hypothesis being more adequate.

We can imagine a similar situation with Fermi's paradox as well. For example, we might have a scenario in which 95 per cent of the sophonts in the Galaxy destroy themselves after inventing nuclear weapons—and the remaining 5 per cent migrate to the Galaxy's rim, where they have a very small cross section for detection and contact.[57] This would violate the non-exclusivity principle—but it is still a viable solution. Contrary to naive expectations, the non-exclusivity principle cannot guarantee a clean-cut solution to a problem as complex as Fermi's paradox is; what it *can* do, however, is to provide us with a rating of solutions *at any given stage of our knowledge*.

We need to resist the temptation—often occurring in philosophical debates—of thinking that our principles actually *do* something in the world itself. No—they just help us organize and manage our knowledge about the world. This clearly applies to non-exclusivity: while it cannot guarantee a single, all-encompassing solution, it can at least keep the number of different viable explanatory hypotheses in the mix low. Two different mechanisms explaining Fermi's paradox are somewhat worse than a single explanatory mechanism; but they are still significantly better than a hundred different explanatory mechanisms. This intuitively obvious state of affairs is made precise with the non-exclusivity principle.

## 3.7 The Continuity Thesis

One of the central issues of astrobiology is the extent to which we can talk about abiogenesis (and, by extension, noogenesis) in naturalistic terms. This issue has been investigated in depth by the Israeli philosopher of science Iris Fry, who showed that a necessary ingredient in any scientific account of abiogenesis is the so-called *continuity thesis*: 'the assumption that there is no unbridgeable gap between inorganic matter and living systems, and that under suitable physical conditions the emergence of life is highly probable.'[58] Adherence to the continuity thesis, as Fry amply demonstrates, is a precondition for the scientific study of the origin of life; contrariwise, the view that abiogenesis is a 'happy accident' or 'almost miracle' are essentially creationist, that is, unscientific.

The taxonomy of solutions to Fermi's paradox suggested in this book relies on this analysis of the continuity thesis and, in part, on its extension to noogenesis.[59] The continuity thesis has been supported by many distinguished scientists and philosophers throughout history, but none did more to promote it in wildly varying contexts than the great British polymath John B. S. Haldane (1892–1964). In both his research writings in biology, mathematics, astronomy, and so on, and in his philosophical essays,[60] he insisted on the continuity between physical (in particular, cosmological), chemical, biological, and even cultural evolutions. Haldane was a co-author of the famous Oparin–Haldane theory of abiogenesis, which emphasized the law-like aspects of the process. This was in complete accordance with his philosophical and methodological principles, which enabled him to lay down foundations of what is today often called future studies as well.

The continuity thesis—and any of its extensions—has nothing to do with metaphysical notions such as *necessity*. It does not state that the emergence of life or intelligence is necessary in any reasonable construal of the word. Nor does it claim that we can *derive* features of emerging life, intelligence, or technological civilizations from the form of the laws of physics (even if they were known fully). One needs to tread very carefully here. As the great Richard Feynman observed half a century ago:[61]

> The next great era of awakening of human intellect may well produce a method of understanding the qualitative content of equations. Today we cannot. Today we cannot see that the water flow equations contain such things as the barber pole structure of turbulence that one sees between rotating cylinders. Today we cannot see whether Schrödinger's equation contains frogs, musical composers, or morality—or whether it does not. We cannot say whether something beyond it like God is needed, or not. And so we can all hold strong opinions either way.

This comes right after a paragraph in which the great teacher emphasizes that both the lack of detected life on other Solar System planets and the enormous diversity of the terrestrial biosphere are consequences of one and the same set of physical laws. So, while the naturalist world view implies that life and observers like ourselves are compatible with the laws of physics (without going into the question of to what exact degree do we understand them), there is nothing in laws themselves which tells us the frequency of such events as abiogenesis or noogenesis. The continuity thesis is a working hypothesis (as well as a heuristic, as previously explained) that the part of the relevant parameter space containing life or intelligence is appreciably large; it might be wrong, but it certainly is not tautologous or vacuous—or non-empirical.

## 3.8 Postbiological Evolution

An important novelty mandatory for any modern-day review of Fermi's paradox—and insufficiently recognized in SETI circles so far—is the necessity of taking into account hitherto unrecognized evolutionary possibilities, especially the Haldanian notion of *postbiological* evolution. The degree to which previous discussions of the status and the importance of the paradox have been framed by the conventional biological thinking would in itself be an interesting topic for the history of science, albeit one into which I cannot enter here in any detail.[62] But the least one can safely do is consider the impact of generalizing the situation to include other options. And this has been recognized rather recently; for instance, the great historian of science Steven J. Dick cogently wrote in 2003:[63]

> But if there is a flaw in the logic of the Fermi paradox and extraterrestrials are a natural outcome of cosmic evolution, then cultural evolution may have resulted in a postbiological universe in which machines are the predominant intelligence. This is more than mere conjecture; it is a recognition of the fact that cultural evolution—the final frontier of the Drake Equation—needs to be taken into account no less than the astronomical and biological components of cosmic evolution.

Before we enter the (dis)contents of the Drake equation, it seems worthwhile to pause and consider the impact of postbiological options on our thinking about SETI. Prompted by Moore's Law and the great strides made in the cognitive sciences, discussions of postbiological evolution of humans (and the related scientific and philosophical questions) have proliferated in recent years.[64] There is much talk now—if not always in scientific, at least in protoscientific terms—about different forms of human enhancement, implants, cyborgs, mind uploading, or even technological singularity. While many people continue to consider these topics as a staple of science fiction, they are fighting the rearguard now: the future is coming and, barring some global cataclysm causing the destruction of human civilization, these and related topics will certainly become more active and important as science and technology progress.

It is easy to understand the necessity of redefining SETI studies in general and our view of Fermi's paradox in particular in this context. For example, postbiological evolution makes behavioural and social traits like territoriality or the 'expansion drive' (to fill the available ecological niche), which are—more or less successfully—'derived from nature', lose their relevance. Other important guidelines must be derived which will encompass the vast realm of possibilities stemming from the concept of postbiological evolution.

Since the 'expansion drive' has often been promoted as the root of all problems with Fermi's paradox, and naive readings of it ('You certainly do not expect advanced extraterrestrial intelligences to behave like the Spanish conquistadors?', with all the appropriate body language and gestures of offence[65]) still occupy a large niche in the literature, it is worth going here into slightly more detail. Without entering the debate on the merits or demerits of sociobiological and evolutionary–psychological explanations, one thing is clear on purely logical grounds: if we accept that *in principle* these disciplines could be satisfactorily grounded in some inclusive form of biological evolution (a very weak assumption), and we accept that postbiological evolution is a serious possibility (another weak assumption), then it is clear that we should not use the same concepts derived from our biological past to explain features of our *post*biological future. Ditto for other hypothetical intelligent species in the universe. In the specific case, we should not expect that postbiological civilizations will have the same evolutionary-conditioned 'expansion drive' as the biological ones. We should not expect other features we observe in human, still largely biologically grounded, society to be truly universal, that is, to apply to postbiological/non-biological societies as well; particularly important among those features are aggression and warfare, both of which are claimed by sociobiologists, with more or less success, to follow from our evolutionary inheritance.[66] We should not consider either the future of humanity or the advanced stages of evolution of other intelligent beings to be bounded by the Malthusian problems, which, as is well known, motivated Darwin himself. In brief: we should avoid biases grounded in our biological origin.[67]

A postbiological approach, therefore, frees us from at least a part of the tyranny of the biological. This conclusion, of course, needs to be coupled with the working assumption that large-scale postbiological evolution is possible. There seems at present to be no conceivable reason why it should not be so. Whether it will come to pass on Earth—or in any other place inhabited at any moment of time by intelligent beings—is a completely different agenda. To claim that humanity is either a very good or a very bad model for possibilities of postbiological evolution is to violate Copernicanism without any compelling reason to do so. However, if some civilizations go along the postbiological pathway and some not, it is reasonable to expect an outcome which can be construed as a monumental irony: the mechanism of *natural selection*, that keystone of all *biological* approaches to explanation, might give a huge advantage to *non-biological* actors on the Galactic stage.

This may happen for some of the reasons outlined by Dick, but can also follow from the sobering fact mundane to anyone acquainted with technology (and history) or astronautics: machines are, simply, much better at tasks relevant for the exploration and colonization of space than people *in their current biological form*.[68] It is exactly for this reason that Tipler's scenario with self-replicating interstellar probes[69] is the strongest form of Fermi's paradox and the biggest obstacle any contact optimist has to face.

While I shall return to some specific scenarios relying in larger or smaller part on postbiological evolution, it is important to keep this option open as we consider each and every one of the explanatory hypotheses. Along the same mirroring principle described in Section 1.9, this applies to studying scenarios for the future of humanity as well. Of course, a huge space of opportunities provided for by the postbiological evolution is a significant factor in the *limitations* of our discourse in both astrobiology and future studies. Here, as everywhere, to ever hope to overcome our limitations, we need the courage to face them first.

## 3.9 The Drake Equation, for Good or Bad

It seems impossible to have a book even tangentially discussing SETI issues without the customary, or even ritual, mention of the Drake equation, which supposedly supplies us, in principle, with a valid measure of the number of technological civilizations likely to exist in the universe. Somewhat paradoxically, this does not have to do with the substance of the discussion; in some cases, it seems to be motivated exclusively by the fear that a discourse without equations will not be taken seriously enough in the 'real' scientific world. Such idolatry of mathematics and numbers is entirely misplaced; there have been many similar grotesque cases of trying to force mathematical language and formalism on fields such as literary criticism, art history, and class struggle. Even in these areas where mathematical expression gradually took roots, such as evolutionary biology, this occurred due to deep methodological and historical reasons, and not because practitioners felt insecure and uncomfortable without repeating ad nauseam some simple piece of mathematical regularity such as, say, the Hardy–Weinberg equilibrium.[70] In the SETI field, invocation of the Drake equation is nowadays largely an admission of failure. Not the failure to detect extraterrestrial signals—since it would be foolish to presuppose that the timescale for the search has any privileged range of values, especially with such meagre detection capacities—but of the failure to develop the real theoretical grounding for the search. This follows from its very structure:[71]

$$N = R_* f_p n_e f_l f_i f_c L, \tag{3.1}$$

where $N$ is the 'predicted' number of extraterrestrial civilizations (usually misinterpreted as the number of SETI targets; see below) in the Milky Way; $R_*$ is the star-formation rate in the Milky Way, appropriately averaged; $f_p$ is the fraction of stars possessing planets of any kind; $n_e$ is the average number of habitable planets per planetary system; $f_l$ is the fraction of habitable planets actually possessing life (either through local abiogenesis or panspermia); $f_i$ is the fraction of inhabited planets developing intelligent life; $f_c$ is the fraction of intelligent communities developing the technology relevant for detection and communication over the interstellar distance; and the famous factor $L$ is the lifespan of the civilization in the detectable mode (often misinterpreted in a variety of ways).

Each fraction term in the Drake equation should, in fact, be explicated in terms of relevant probability distribution functions, which should be integrated over the relevant region in the parameter space. Roughly speaking, there should be an integro-differential equation—in the most general case—for each of the probability terms in the equation. For instance, the average number of habitable planets is an integral over the rate of planets becoming habitable minus the rate of their ceasing to be habitable by various processes (runaway greenhouse effect, stellar evolution, and so on); this should be integrated over spatial volume of the Galaxy over the course of Galactic history. In this manner, we could finally make the transition between guessing (even educated guessing!) and computing within the context of quantitative astrobiological models. In fact, all individual terms, with a partial exception of the average lifetime of technological societies, belong squarely and without any doubt in the domain of astrobiological research, which becomes more and more sophisticated and precise as we speak, write, or read these lines. So, to those who continue to rewrite, cite, and re-cite the Drake equation without trying to get a deeper theoretical insight, one is more and more justified to ask to 'put up—or shut up!'

Those cases are legion, however; it is particularly discouraging to find even (section) titles such as 'The Drake Equation: *The Theory* of SETI' or a labelling of the equation as a 'SETI creed',[72] when it is exactly the Drake equation which is used *instead of* any real SETI theory, and when it is exactly the reputation of the entire field which too often suffers from quasi-religious associations. Even worse, that simple fact—that SETI issues have been historically considered too complex or too unworthy of effort for the development of a real theoretical framework—fuels scepticism of those who are opposed to SETI on other grounds.[73] The rhetoric here is twofold: either the lack of a proper theory is used to demonstrate that SETI is not a 'real science', or it is claimed that some theoretical underpinnings of SETI do exist, but they are all contained in the

Drake equation with its meagre informational content and, at best, vague predictions. The latter strategy is particularly insidious, since it claims real insight into SETI research, albeit one which is outdated and prejudiced. In both cases, a reader is led to the impression that SETI should not be taken seriously.[74]

To the extent that the Drake equation represents any elements of SETI theory (in contrast to using it as a fig leaf instead of a real theory), it does so in a manner which is constraining and impractical. This has been noted even by researchers otherwise quite sympathetic to it. As a typical example, consider that the equation prima facie rejects interstellar colonization and the emergence of independent SETI targets by such spatial expansion. Walters and co-workers have tried to correct for that by offering a quantitative model for modifying the Drake equation, but the predominant state of mind in the majority of the SETI community has been amply demonstrated by the sad fact that this study has not been widely cited or employed in justifying practical searches.[75] Let us consider this example in a bit more detail from a methodological point of view. If the cornerstone of a theory in science is modified in such an important way, it is only rational and, indeed, mandatory to use the modified version from that moment on. It must not be regarded as just 'an option', something to mention in footnotes—or not at all. When Fischer, Haldane, and Wright modified the old Darwinian theory, thus creating what subsequently came to be known as the Modern Synthesis, anyone interested in evolution from that epoch on was *obliged*, if intellectually honest, to use the synthetic theory and the explanations based on it. There is no disrespect towards Darwin and Wallace in this; if anything, modified insight of the Synthesis leads to a better appreciation of the achievement of the old masters. It makes no sense—except when dealing with history of evolutionary biology or in searching for an inspiration—to revert to what Darwin, Weissman, Haeckel, Thomas Huxley, or any number of distinguished pre-Synthesis authors considered the best explanation or the best way to proceed.

Similarly in other fields: after 1998, almost all cosmological formulae have been modified to take into account the large contribution of dark energy to the total energy budget of the universe. Such high methodological standards have not been seen in the SETI studies. The study by Walters et al. should have permanently altered the understanding and citing of the Drake equation; the fact that it was largely ignored tells us something important about the immaturity of the field.[76] (A closely related reason why the Drake equation is irrelevant from the point of view of the present book is that it is essentially incompatible with the Kardashev classification of the technological civilizations, especially Types 2.x and Type 3. Space colonization—direct or indirect via von Neumann probes—causes the emergence of detectable locations which are entirely unconnected with the probabilities that a star has habitable planets, that the planet experiences abiogenesis, etc. As mentioned in Chapter 1, it is exactly Type 2 and higher civilizations which are the worst offenders in this respect, i.e. the sources of the paradoxical conclusion in the strongest versions of Fermi's problem.)

Even without the specifically astrobiological arguments discussed above, this tendency of portraying Drake equation as a cornerstone—or, even worse, *the* cornerstone—of SETI by both proponents and opponents alike should give one a pause. After all, in other grand controversies in the history of science, a particular piece of theoretical apparatus was invoked by either critics (as a weakness) or defenders (as a strength), but not by both alike. Consider, for instance, epicycles in the old geocentric cosmology of Ptolemy: they were introduced to 'save the phenomena', and no rhetorical manoeuvre could make them a *virtue* of the theory. The Copernican opponents charged that excessive epicycles are one of the major *weaknesses* of the old theory; supporters such as Clavius or Riccioli at best muttered something along the line of 'they enable us to make accurate predictions'. It would be a strange situation indeed if geocentrists were to emphasize epicycles as one of their *achievements*!

As above, so below. Ludwig Boltzmann was correct in emphasizing that his new statistical mechanics could explain many observed thermodynamic regularities such as Boyle's law or Charles's law. He was not that sure, mildly speaking, that the same holds for the second law of thermodynamics and the accompanying arrow of time, since his theory was essentially time symmetric.[77] Thus, his opponents, like Ernst Zermelo, used this particular feature of the theory—temporal symmetry—to argue against it. And Boltzmann himself did not try to 'fig leaf' the problem, but faced it courageously, pioneering an important anthropic-style explanatory hypothesis in the process, which still might have something to teach us.[78] How strange would it be if *both* Boltzmann and Zermelo were to extoll the virtues of temporal symmetry of the underlying dynamics! Such examples abound, and all contrast sharply to what we have seen in more than a half-century of SETI debates.

Therefore, it is high time for the practice of the ritual invocation of the Drake equation to stop.[79] Exceptions are, of course, those innovative works which use the equation effectively as a *provocation* to reach deeper theoretical insights.[80] While the SETI research community cannot, of course, prevent abuses of history and its 'Whiggish' reinterpretation, what it can do is to start insisting on building a more serious and strict theoretical scaffolding for its enterprise.

Since the Drake equation is a rule of thumb which should be *derivable* from any such real SETI theory by a chain of approximations, integrations, and averaging, it should not be advertised as anything more than that. Instead, we should strive to reach a deeper understanding through more precision, more numerical models, more simulations, more specific scenarios subject to quantification, and so on. Fortunately, the winds seem to be changing in this area as well.

## 3.10 Let the Games Begin!

We have so far formulated the problem (Chapter 1) and given its astrophysical (Chapter 2) and philosophical (the present chapter) background. Now, the search for the culprit—the best explanation—may begin in earnest. In the rest of this book, I shall discuss and compare specific explanatory hypotheses, based on the taxonomy imposed by the philosophical assumptions sketched above. Specifically, if we abandon each of the major three philosophical assumptions (realism, naturalism, gradualism), we obtain one of the major types of solutions (which I dub *solipsist*, *rare Earth*, and *neocatastrophic*, respectively). A family of explanatory hypotheses corresponds to each of these types. In addition, another type of solution emerges in cases in which one or more of the philosophical

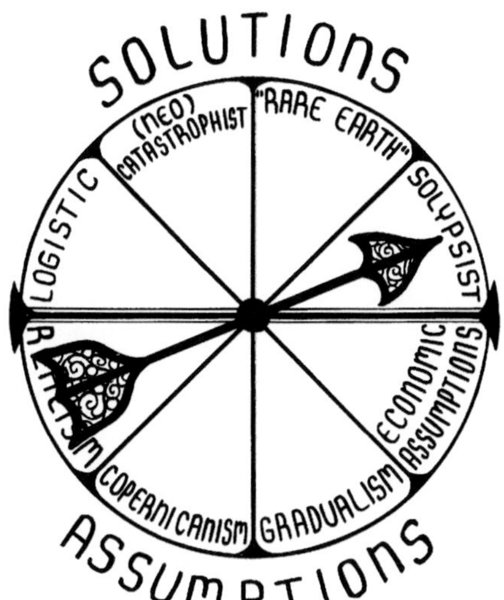


**Figure 3.1** The proposed high-level classification of the solutions to Fermi's paradox. In an extremely simplified form, the respective replies to Fermi's question *Where is everybody?*, by proponents of solipsist, 'rare-Earth', (neo)catastrophic, and logistic hypotheses are 'They are here', 'They did not evolve', 'They have been destroyed or transformed', and 'They have been prevented from coming/ developing yet', respectively. By negating each philosophical assumption (or, in the case of logistic hypotheses, economic assumptions), we obtain a wide class of solutions to the puzzle Symbolically, our arrow obscures ('rejects') an assumption while pointing at the corresponding category of solutions.

*Source*: Courtesy of Slobodan Popović Bagi

assumptions are only weakly violated, with violations of additional economic assumptions as well; these are *logistic* solutions, to be considered in Chapter 7. Schematically, such fourfold taxonomy has been shown, with some characteristic examples, in Figure 3.1. We shall return repeatedly to this scheme, especially in comparing different classes of hypotheses in Chapter 8.

The game is afoot!

CHAPTER 4

# L'Année dernière à Marienbad

## *Solipsist Solutions*

In the beautiful 1961 movie directed by Jean Resnais, *Last Year in Marienbad* (*L'Année dernière à Marienbad*), a group of characters are, apparently, holidaying in a spacious baroque castle. The protagonist, unnamed in the movie but called 'X' in Alain Robbe-Grillet's screenplay, desperately tries to convince an unnamed woman, 'A', that they had known each other since the summer before, when they spent holidays in Marienbad (an old Austro-Hungarian resort, today Mariánské Lázně in the Czech Republic). She doesn't remember him at all, but what he tells her has the power to create a past for her and blend it into her present. A second man ('M'), who may be the woman's husband, appears and repeatedly asserts his dominance over X, including beating him several times at a strange mathematical game. But no real resolution of their conflict is in sight.

Is there anything real about their situation at all? Everything in the film looks *just apparent*. Mirrors, mysterious games, labyrinthine corridors of the castle, the puzzling selective lack of shadows (see Figure 4.1), the seemingly eternal sunshine outside, repetitive conversations, ambiguous voiceovers—all of those just strengthen the effect. The characters move like somnambulists through a hermetically sealed world that seems totally surreal; at moments they seem all caught up in a Gödelian loop of disjointed time. The world as captured by the camera is subtly incoherent; but is it a consequence of the erroneous subjective perceptions of our characters, who are frozen in the timestream like insects

**Figure 4.1** The famous courtyard scene from *L'Année dernière à Marienbad*; which is more real, the shadow-casting protagonists or the shadowless scenery?

*Source*: Courtesy of The Ronald Grant Archive/Mary Evans

captured in a drop of amber? Or is it a 'God's-eye' view of a truly incoherent, chaotic, decaying world? As a reviewer poignantly asks, 'Who are these people? Are they people? It's *civilization as a hallucination*.'[1]

In the literary inspiration for Resnais's movie, *The Invention of Morel*, an under-appreciated, brilliant novel by the Argentinian writer Adolfo Bioy Casares, first published in 1940, there is a similar effect of blurring the distinction between apparently real and simulated features of the environment.[2] The protagonist of this strange and sorrowful story finds himself on an apparently inhabited island in the Pacific, whose inhabitants have sophisticated habits and obey complex social rituals and motions. Not only does the protagonist come to 'understand' them, but they also have a deep emotional impact on him and his entire view of the world. However, there is much more on the unnamed island than meets the eye... *and much less in the same time*. Without any obvious spoilers—this masterpiece of a novel is still insufficiently read and recognized—it is enough for our present purpose to note that the observations of the protagonist are in a dramatic conflict with the conventionally understood 'reality', although he is in fact a quite *rational* and *perceptive* observer. That such a state of affairs is possible and logically consistent, at least within the framework of the novel, is testimony not just to Bioy Casares's literary mastery but also to an inherent property of complexity: above some threshold, complex systems are capable of simulating one or more aspects of reality in any number of ways, including Turing-complete simulators themselves.

These artistic works—together with fiction by authors like Philip K. Dick, Italo Calvino, Thomas Pynchon, and Julio Cortazar—precede most of this chapter's hypotheses for resolving Fermi's paradox, but do possess some important features which are unavoidable in this category of hypotheses. As Bioy Casares's close friend and occasional collaborator Jorge Luis Borges famously stated, 'the present is always anachronistic.'

The ideas investigated in this chapter, as well as in Chapters 5–7, are often not formulated in the form of clear scientific hypotheses. Occasionally, they lack the necessary generality, precision, specificity, or some combination of these ingredients to be treated as such. This should not be a reason for their outright rejection if some intrinsic merit of the ideas is perceived—instead, it should be understood as a call for more work on their better formulations. To that end, I have tried to formulate some of these in the form of scenarios, thought experiments, or, to use a much abused term, 'just-so stories.' There are many justifications for such an approach when a problem as complex as Fermi's paradox is concerned. Even much more mature fields use storytelling as a fruitful modus operandi.[3] Also, the history of science clearly shows that some of the most dramatic theoretical steps in our understanding have been prefigured in other discourses and even other media.[4] Finally, we need to keep in mind at all times how limited our perspective is within a general framework, such as the astrobiological landscape of the Milky Way: the argument that we do not have experience with X, Y, or Z and that *therefore* X, Y, or Z is non-existent or unimportant should be consistently and relentlessly opposed.

After we consider some of the 'devil's-advocate' reasons for rejecting scientific realism (Section 4.1), various hypotheses and approaches based on the remaining set of assumptions will be discussed (Sections 4.2–4.7) in a *rough* and admittedly subjective order of seriousness. A summary of these is given in Section 4.8, and further comparative discussion is left for Chapter 8.

## 4.1 Down with Realism!

The label 'solipsist' refers to Sagan and Newman's classic 1983 paper criticizing Frank Tipler's scepticism towards SETI studies, based on Fermi's paradox and strengthened by the idea of colonization via von Neumann probes.[5] Here, however, we would like to investigate solipsist explanatory hypotheses in a different—and closer to the vernacular—meaning. The latter is usually interpreted as the idea that only one's own mind is sure to exist; the rest of the empirical universe is at best uncertain and possibly illusory. The doctrine originated in ancient Greece and has had a long and not really illustrious history, mostly lurking in the dark corners of philosophical thought for centuries. In the last several decades, due to the rapid development of neurosciences, ideas and thought experiments of the 'brain-in-the-vat' type have again become fashionable in circles of cognitive scientists and neuroscientists, as well as metaphysicists and philosophers of mind. As we shall see in this chapter, there are important recent developments which have rekindled interest in these ideas.

In our present context, solipsist solutions reject the main empirical premise of Fermi's paradox, namely that there are no extraterrestrial civilizations either present on Earth or detectable (through our observations) in the Solar System and the Milky Way so far. On the contrary, such solutions usually suggest that extraterrestrials, indeed, are, or have been, present in our vicinity, but that the reasons for their *apparent* absence lie more with our cognition, our observations, and even our politics (and their limitations), than with the true state of affairs. The analogy with vernacular solipsism is, then, clear: solipsist explanatory hypotheses submit that those aspects of the external world dealing with extraterrestrial intelligence are in part or entirely illusory. Thus, their proponents deny the assumption of scientific realism.

Why is such an outrage permitted at all? Why don't we immediately realize that only crackpots, pseudoscientists, and a deluded philosopher or two might reject realism? Why don't we pass on to other categories of hypotheses which are—at least from afar—much less controversial? Hasn't this been a province of the lunatic fringe long enough? Such a commonsensical position is, indeed, taken by many astrobiologists and students of SETI so far. There are, however, at least two reasons of principle to justify considerations of the rest of this chapter—while remaining healthily sceptical. The first is best expressed by Brin in his famous sentence: 'Aversion to an idea, simply because of its long association with crackpots, gives crackpots altogether too much influence.' Part of the same line of argument is the well-documented historical fact that major explanatory hypotheses put forward in the past have often been labelled as crazy or crackpot by the best authorities of the ages past—and do note that I have referred to *hypotheses* and not to the individuals who put them forward. The second principled reason to consider solipsist hypotheses in this framework is the fact that we have very little experience with the systems of such high complexity like hypothetical extraterrestrial sophonts and their presumed artefacts, systems which—like ourselves and our artefacts—are characterized by *intentionality*. The best analogies drawn in popular culture would be dealing with supernatural mythological creatures, like gods, demons, God, spirits, angels, and so on—but there can be no *scientific* knowledge of such systems, of course. Rather unscientific, but very common and sensible pop-psychological inference from our experience with other humans is that other humans (i.e. complex physical systems with intentionality) tend to *cheat* in a sufficiently generalized sense. That is, some evidence about humans and their relationship with their environment is often either hidden or tampered with. Scientists who are researching humans and their artefacts—people like psychologists, anthropologists, historians, and archaeologists—are very well aware of this situation, and a large chunk of their working methodologies is geared towards correcting for the consequent distortions of knowledge about these subject matters.

This situation differs dramatically from the one encountered, for example, by researchers studying elementary particles and fields. While the Higgs boson was quite elusive until recently,[6] no sane physicist would ever ascribe that state of affairs to higgson's *intention* and *capacity* to hide from human researchers. In contrast, archaeologists studying the tomb of Tutankhamen are not surprised to learn that there is little direct evidence that he was assassinated; even if that were the case, the traces of the crime would have been *intentionally* concealed by human beings with *capacity* to do so, and thus distort or falsify the historical record.[7] The creation of more or less artificial (or 'virtual') reality for moral, religious, economic, or political reasons has accompanied human societies in the entire history of our civilization. It would be unreasonable to expect in advance something completely different from other Galactic sophonts. As have already been mentioned, the quest for resolving Fermi's paradox has some elements of 'interstellar archaeology'; one of those elements is certainly the need for interpreting evidence some degrees less literally than it is the case in natural sciences such as particle physics or geology.[8] While this example is extreme on purpose, one should keep in mind that such distortions of evidence might arise either from different cognitive capacities of extraterrestrial sophonts or from the huge temporal and spatial scales involved (this particularly pertains to the artefacts, or cases such as directed panspermia, which will be discussed in Section 4.6). In brief, while bosons, quasars, and bacteria cannot use resources in order to intentionally hide from our observations, intelligent extraterrestrials could.

To these key reasons of principle, one might add two minor reasons for discussing 'crazy' solipsist hypotheses here at some length. One is the fact that at least some of them are powerfully present in the popular culture—and have been so from the very beginning: remember that UFOs were the main motivation for Fermi's small talk at the famous lunch! To entirely disregard them would not only be bad outreach strategy, which is something astrobiology and SETI studies cannot afford and do not really need, but also would be unfair towards the history of science. The tradition of the 'alien hiding in our midst' is long and venerable in the cultural discourse. Starting with E. A. Poe and H. P. Lovecraft, the common denominator is that the alien—not necessarily of the kind we would today associate with SETI studies, but we should be wary of anachronism and a 'Whiggish' interpretation of the history of ideas—is hidden but exercises some (usually malign) influence upon human comings and goings. As Jason Colavito argued, the work of Lovecraft in particular influenced much of the pop-cultural and pseudoscientific thinking about advanced aliens, although the influence is rarely, if ever, admitted; this should be tackled head on if we wish astrobiology and SETI studies to gain wider respect in society. The other auxiliary reason is that some issues tightly related to some of the solipsist ideas (like the possibility of our living in a large-scale computer simulation) have recently gained traction in both science and philosophy.

The extent of the denial of scientific realism varies a great deal in this category, as we shall see. Some proponents of the solipsist hypotheses are content to reject the conventional scientific accounts of the independently contentious issues—like the origin and extent of technological knowledge in ancient human civilizations, as in the case of followers of the 'ancient astronauts' quasi-mythologies—while others pose more sweeping challenges. The latter, like the newly refashioned **Simulation Hypothesis** (Section 4.5) or **directed panspermia** (Section 4.6) are generally much more interesting and may indeed eventually become testable. Instead of rejecting solipsist options by fiat, we need to consider all the alternatives, and some of these clearly form well-defined, albeit often provably wrong or undeveloped, ideas. Hypotheses in this class serve another important role: they remind us of the magnitude of the challenge posed by Fermi's paradox to our naive world view—and they should be evaluated in this light. While the rejection of scientific realism may look and sound like too steep price to pay, part of our revulsion at the price comes from downplaying the magnitude and difficulty of the puzzle itself, as discussed in Chapter 1. The distinguished palaeontologist Simon Conway Morris thus in a recent paper argues that we should endorse what he calls a 'mad' view of the problem—that our understanding of our own existence is wrong:[9]

> This proposal, of course, is totally mad but unfortunately it is the correct solution to Fermi's paradox... we happen to live in the 'wrong sort of universe' or to be more precise in a universe where the question of extraterrestrials will have to be entirely re-formulated. Under this banner there are a gratifying range of alternative possibilities, but they fall broadly into three categories... The first possibility is that sentient extraterrestrials remain in the universe, but in one way or another become 'invisible' to our current technologies... The second category of explanation is that the universe we think we live in is virtual, constructed by 'people' who may (or more likely do not) have our interests at heart... The third alternative, which I think is more open to demonstration, is that the universe we live in is not in a strict sense 'virtual', but nevertheless is not at all as we imagine it to be. Rather than proposing a Matrix-like solution to our 'existence'... our Universe consists of a series of intersecting orthogonal realities.

Some of Conway Morris's categories will be easily recognized in what follows. In trying to digest the rather extreme point of view of as quite an established scientist as Conway Morris, we need to be constantly reminded that, although Fermi's paradox is a scientific problem, this does not imply that all possible—or even all actually suggested—solutions are scientific *as yet*. This is particularly transparent when we consider solipsist solutions. To reiterate, I use 'solipsism' in somewhat more relaxed manner, meaning not necessarily the extreme position that

**there is no external reality**

but rather the weaker statement that

**the external reality is not adequately described by our conventional scientific insights.**

Note that this does not reject *all* scientific insight, just those which are today available and widely—or even consensually—accepted. In other words, while consideration of the solipsist hypotheses might give boost to pseudoscientific and/or postmodern nonsense, this is not *necessarily* so. In a sense, since Fermi's paradox is such a wide and difficult problem, we are obliged to try to find nuggets of worthy ideas even in a huge pile of garbage.

Well-defined examples of limits of our scientific knowledge exist which could be listed under this heading, and motivate taking solipsism more seriously. Some philosophers and physicists have discussed in quite a serious manner whether the complexity of the universe could ever be entirely cognizable to the human mind.[10] Suppose that one day we all agree on a unique metric for measuring the complexity of any one particular thing $X$, and let us call this metric $\mathbb{C}(X)$.

It gives us a number (in any convenient unit; in this case, let us say bits, as in the algorithmic information theory[11]) for each investigated thing, and this number can be then compared with the numbers obtained for other things. We assume that the complexity measure gives us intuitively plausible results, for example $\mathbb{C}(A \text{ and } B) \geq \mathbb{C}(A) + \mathbb{C}(B)$, where $A$ and $B$ are any objects—including objects of the mind, like ideas or thoughts—or that

$$\mathbb{C}(\text{human}) > \mathbb{C}(\text{insect}) > \mathbb{C}(\text{bacterium}) > \mathbb{C}(\text{crystal})$$

as we would expect from structural insights. If such ordering of things by complexity is possible, we are entitled to ask whether there is a maximal complexity (say, $\mathbb{C}_{max}$) which is intelligible to human mind; of course, this pertains not to an individual human mind but to the sum of all minds of humanity at any given time. Suppose that such a limit exists. Perhaps the solutions of big unsolved problems in science and philosophy are of this variety, which is exactly why they remain unsolved (the solutions to problems such as the Riemann hypothesis, the problem of evil, why is there something rather than nothing, and other, similar problems might qualify as such). Could extraterrestrial intelligence or its artefacts (including messages) be of this type?

This needs not be so extravagant or cynical as it might sound. As previously discussed, Lineweaver's results coupled with the naive application of Copernicanism suggest the existence of giga-anni-old supercivilizations. The gulf between capacities of such an old community and what we can do or imagine now could be so vast that their products might be such $X$ for which $\mathbb{C}(X) > \mathbb{C}_{max}$. There is no a priori reason for rejecting this possibility (except, of course, misguided anthropocentrism in using the human mind and its properties as the 'measure of all things', an attitude rightly criticized by philosophy since Socrates). Charles Darwin thought about such a possibility—although he wrote about it only informally, in a letter—as we shall see in Section 4.4. Unless further research shows that there is an absolute upper limit to the chosen complexity measure—and such a conclusion is not immediately forthcoming in contemporary physics, computer science, or cosmology—we could, in contrast, expect ever-higher peaks of complexity to exist somewhere in the universe. Per analogy with Earth and the local universe, we expect those peaks to be associated with outcomes of biological and cultural evolution. If it is so, then such supercomplex artefacts could *never* be adequately described and explained by human science.[12] This situation would violate the usual construals of scientific naturalism (as presented in Section 3.2). It would not necessarily entail supernaturalism, although the tendency needs to be accounted for, but just another level or tier of scientific treatment.

For instance, consider some artefacts of very advanced civilizations described in fictional discourse.[13] Such things could be unintelligibly complex, but we'll still be able to perceive them, and perhaps give some partial description of them. After all, Dr Floyd and his colleagues were able to precisely measure the *size* of the black monolith in *2001: Space Odyssey*, with not the faintest idea about its purpose or way of functioning. Stanislaw Lem in his famous metaphor does indicate that ants have *some* use for a dead philosopher they have stumbled upon in the woods.[14] While that use is not in any way commensurable with the content of philosopher's mind while he was alive—even if we accept the strictest physicalist view that everything is encoded somewhere in brain's physical structure—it still is of some use. Ants could, in principle, distinguish between the day or season in which they found a dead philosopher and some other day/season in which no such nutritious occurrence took place. Cognitively advanced ants, like us, might hope to advance their insight even a tiny bit more.

## 4.2 Saucers, Utensils, and Other UFOs

For very unfortunate and historically contingent reasons, it seems impossible to discuss extraterrestrial intelligence without making even a token mention of the so-called unidentified flying objects (more commonly referred to as UFOs) and the extraterrestrial hypothesis for their origin. Briefly, hundreds of sightings of UFOs in the period after the Second World War, all over the increasingly globalized and interlinked world, led to speculations about their extraterrestrial origin.[15] In 1969 the physicist Edward Condon defined the extraterrestrial hypothesis as the 'idea that *some* UFOs may be spacecraft sent to Earth from another civilization or space other than Earth, or on a planet associated with a more distant star.' Somewhat trivially, the corresponding 'solution' to Fermi's paradox—and alleged motivation of the celebrated lunch debate—can easily be concocted:[16]

> **Fermi's flying saucers:** Extraterrestrial intelligent beings are, indeed, here on Earth, flying around in UFOs, and are engaged in various covert activities. While we do not know when they first arrived, it could be the case that Earth has been surveyed for quite some time, so the visitors must be much older from us, even comparably to Lineweaver's scale in Eq. (2.4). Since they exist, are older than us, and are here as expected, there is no Fermi's paradox.

Those who believe UFOs are of extraterrestrial intelligent origin clearly do not have any problem with Fermi's paradox. This is a direct consequence of abandoning the assumption of (scientific) realism—one rejects the premise of absence of extraterrestrial intelligence from the local universe. Instead, 'ufologists' are confident that extraterrestrial intelligent beings are present not only in the local universe but on Earth as well, and usually in contact with at least some humans. Therefore, there is no paradox. For instance, the (in)famous UFO researcher Stanton Friedman claims exactly that.[17]

There are all sorts of problems with such an explanation, however. Since the weight of scientific evidence obviously tells us that UFOs are not of extraterrestrial origin, proponents of this hypothesis tend to reject scientific evidence and even the entire scientific method. Since this is a non-starter, it is difficult to say anything further about this hypothesis. Still, a 'rational ufologist' (insofar as such an animal exists) might face a host of obstacles in justifying the extraterrestrial hypothesis, as long as some vestiges of scientific method are retained.

For instance, apart from bizarre modus operandi of the alleged extraterrestrial visitors and their spacecraft, one might question the timing of alleged visitations. Since most of the known UFO activities have been recorded after the Second World War, it would be a strange coincidence that extraterrestrials should have arrived on Earth exactly—from the point of view of astronomical and evolutionary, even historical timescales—when space flight has become a reality within human civilization. If not premeditated, this would be an example of truly extreme fine-tuning, something which in most of the sciences would require an explanation in itself. The response sometimes given by ufologists that the extraterrestrials actually reacted to the emergence of human technology does not really help, unless the visitors originate from some nearby location—or possess faster-than-light travel. Faster-than-light travel would undoubtedly create much *stronger* forms of Fermi's paradox than those discussed in Chapter 1.[18] On the other hand, a nearby origin is incompatible with everything we know from both conventional astronomy and SETI projects undertaken so far. In addition, a marked decrease in the frequency of UFO sightings after the end of the Cold War is completely inexplicable based on the extraterrestrial hypothesis for their origin.[19] As Robert Sheaffer writes in the classic anthology of Zuckerman and Hart:[20]

> We find that, despite the 788% increase in reports from 1951 to 1952, the percentage of the supposed unidentifieds was essentially the same. This is a most puzzling factor for the following reason. Suppose that the unidentified UFO

> reports represent sightings of alien spacecraft. Then, when the number of genuine UFO sightings triples in a given year, it is presumably because the ones that are here have become three times as active. One would expect the percentage of supposed unidentified to go up dramatically, as the signal-to-noise ratio improves, but they do not. Why should a three-fold increase in extraterrestrial activity cause people to also report Venus and weather balloons as UFOs at three times the previous rate?…I suggest that the most straightforward explanation of the above dilemma is that the signal-to-noise ratio in UFO reports is exactly zero, and that the apparently unexplainable residue is due to the essentially random nature of gross misperception and misreporting.

Some further arguments against the extraterrestrial origin of UFOs are discussed in detail by authors such as Carl Sagan, Philip Klass, and others.[21] There is, however, one unlikely way of framing the extraterrestrial UFO hypothesis to be an accidental consequence of a different explanatory hypothesis—I shall return to this issue ('**LEAKY INTERDICT**') in Section 4.4.

One should be careful to delineate here between ufology and the *Search for ExtraTerrestrial Artefacts* (SETA) as mentioned in Chapter 1, and as will be discussed throughout this book, especially in Chapter 8.[22] The actual boundary line might seem blurred from outside science, but effectively, the two are clearly separated by their approach and justification strategies. While paranoid mind frame of most—although not all—ufology tends to rant about NASA's censorship of pictures of the 'face on Mars' and of alleged UFO sightings by space probes, SETA is the only way of systematically and scientifically testing our assumption of scientific realism vis-à-vis the absence of an extraterrestrial presence in the Solar System. While the absence of any artefacts of an advanced Galactic civilization in the Solar System can hardly ever be decisively proved, the relevant parameter space of such artefacts can—and will be, as long as human or post human civilization survives—be more and more constrained.[23]

Robert Freitas, one of the most distinguished SETI researchers of the 1980s, has tried to argue against the sceptical outcome of Fermi's paradox ('we are alone') on the basis of insufficiency of conditional probabilistic logic coupled with uncertainty as to the absence of extraterrestrial artefacts from the Solar System.[24] While he correctly points out that, due to a paucity of SETA activities and data, our empirical evidence for such an absence is weaker than is usually assumed, his argument is of limited value in the general astrobiological context. Notably, it uses only **WeakFP**, not offering any reply to the stronger versions of the paradox, which include detectability over interstellar distances. In addition, the 30+ years which have passed since Freitas's study have brought about several revolutions in observational astronomy and in our understanding of the Solar System. The failure of our advanced astronomical tools to detect any manifestations or traces of extraterrestrial intelligence adds further weight to the probabilistic version of **StrongFP**.

SETA is a useful, although very minor, part of the SETI research. As much else in the field, it is largely a parasitic programme, the virtue here being that it may parasitize on very broad scope of research, including multibillion-dollar space missions. There are two important aspects to consider and balance here: (i) the all-too-often encountered public image of SETI as something 'far out', 'science fictional', and unscientific, and (ii) the safety concerns following from the existence of any hypothetical extraterrestrial artefacts. The latter should be obvious, but is often not; it has nothing to do with any volition—it is essentially the same reason why children or members of a primitive tribe are at risk when coming upon live power lines.[25] Unfortunately, the first aspect has been much more present thus far in the overall downplaying of SETA research thus far. If humans are to survive as species—which means expansion into other locales in the Solar System, and utilization of resources available there—the second aspect will unavoidably become more important, even acute. Prima facie, there is no essential difference between SETA and the archaeological search for artefacts of long-extinct human cultures at unexpected places. In at least some cases, we had no prior historical evidence for the very existence of some cultures before their artefacts were serendipitously found.[26] If anything good is to follow from the UFO phenomenon in general, and **Fermi's flying saucers** in particular, it is to obtain some scientific respectability for SETA/interstellar archaeology.

Of course, the lack of imagination is not in itself an argument against extraterrestrial interpretation of UFOs, but it still quite indicative. In contrast to the tabloid picture of extraterrestrial visitations, it is enough to study, for instance, a truly non-anthropocentric scenario of alien visitation, as described by the Strugatsky brothers in their masterpiece *The Roadside Picnic*.[27] *That* is how a visitation might realistically look like—puzzling, mysterious to the degree that it literally makes Nobel laureates scratch their heads in disbelief and puzzlement (and is likely to have quite an unexpected societal impact on human society as well).

A closely related hypothesis is that extraterrestrial visitation occurred in recent (astronomically speaking) past:

> **Ancient flying saucers:** Extraterrestrial visitations occurred in humanity's historical past, and these early contacts influenced the development of human

cultures, technologies, and religions. We can find traces of these influences in myths about visitors from the sky and in archaeological artefacts which cannot be explained solely as products of human societies. Since the extraterrestrial beings were here, the premise of Fermi's paradox is false and the problem dissolves.

Apart from the colourful notoriety of some of its proponents, like Peter Kolosimo or Erich von Däniken,[28] this hypothesis brings hardly anything new in comparison to **Fermi's flying saucers**, since the timescale of human history is so minuscule in comparison to $\tau_L$ or even $t_{FH}$ that we can effectively compress it to a single point in time: *now*! Of course, it is much more difficult to *falsify* a hypothesis pertaining to the past by empirical evidence, which is the reason their proponents feel more comfortable. In fact, past visitations aggravate Fermi's paradox, since we are even more entitled to ask '*Where is everyone?*' if there is some evidence of extraterrestrial civilizations possessing the technology for interstellar flight at some point in the very recent past. The same applies to traces and manifestations of astroengineering, which we shall cover in detail in Chapter 8.

## 4.3 Special Creation

As far as they can be formulated as hypotheses, traditional views of special creation (of Earth, life, and/or humanity) belong to this category. Such views enjoy a long history but lie, *in general*, outside science, with a few marginal cases deserving mention here. The first and perhaps most valiant of such marginal cases has been made, as already mentioned, by Alfred Russel Wallace, who in 1903 argued for the key role of a 'cosmic mind' in the grand scheme of things on the basis of (i) the alleged incompatibility of the 'higher' properties of the human mind with natural selection, and (ii) a teleological (mis)interpretation of the then fashionable model of the universe, similar to the classical Kapteyn universe.[29] As discussed in detail by Michael Crowe and others, such views were occasionally dressed in the garb of traditional theology (especially of Christian provenance), but the association is neither logically nor historically necessary.[30] The essential points of this view can be summarized as follows:

**Special creation**: Some or all of the prerequisites logically or physically necessary for the present-day human intelligence and civilization are either impossible or highly improbable to have emerged in a naturalistic manner.

> Instead, such a prerequisite or prerequisites have been created in a supernatural way by a supernatural agency. Therefore, there is no reason to expect the existence of other intelligences elsewhere in the universe, and there is no Fermi's paradox.

Obviously, this formulation lends itself to multiple versions, depending on the prerequisite in question: it might be Earth as a habitable planet; abiogenesis; the emergence of complex life; noogenesis; or any such individual item or a set of items. Of course, this hypothesis (let us keep the label, while bearing in mind that it is not necessarily a scientific hypothesis) does not offer a clear reason *why* the same supernatural agency, or a different such agency, would not have created other sophonts and their societies. In the conventional theist discourse of centuries past, it has been postulated as expression of a particular *preference*—usually deemed *love*—of the said agency for humans, and not arbitrary other sophonts. In the context of Western monotheist traditions stemming from Judaism, it has even been expressed in a morphological manner: the Creator has created humans 'in his own image'. Even if anthropocentrism is not so overt, but rather implied, as in some other religious traditions, this shows that, apart from naturalism and scientific realism, this kind of 'explanation' violates Copernicanism as well.

To some mild relief, thus, this way of looking at the problem of life and intelligence beyond Earth has today been abandoned in most mainstream theologies.[31] Old-fashioned qualms about the 'uniqueness of salvation' are, seemingly, retreating before more ecumenical and liberal views. But, if we do not need to understand alleged supernatural Creation in anthropocentric terms, we cannot use it to resolve Fermi's paradox either! This blade cuts both ways: as the extent to which the anthropocentrism of most special design ideas sounds ridiculous increases, the hope of resolving the problem in this manner recedes, to the same extent. It is not without historical irony that even sincere theists among the scientists of today are likely to recognize Fermi's paradox as a *scientific* problem that is *unresolved* as of today.

Special creation possesses some methodological similarities with the 'rare-Earth' hypotheses as well (see Chapter 5). In conventional creationist accounts, at least one of the key steps of the evolutionary chain leading to the present situation is either flatly impossible or so improbable that supernatural intervention is the only intelligible way of explaining why it has actually happened in our past. This is similar, as we shall see, to the variety of highly improbable requirements postulated by the 'rare-Earth' theorists. Of course, a 'lucky accident' is still different—and certainly more scientific, insofar as a comparison is possible at all—from a 'miracle'. Therefore, 'rare-Earth' hypotheses are still preferable to **Special Creation**. This is especially so if you take into account philosophical arguments to the effect that scientific methodology is not—contrary to the conventional assumption often abused by creationists and supporters of the 'intelligent-design' movement—metaphysically neutral.[32] After all, **Special creation** is a weapon of last resort—we may always return to this if everything else fails; clearly, with any version of Fermi's paradox, we are still very far from that kind of quandary. We are dealing with too many hypotheses, not too few.

Surprisingly enough from the point of view of a conventional scientist, if **Special Creation** is unpalatable to our naturalistic assumptions, there is a *naturalistic version* of the very same explanatory hypothesis—or at least something which looks like a naturalistic version. Consider, however: even if such a possibility were real, would it be a better offer? Perhaps, to a scientist aware of the philosophical and methodological issues surrounding the central tenet of realism, it might be, which is the reason why we shall encounter it in this chapter (see Section 4.7). Most people in this still-largely religious or religion-influenced world would not concur. But in another expected twist, it might not be appealing to many scientific naturalists as well since it... sounds too much like **Special Creation**.

## 4.4 Zoos, Interdicts, Dogs, and the Mind of Newton

In his 22 May 1860 letter to the Harvard naturalist Asa Gray, the father of evolutionary theory discussed metaphysical questions of the 'meaning-of-life' type, concluding resignedly that[33]

> I am inclined to look at everything as resulting from designed laws, with the details, whether good or bad, left to the working out of what we may call chance. Not that this notion at all satisfies me. I feel most deeply that the whole subject is too profound for the human intellect. A dog might as well speculate on the mind of Newton.

One wouldn't bet on the dog doing that successfully, would one? And yet, there is no a priori reason to consider the gap in intelligence and other cognitive abilities between humans and hypothetical members of an advanced technological civilization or a hypothetical Kardashev Type 2.x civilization to be any less than that which separates your average Snoopy or Ace from the founder of classical physics (or Darwin himself!). After all, Snoopy's ancestors were our ancestors within the *Eutheria* clade of mammals as recently as about 85 Ma ago, in the Cretaceous. Compare this with the flabbergasting 1.8 Ga in the Lineweaver's scale (Eq. 2.3). Even the ants of Lem's parable are evolutionary closer to the dead philosopher, since both arthropods and chordates originated together in the Cambrian explosion, about 540 Ma ago, still less than a third of Lineweaver's scale. So, if Snoopy is extremely unlikely to ponder the laws of mechanics (or alchemy, all the same), and the ants are in no position to extract deep ethical reflections from philosopher's delicious brain tissues, how minuscule are our chances to even remotely and vaguely understand the mind of a really advanced sophont?

Some people have been wildly optimistic on this issue. The **Zoo hypothesis** of John A. Ball and the related **Interdict hypothesis** of Martyn J. Fogg suggest that advanced extraterrestrial civilizations have a uniform policy of avoiding any form of contact (including visible manifestations) with newcomers to the 'Galactic Club':[34]

> **Zoo hypothesis**: Advanced Galactic civilizations intentionally refrain from contacting newcomers for ethical reasons, reasons to do with security, or some other reasons (which would be incomprehensible to newcomers). We are located in a Galactic analogue of a zoo or a wilderness preserve—a chunk of space set aside for the low-level civilizations to evolve without interference. This no-contact policy extends to hiding traces and manifestations of their existence. We may be confident that they observe us, as we observe animals in a zoo, a lab, or a wilderness preserve, without us being aware of the fact.
>
> **Interdict hypothesis**: Advanced Galactic civilizations did actually explore and colonize almost all of the Milky Way long time ago and achieved a high level of technological and cultural uniformity in the process. Due to information being the most important non-renewable resource, individual planets likely to evolve intelligent life (as the best source of new information) in the fullness of time were placed under interdict upon the formation of this global, Galactic equilibrium state. Therefore, local 'zoos' are set up around any habitable planet with an indigenous biosphere—and we may assume that the really advanced Galactic sophonts have long ago moved past using Earth-like planets as main habitats; therefore, the loss of space and resources for them would be negligible in comparison to the Kardashev Type 3 (or Type 2.x) resources they possess. Since they are fully capable of remaining hidden from newcomers like us, there is no Fermi's paradox.

Two important aspects of these should be kept in mind: the *motivations* of the hypothetical advanced sophonts, and their *capacities*. As far as motivations go, the reasons behind the behaviour of 'zookeepers' may be those of ethics, prudence, or practicality.[35] These are often interpreted in analogy with the *Star Trek* 'Prime Directive', which forbids interference and intrusion by the United Federation into the affairs of inhabited, but unaffiliated planets, but they also could be understood as being analogous to some of the ecological policies implemented by modern-day humans, such as setting up national parks and wilderness preserves. Advanced civilizations might find it immoral to influence young civilizations, at least before they reach some level of technological maturity. Obviously, if the history of human cultures is any guide, such ethical motivation seems justified.[36] Another motivation often found in the pop-cultural context is the risk aversion of a more advanced society confronted with primitive but potentially disruptive ideas and practices.[37] This does not mean that a less-advanced civilization might threaten the existence of their more advanced Galactic neighbours—but there are other, more subtle dangers and uncertainties lurking in the cultural contact of very different entities. We simply do not know enough to decide one way or another.

(The situation can be compared to the power relations in real zoos on present-day Earth. While tigers do not present any existential threat to humans—quite the contrary—this circumstance does not mean that freeing tigers from a zoo is not an *undesirable* occurrence. It is, in fact, so undesirable that resources are invested on a routine basis for the prevention of any possibility of such an accident. Even the most animal-loving group of humans will not tolerate any large carnivores moving freely in their midst—and the force, even if non-lethal, used to mitigate such an occurrence might be regarded as desirable from a moral point of view. Also, one might speculate as to what extent morally superior communities might be at risk of *memetic* contamination by their less developed neighbours.)

In each case, these do not really offer testable predictions (if the extraterrestrial civilizations are sufficiently powerful, as suggested by the magnitude of $\tau_L$), for which they have been criticized by Sagan, Webb, and others. As a consequence, a **'LEAKY' INTERDICT** scenario is occasionally invoked to connect with the alleged extraterrestrial origin of UFOs, which is clearly problematic:[38]

> **LEAKY INTERDICT**: While the general policy of extraterrestrial intelligent communities follows the guidelines/regulations described in the **ZOO/INTERDICT HYPOTHESIS**, there are in those communities some free agents capable of announcing their presence to other Galactic cultures and violating the guidelines/regulations. Even then, the violations would be sporadic, unpredictable,

> and clandestine—and the violators would tend to leave the *mainstream* of the newcomer civilization ignorant of their activities. UFO sightings are manifestations of such free agents on (and near) our planet.

Unfortunately, the main idea here is to explain the unsystematic nature of the UFO sightings. But **Leaky interdict** is still important as a reminder of a wider problem that a large chunk of the solipsist class of hypotheses faces. The **Zoo** and **Interdict** hypotheses share with some of the other solipsist hypotheses a typical weakness which can be called the *Ashurbanipal-of-our-street* problem: how to prevent individual actors within the containing civilization from violating the regulations and initiating contact with us? In more general terms: how can *leakage* of any information related to the containing civilization be prevented?

Consider one of the most pronounced historical trends since the dawn of human civilization: the increase of personal control over the environment. In the present-day world, an average person often—at least in the northern hemisphere—has capacities over both his or her physical environment and his or her personal well-being which are larger than those enjoyed by the most powerful and privileged members of ancient or medieval societies. Not only is the life expectancy for Average Joe or Median Chang dramatically larger than that of the most powerful aristocrats of past eras; the capacities for communication, transport, exhibition of creativity, material prosperity, and even pure physical control of the environment (like lighting or air conditioning) are, due to the effects of our technology and education, orders of magnitude larger than the corresponding values for Alcmeonids, Grakhi, or Sforzas. If such a trend continues, even at an attenuated pace, it is to be expected that by the time humanity or post humanity rises to the level of an advanced technological civilization, even a future analogue of a village idiot will have capacities over his or her environment larger than even the most powerful despots and monarchs of the pre-industrial age. Even a lowly member of an advanced technological civilization would be immensely more powerful than the absolute rulers of oriental empires or modern totalitarian states. If those capacities are to be freely exercised without strict control by the government or its analogues, it is hard to see how global policies such as the Prime Directive or any analogue could be implemented and enforced.[39] Preventing individuals and groups from violating **Zoo/Interdict** regulations would be possible only if such societies were tightly controlled and regulations enforced with an efficiency unknown in human history so far; one is tempted to say that such societies must be *totalitarian*. Even more, in the known examples of totalitarian societies, there were dissidents in many spheres of human activity, in spite of rather small spatial and temporal scales involved. In order to efficiently enforce all details of any policy, including supposed prohibition on contacts (and, as we shall see in Section 4.6, seeding of life on other celestial bodies), such an advanced society would need to be what Nick Bostrom calls *singleton*; this does not necessarily mean that it has to be coercive or violent, since advances in both social and cognitive sciences might enable morally acceptable forms of unanimity.[40] But, in any case, *it will be very different from what futurists have—usually—imagined thus far.* This might be overkill in searching for the resolution of Fermi's problem; I shall return to this issue in Chapters 6 and 8.

## Alien Motivations and Vico's Argument

If speculations on the alien motivations seem too outlandish from the start, one should keep in mind several sobering facts. First of all, opponents of such speculation—even those generally sympathetic to SETI science, such as Steven Jay Gould—emphasize the fact that we do not understand even remotely adequately the social behaviour and motivations of *humans* in the first place, in contrast to our, allegedly superior, understanding of the natural, non-anthropogenic world:[41]

> I don't mean to be a philistine, but I must confess that I simply don't know how to react to such arguments. I have enough trouble predicting the plans and reactions of people closest to me. I am usually baffled by the thoughts and accomplishments of humans in different cultures. I'll be damned if I can state with certainty what some extraterrestrial source of intelligence might do.

This is often manifested in classifying hypotheses such as the ZOO/INTERDICT HYPOTHESIS as 'soft', meaning, foremost, that they originate with 'soft sciences' such as sociology. Interestingly enough, very distinguished minds used to argue for exactly opposite conclusion—that, in fact, the social sciences have a better chance of achieving a satisfactory explanation than the natural sciences do! Thus, Giambattista Vico writes:[42]

> Whoever reflects on this cannot but marvel that the philosophers should have bent all their energies to the study of the world of nature, which, since God made it, He alone knows; and that they should have neglected the study of the world of nations, or civil world, which, since men had made it, men could come to know.

*continued*

Although, since the Enlightenment, science has rejected the idea of a deity being the ultimate cause of phenomena in the natural world, note that this idea is actually rather peripheral to Vico's main argument. The same conclusion would hold if we simply stated that the natural world is huge, complex, and ancient and gives us—when we account for observational selection effects—little prior reason to assume that it is comprehensible to human intellectual capacities. The fact (unknown in Vico's time) that such human intellectual capacities have emerged through evolutionary processes such as natural selection and genetic drift should just strengthen this part of the argument.[43]

Now, Vico's argument ought not to be used to justify just any speculation about the motivations and behaviour of extraterrestrial intelligent beings; instead, it should be used against blithe and unthinking rejection of any such speculation as unworthy of discussion. The **Zoo** and **Interdict** hypotheses might not be falsifiable, as many have pointed out,[44] but the extraordinary nature of Fermi's puzzle is such that conventional epistemology is perhaps too narrow. As previously mentioned, SETI studies share with archaeology and similar historical disciplines a feature lacking from most of sciences: in contrast to Higgs bosons, granite, or chromosomes, sophonts (both terrestrial and extraterrestrial) can be intentionally deceitful. The same idea goes some steps further in the next two hypotheses we consider.

## 4.5 Living in a Planetarium—or a SimCity

**Planetarium hypothesis:** Our astronomical observations do not represent reality but a form of illusion, created by an advanced technological civilization capable of manipulating matter and energy on interstellar or Galactic scales. While Earth and the Solar System, as visited by our interplanetary probes, are real enough, the universe at large is an illusion similar to the planetarium projectors or 3-D movies. Since the 'real' universe outside of our planetary theatre could be much different—and is, indeed, inhabited by at least one immensely powerful intelligent species—let us call them Directors—there is no Fermi's paradox.

The **Planetarium hypothesis** was suggested by the renowned British science-fiction author and engineer Stephen Baxter in 1999.[45] This hypothesis can be regarded as a special case of the **Zoo hypothesis**, substituting passive 'hiding' of the advanced civilization for 'active screening' of the ultimate reality. Whether one option is more or less plausible than another is an interesting engineering and technical issue; one should keep in mind that, even if we entirely agree on the objective content of scientific knowledge, there is much more freedom for action of historical contingency in the development of technology.

The eponymous Planetarium will, therefore, be the *ultimate* alien artefact (as artefacts are usually understood, as 'intentionally created things'; we shall see a more liberal understanding of the term in Sections 4.6 and 4.7), encompassing all of observable 'nature'. This is a bit paradoxical in itself, since, in this chapter, we reject broadly construed realism, not naturalism. Bear in mind, however, that Baxter's idea is that only large-scale astronomical features of the natural universe are illusions: the Solar System, Earth, our laboratories, atoms, particles, and so on, are still considered real in the usual sense. Moreover, we could use their properties to infer that the 'real' universe, behind the Planetarium, is actually very similar to the one we see through our telescope: not necessarily the *very same* stars, gaseous clouds, or galaxies, but similar stars, gaseous clouds, and galaxies. The only really important difference in the context we are interested in is that, most probably, the universe behind the Planetarium is inhabited by the Planetarium Builders (and/or other kinds of sophonts). So, the artefact is, in fact, just a veil or a screen—Magritte's man with a newspaper is indeed there in all panels, but veiled by subtle and powerful stealth technology. Insofar as we stick to common physical reductionism, we can even infer important properties of Planetarium Builders, since we share a common physics (and possibly chemistry) with them.

What if the entire cosmological domain (and not just our immediate Solar System surroundings) is a product of intentional intervention? This is the subject of Paul Birch's 'peer hypothesis'. Birch, a British scientist, engineer, and author, particularly interested in astroengineering projects, correctly understood that such large-scale achievements exacerbate Fermi's paradox and, in an unpublished manuscript, suggested a radical solution:[46]

> **Peer hypothesis**: A supercivilization has created our entire cosmological domain as an experiment, presumably using processes indistinguishable from those 'conventional' cosmology suggested took place in the very early universe, at the time of inflationary expansion. The universe is not only habitable, but actually inhabited, since the research or creative goals of the experiment is to create 'interesting history' by having many biospheres of roughly similar properties and ages (our eponymous 'peers'). Since other such civilizations are not *significantly more* advanced than us, the concerns of Fermi's paradox are invalid, and we might expect contact with some of our peers very soon.

Unfortunately, it seems that the **Peer hypothesis** cannot cope with present-day versions of Fermi's paradox. In particular, if the 'basement-created universe' evolves according to preset physical laws,[47] then the magnitude of $\tau_L$ in Eq. (2.4) still requires—naturalistic—explanation. We still are not typical by the chronological order of appearance, unless the evolution on older planets, those closer to Lineweaver's median age, was not intentionally slowed down by the Creators. But if it was intentionally slowed down, the last tangible difference between the **Peer hypothesis** and **Special Creation** disappears! There is no real explanatory advantage in Birch's hypothesis.

But why shouldn't we dispense entirely with the substrate and its conventional ontology? The **Simulation hypothesis** of Bostrom, although motivated by entirely different reasons and formulated in a way which seemingly has nothing to do with Fermi's paradox, offers a framework in which the puzzle can be naturally explained:[48]

> **Simulation hypothesis**: Physical reality we observe is, in fact a simulation created by Programmers of an underlying, true reality and run on the advanced computers of that underlying reality. Due to a form of principle of indifference, we cannot ever hope to establish the simulated nature of our world, provided that the Programmers do not reveal their presence. As a parenthetical consequence, the simulation is set up in order to study a rather limited spatio-temporal volume, presumably centered on Earth—there are no simulated extraterrestrial intelligent beings, so there is no Fermi's paradox.

Bostrom offers a Bayesian argument for why we might rationally think we live in a computer simulation of an advanced technological civilization inhabiting the 'real' universe. This *SimCity*-kind of argument has a long philosophical tradition, going back at least to Descartes's celebrated second *Meditation*, discussing the level of confidence we should have about our empirical knowledge.[49] Novel points in Bostrom's presentation include the suggestion that we might be technologically closer to the required level of computing sophistication than we usually think (invoking Moore's law), as well as the addition of Bayesian conditioning on the number (or sufficiently generalized 'cost' in resources) of such 'ancestor simulations'. It is trivial to see how any version of Fermi's paradox is answered under this hypothesis: extraterrestrial civilizations are likely to be simply beyond the scope of the simulation in the same manner as, for example, present-day numerical simulations of the internal structure of Sun neglect the existence of other stars in the universe. Thus, the **Simulation Hypothesis** is not that different from **Special Creation**: in both cases, the teleological element is what resolves the problem—the observed reality, in some sufficiently relaxed sense, truly exists because of us, humans on Earth.

The ontological turnround indicated by both the **Planetarium hypothesis** and the **Simulation Hypothesis** could, of course, be construed as answering the Fermi question: the Builders (of the planetarium, or of our entire universe as per the **Peer hypothesis**) or the Programmers are indeed those old aliens which we seek. For reasons which are inscrutable to minors like us, they have or have not included other comparable sophonts, our peers, in our environment, physical or simulated.

The **Planetarium hypothesis** and the **Simulation Hypothesis** are similar insofar as the largest part (99.9999 . . . per cent) of everything we observe is not 'real', but is a form of illusion. So, our *Marienbad* trees do not cast shadows as well—these two hypotheses are paradigmatic examples of solipsist solutions, which deny the scientific realism assumption. But they are both *naturalistic*: it is assumed that both the large-scale astronomical illusion in the **Planetarium** and the simulated reality of not-quite-specific spatial extent in the **Simulation hypothesis** are achieved through advanced technology, not supernatural miracles. The type of technology is different, though; while the **Simulation hypothesis** does not require more than sufficiently advanced computer science, the **Planetarium hypothesis** requires something more akin to astroengineering (and it is not immediately obvious that such a mega-artefact is possible at our current level of materials science). On the other hand, the **Simulation hypothesis** has its own special assumptions, notably physicalism about observers, and the validity of Leibniz's principle of indifference.[50] In all likelihood, if physicalism about observers and mental phenomena in general is true, the **Simulation hypothesis** is more plausible than the **Planetarium hypothesis**, since the resource expenditure is much smaller.

On the other hand, there might be ethical reasons why very advanced sophonts could refrain from large-scale simulations such as Bostrom's ancestor simulations: if *our* simulated history so far is an average one, such simulations must contain a huge amount of pain and suffering inflicted on sentient beings with moral standing. A humorous perspective on this issue is offered by Scott Adams in Figure 4.2. In contrast to our ordinary metaphysical understanding of our history as the 'real thing' where bad things such as wars, totalitarian regimes, pandemics, terrorism, and intolerance cannot be avoided in any simple manner, such things in simulated histories are consequences of the decisions made by the Programmers. If the latter are highly moral beings, they might find whatever research or amusement benefits created by ancestor simulations simply

**Figure 4.2** *Dilbert* strip from 17 June 2013 (http://dilbert.com/strip/2013-06-17). While Dilbert wisely pinpoints the key *ethical* issue with the **Simulation Hypothesis** (and its versions), the comic effect is achieved by the anthropocentric assumption that the 'Programmers' are indeed similar—although more advanced—beings to us. This is unwarranted.

*Source*: Courtesy of Scott Adams

insufficient to compensate for the negative value accumulated by pain and suffering of the simulated beings.[51] In contrast, a similar ethical calculus is not involved in the **Planetarium hypothesis**, since the Planetarium Builders have just created a particularly funny **Zoo** for *independently*—and non-intentionally—evolved sentient beings, for which they cannot be considered responsible in the same manner as the Programmers are for those simulated beings.[52]

A 'lite' version of the **Planetarium hypothesis**, and a further step towards the final hypothesis in this chapter (see Section 4.7), is the idea that we could be actually receiving intelligent signals all the time, hidden in the 'normal' noise of astrophysical sources. In other words, the large-scale cosmological scaffolding is real enough; only the parts pertaining to the existence of sophonts are censored/unreal. The idea was recently suggested by no less a controversial source than the famous NSA whistleblower Edward Snowden in a September 2015 interview given from his exile in Russia:[53]

> **The paranoid style in Galactic politics**: The intentional signals of other Galactic sophonts cannot be distinguished from the radiation background, due to highly sophisticated and ubiquitous encryption. Although we could mistake a particular intentional signal for noise a couple of times, there is no a priori reason to think that the intentional emissions would be indistinguishable from noise *all the time*—unless we are dealing with systematic hiding efforts. Advanced encryption could be essentially unbreakable for younger and more primitive civilizations, since it supposedly scales with the computing power available to a civilization. Galactic sophonts typically refrain from direct visitations, and their communications are indistinguishable from natural noise, so there is no Fermi's paradox.

The title of this hypothesis paraphrases the evergreen (sadly!) essay and the eponymous book of the historian Richard Hofstadter, *The Paranoid Style in American Politics*, originally published in 1964.[54] While the idea is somewhat tongue-in-cheek, we are here considering all possibilities, so that even spy concerns should get a hearing! Keep in mind that in this chapter we study hypotheses which reject scientific realism from the get go—we have already bitten the bullet of strangeness. As Kevin Costner's character in one of the most paranoid movies in history, Oliver Stone's *JFK*, exclaims: 'Y'all gotta start thinking on a different level—like the CIA does. We're through the looking glass. Here white is black and black is white!'

Physics tells us that natural objects and processes produce noise—that much is uncontroversial. Noise is usually a consequence of chaotic thermal motion of atoms and other constituents in both the system under study and the detector. But physics cannot tell us that *all* noise is produced in this way; this is another instance of classic Newtonian distinction between the general dynamical laws and the specific boundary conditions. Part of the actual noise could be generated by Galactic sophonts and just *sound* like noise, just as any good cryptographic (or more appropriately, *steganographic*) work looks like a garbled series of senseless letters or other symbols to the uninitiated. In terms of plausibility, **The paranoid style in Galactic politics** certainly is plausible, unlike some of the contenders already discussed in this chapter. *Some* sophonts in a sufficiently large set would certainly go far to remain hidden.

It might even have more than mere plausibility, if Copernicanism is to be taken very strongly in the sense that not only our species but also the historical development of our culture is considered typical. The dramatic expansion of information technologies in the last couple of decades brought about, among other things, a veritable explosion of paranoid memes.[55] If this trend is common to all technological societies above some threshold, and if it pertains to interstellar contacts, as there is some indication in the current debates about the risk of messaging, we might be on our own way to **The paranoid style in Galactic politics**.[56] And there is an argument for this hypothesis based on energy efficiency: as demonstrated in 2004 by three researchers associated with the Santa Fe Institute, 'the most information-efficient format for a given message is indistinguishable from blackbody radiation', which astronomers tend to discard as noise or background.[57] This crucial result has been largely ignored by the SETI community.[58]

What manifestly has *not* been ignored in the mainstream SETI and even the news media is the issue of the possible dangers of our own messaging to hypothetical other Galactic sophonts, and the somewhat oversold and overheated debates related to this.[59] While the problem of detectability of Earth and the human civilization in various wavelength bands and on various observational methods is certainly an important research topic for SETI studies, the emphasis on limited-showtime messaging projects is misplaced. By the mirroring effect, however, if **The paranoid style in Galactic politics** is even remotely true, it will tell us something about our future—and it is not pretty. It might be the case that a future of global encryption paranoia and close-mindedness about all things cosmic awaits us. It might even pave the way for new forms of totalitarian control; there will be more on this in Section 6.5.

As with so many other things, this has been prefigured by Stanislaw Lem. In *Fiasco,* future human explorers observe strange things about Quinta, the planet indubitably inhabited by intelligent beings:[60]

> But the most astounding thing was the result of the Fourier analysis done on the entire radio spectrum of Quinta. All trace of modulation disappeared, while at the same time the power of the transmitters increased. A radiolocation map of the planet showed hundreds of transmitters of white noise, which merged into shapeless blotches. Quinta was emitting noise on all wavelengths. The noise was either a scrambling of the broadcast signals or a kind of coded communication concealed by the semblance of chaos—or else it was chaos indeed, created intentionally.

Of course, Lem's interstellar humans of the far future are in much better position than we are, since they know for certain that their target planet *is* in fact inhabited by technological civilization. Thus, they implicitly discard the option of natural chaos. Unfortunately, we must consider this third option as well.

Lem's epic story in *Fiasco* highlights another link in the space of explanatory hypotheses for **StrongFP**, one which could have crucial impact on future of humanity as well. Not all societies—both Hofstadter and Snowden would have agreed!—are equally prone to the paranoid outlook. Our current fight for basic civil rights, including the right to privacy and freedom from government intrusion and Orwellian oversight, makes sense only insofar we are aware of a better alternative. Totalitarian societies, on the other hand, thrive on secrecy and paranoia; there will be more on this in the crossover with the **Introvert Big Brother** scenario discussed in Chapter 6.

The problems with **The paranoid style in Galactic politics** are rather straightforward: implausible cultural convergence violates the non-exclusivity principle, and the lack of manifestations and artefacts is in tension with the huge Lineweaver timescale. Since we are 'through the looking glass' here anyway, before we consider the longest-ranging idea of this category, we might ask whether there are cheaper prices to pay on the market of ideas.

## 4.6 Directed Panspermia: Are *We* the Aliens?

An apparently extreme violation of realism vis-à-vis absence of extraterrestrials or their manifestations was suggested by the two giants of biochemistry, Francis Crick and Leslie Orgel, in 1973 in order to half-seriously account for the origin of life on Earth: maybe our planet was seeded with early life forms from outer space—and intentionally so.[61] The source of the terrestrial life in this scenario is, therefore, extraterrestrial *intelligent* life, since we have intentionality as the central element of the explanation. This hypothesis is known as *directed panspermia*. In essence, it offers a shocking response to Fermi's question: *they are here, since we are the aliens!* Or, at least, we are the descendants of simple alien life forms which landed on our planet in the distant Precambrian, thus establishing a strange form of 'contact'. This pertains to the entire terrestrial biosphere, since it shares a common ancestor.

> **Directed panspermia:** Our planet has been intentionally seeded with simple life forms originating elsewhere. Thus, intelligent beings did visit the Solar System (if only by proxy) and left a visible artefact: the terrestrial biosphere itself, including us. Since we indirectly recognize the existence of creators of this artefact as old Galactic sophonts, there is no Fermi's paradox.

This motive of intentional seeding of Earth in distant past has been extensively used in fiction.[62] As a character in Ridley Scott's *Prometheus* summarizes it: 'We're just some experiment. And the Earth was a goddamn petri dish.'[63]

Any idea published by a Nobel Prize winner, such as Crick, and one of the pioneers of prebiotic chemistry and abiogenesis studies, such as Orgel, is likely to be taken seriously. How seriously, though, depends on how desperate are we in looking for other explanatory roads towards our goal. The present state-of-the-art research in abiogenesis offers much hope that shortcuts such as proposed by Crick and Orgel are unnecessary: great strides have been made towards completely naturalistic, terrestrial, and 'regular' explanation of the emergence of life on Earth. In this manner, radical 'explanations' are nowadays not truly needed, in contrast to the situation in the 1970s, when Crick and

Orgel wrote their paper. In itself, this is not tremendously important for our purpose in this book: even if unnecessary for explanation of abiogenesis, **Directed panspermia** could be useful as an explanation of Fermi's paradox. Whether it is a remotely plausible explanation is an interesting question in itself, tangential to several key issues in modern astrobiology.

The first key issue is the timescale. Earth was uninhabitable in the course of roughly the first 0.5 Ga of its history. The first traces of life on Earth appeared about 3.8 Ga ago, although there are tantalizing new indications that life had been present earlier, even at 4.1 Ga ago.[64] Since then, all known life forms share the same biochemical basis. Therefore, if **Directed panspermia** occurred and was successful, it had to occur in the window spanning about 200 Ma at the end of the Hadean and the beginning of the Archaean. If it occurred outside of this window, it would probably have been unsuccessful, due to either hostile physical conditions (before the window) or incompatibility with the existent terrestrial biosphere (after the window). Since no traces of what astrobiologists nowadays dub 'shadow biosphere' have been found so far,[65] we can provisionally assume that the seeding occurred within the window. This would, in turn, imply that a Galactic civilization was capable of interstellar flight at least 3.8 Ga ago. (Note that this is entirely consistent with the Lineweaver timescale in Eq. 2.3, which represents only the median of the age distribution—some habitable planets are much older. When talking about Fermi's paradox, we are, to reiterate the conclusions of Section 2.3, primarily interested in that oldest subset.) So, where are the Seeders—and/or their other artefacts—now?

The second issue is the spatial one, although the situation here is less clear. A problem *any* type of panspermia faces is the one of spores' surviving long transit times between the originating habitat and the destination one. One major limiting factor is presented by cosmic rays and their damage to biochemically important compounds, such as nucleic acids. If spores were inactive, in hibernation, or in some other form of suspended animation, the cosmic ray damage would be cumulative and roughly proportional to the distance traversed; local enhancements such as spiral-arm crossings or star-forming regions will magnify the problem. There is no effective shielding from high-energy cosmic rays, some of which penetrate kilometres of hard rock; even worse, after prolonged exposure, shields create secondary radiation, which can worsen the damage to biotic structures. The only realistic option for overcoming the inexorable degradation of such structures is to rely on massive redundancy (in the case of **Directed panspermia**, to try to send as many seeding probes to the same destination as possible) and try to shorten the transit period as much as possible.

The latter branches into two alternatives: (i) either the Seeder civilization was close to the Solar System at the epoch of seeding, or (ii) the Seeder civilization was willing to spend an enormous amount of energy to accelerate the seeding probes to high velocities in order to reach much larger volume of the Galaxy. For several reasons, the second alternative does not seem promising: not only would the required high velocities exacerbate the cosmic ray problem (at relativistic velocities, normal interstellar matter effectively becomes an incoming stream of cosmic rays), but the energy and resource cost for seeding would approach the one needed for direct colonization, and so leads us back to Square One: why was the Solar System—under this hypothesis—just seeded and not colonized?[66]

A plausible conclusion, strengthened by ideas about *our* possible attempts at seeding, is that the first alternative is correct and that seeding originated somewhere nearby in Galactic terms. Unfortunately, due to the differential rotation of the Milky Way, as well as other dynamical mechanisms acting on stars in the disc, even if the home system of the Seeders was very close 3.8 Ga ago, it surely is far away now. All is not lost, however, since *radial* mixing of stellar populations is not so pronounced as the azimuthal one, even on giga-anni-like timescales: while the Seeders' system might be on the other side of the Galaxy at this epoch, it is probably not much closer or much farther away from the Galactic centre than the Sun. This is still a very weak constraint, however (especially since we anyway wish to search for sophonts in the GHZ, which encompasses the Solar circle).

In addition, **Directed panspermia** is interesting as another crossover between astrobiological and futurist themes. There have been speculations and suggestions from time to time that humans could and indeed should seed other planets with the terrestrial kind of life. The reasoning is often based on a form of biocentric ethics: since planet with life is inherently more valuable than a dead planet, we ought to intervene when we encounter a dead—but potentially habitable—planet. The strongest contemporary proponent of directed panspermia and our ethical obligations of seeding the universe with life has been the American physical chemist Michael N. Mautner. In a series of papers and books, he has promoted the view that we have a duty to spread life throughout (mostly) dead universe.[67] Even more, he founded *The Panspermia Society* in 1995, with the explicit goal of bringing this moral imperative to practical fruition. Since the astrobiological revolution started about the same time, all this has resulted in much more vigorous discussions of the relevant technical and bioethical topics.[68] For instance, it is intriguing to consider whether, in spite of all our efforts at preventing planetary contamination, we could actually stop a form of directed panspermia in cases of our spacecraft—and eventually humans—visiting Mars, Europa, or similar potential habitats. Even the simplest space probes are difficult to entirely and reliably sterilize, and it is plainly impossible to do the same with living astronauts! While it is certainly an interesting question in the domain of moral philosophy in general, and bioethics in particular, as to whether there is a moral value in spreading life throughout the universe, this point is still too contentious and unclear to base any serious conclusion—not to mention policies—on it.[69]

If we are, wittingly or not, *that* close to some form of seeding other places, this can only exacerbate the Ashurbanipal problem, which **Directed panspermia** shares with the **Zoo hypothesis** and the **Interdict hypothesis**: in a (near?) future advanced human society or in a past or current advanced extraterrestrial society, it is likely to be individuals and groups willing to engage in directed panspermia and certainly possessing the power to do so (since doing so is rather inexpensive even by today's standards, as Mautner demonstrates). So, why aren't there many more seeded planets, some of which could be expected to evolve creatures capable of sending conventional SETI signals? If seeding is easy and frequent, the 'solution' to Fermi's paradox becomes worse than the problem—a *cane toad solution*, as defined in Section 3.6. If seeding is easy and rare, something is disturbingly wrong in our usual image of advanced technological societies; they might all be hive minds or totalitarian dictatorships. I shall return to this important point in Chapter 6.

Obviously, **Directed panspermia** (as well as any other solipsist hypothesis) has a serious problem with testability, even if we stretch the meaning of the latter. Finding an adequate naturalist theory of abiogenesis will not directly disprove **Directed panspermia**, unless it could be proven that the parameter values correspond exactly to those of primordial Earth and are sufficiently unlikely to be replicated elsewhere. Since the latter condition is impossible to certify, such a theory is likely to remain just 'the best (inferred) explanation'. If the seeding took place 4 Ga before the present, one can hardly hope to find any direct evidence for it. If it occurred in a manner similar to what Mautner and other futurists imagine humankind can do in the near future—essentially by launching cans filled with terrestrial bacteria at various planetary bodies in hope of enough of them surviving to take root and create a new local biosphere—any hope of finding remnants of the seeding probes is unrealistic, since their necessarily low price and expendability will doubtless lead to their ephemeral nature.

On the other hand, the hypothesis could presumably be *corroborated* by finding the civilization responsible for seeding of Earth or finding its artefacts, although such a situation would lead to a host of additional problems. Obviously, this would be a giga-anni-older civilization, which would immediately provoke the Fermi-like question, why haven't we noticed it and its activities before? And this could, in turn, lead us to consider another related radical hypothesis—the one of Lem's **New cosmogony**, to be discussed in Section 4.7. Before we do that, let me just mention that, if the seeding civilization in the **Directed panspermia** scenario turns out to be extinct, it would still be possible, at least in principle, to establish their biochemical kinship to us via their artefacts, *including other biospheres seeded from the same source.* Of course, we would need to falsify the hypothesis that conventional, non-directed panspermia was responsible for multiple biochemically similar biospheres (in which case **StrongFP** remains as big a problem as before). There are contemporary models for interstellar panspermia by quite 'natural' means (i.e. non-intentional and not including sophonts of any kind), but it is still questionable, to say the least, whether they are really operational in the actual Milky Way.[70]

Some attempts have been made to obtain corroboration of **directed panspermia** through finding alleged regularities in the universal genetic code. Under **Directed panspermia**, the genome of particular organisms, as well as the genetic code itself, would be an artifice, so, in principle, it could be expected to contain evidence of its intentional origin, similar to SETI or METI (*M*essaging to *E*xtra-*T*errestrial *I*ntelligence) signals. An early and rather naive attempt in that direction was made by Nakamura, with the alleged pictorial message in a simian virus SV40.[71] Recently, two Kazakh researchers, Vladimir shCherbak and Maxim Makukov, have claimed that arithmetical regularities in the universal genetic code give strong support to its artificial origin, that is, to the hypothesis of **directed panspermia**.[72] However, this sort of numerological argumentation has the same underlying problem as the (in)famous Boeing-747-out-of-junkyard argument for the improbability of abiogenesis: we need not presume that the alternative to design is purely random origin. Quite to the contrary, processes which are 'neither chance nor design' are crucial for a modern understanding of abiogenesis,[73] and we can hope they will help in understanding the origin of the genetic code. Like countless other design arguments (even if naturalistic, as is here the case), the argument of shCherbak and Makukov is not substantially different from the conclusion of Athanasius Kircher who, in the seventeenth century, proclaimed that he had found proof of a supernatural Creator by finding a set of marks similar to the

Latin alphabet in an untouched block of calcite millions of years old. Since there are many hypotheses on the origin of the genetic code, such latitude is allowed; one recent review concluded with the sobering statement that 'the general arguments settled for adopting this or that hypothesis on the origin and evolution of genetic code are strongly based on personal sympathies of authors, but not on strict scientific reasoning.'[74] Until we understand the topic much better, highly speculative hypotheses with a whiff of numerology will appear from time to time, as in all immature scientific fields.[75] But the natural course of events is that this will be the case less often as our understanding progresses (e.g. as was the case with physical cosmology in its early decades[76]).

Why use matter at all, though? The Armenian physicist Vahe Gurzadyan and his collaborators suggested the radical concept of *information panspermia*: life forms existing and travelling through the universe as pure encoded information.[77] In such a picture, 'they might be here' without our noticing; hence, we have another solipsist solution to Fermi's puzzle:

> **Bit-string invaders:** Advanced sophonts travel and/or exist (ambiguity remains on both ontological and technological level) through the Galaxy in the form of encoded bit strings. According to popular algorithmic information theory, the (Kolmogorov or bit-string) complexity of any entity is given by the length, in bits, of the most compact algorithm capable of generating/reconstructing such an entity.[78] Biological information is encoded in genomes which, due to deep redundancy, have much smaller Kolmogorov complexity than naively expected and could be transmitted over interstellar distances and decoded at modest costs. Such bit strings encoded in radio or some other kind of electromagnetic waves could be a much preferred form of interstellar travel for advanced sophonts—provided that decoding and reconstructing devices are present at the destination. Since such 'ghostly streams of information that have the potential to become living'[79] could be present on Earth and in the Solar System at any epoch, there is no Fermi's paradox.

As in **The paranoid style in Galactic politics**, we are dealing with deeply encoded information which looks like noise, but in this case it could be *sophonts themselves*, not just their messages. Note that **Bit-string invaders** can be regarded as a solution to even weaker forms of Fermi's paradox only if supplied with a bunch of additional assumptions. A network of detection and decoding devices has to exist, implying a civilization capable of interstellar flight and potentially detectable in the direct manner. An especially egregious violation of the non-exclusivity requirement is necessary to account for such a uniform cultural evolution: for some unfathomable reasons, all diverse potential sophonts in the Galaxy conclude that they should transmit themselves in the form of bit-string descriptions. If sending such quasi-living bit strings is indeed as cheap as Gurzadyan argues, an advanced technological civilization could employ it—without ceasing all its other potentially detectable activities, including astroengineering feats. Therefore, even accepting all premises of the **Bit-string invaders** may not warrant any explanatory bonus. On the plus side, Gurzadyan's idea offers a straightforward practical consequence: we should study alleged SETI signals from the point of view of the algorithmic information theory and we should try to identify and decode possible bit strings hidden in the noise.

Probably the weakest aspect of **Directed panspermia** and its variations as answers to Fermi's question lies in their evasive nature. Besides a warm and cosy mysticism of being 'descendants' of presumably superior beings like the hypothetical Seeders, it does not give us a meaningful clue about the astrobiological evolution of the Galaxy. Besides expecting some of our siblings (sophonts evolving on other seeded planets) to be around and probably trying to communicate through their own SETI efforts, a truly mystifying question of the ultimate fate of the Seeders or Emitters remains. Where are they, really? If they are extinct, what about their artefacts? **Directed panspermia** does not offer further answers; in addition, **Bit-string invaders** needs to explain the reason for such an exclusive cultural convergence. There is, however, a radical and somewhat disturbing scenario that does try to answer the question about the fate of billion-year-old (super)civilizations.

## 4.7 A New Cosmogony?

A form of intermediate solution between the **planetarium/simulation** hypotheses and **Directed panspermia**, with conclusions arguably more radical than either, is expounded by Stanislaw Lem in his idiosyncratic anthology, *A Perfect Vacuum*.[80] Lem, perhaps unjustly known more as a writer than as an astrobiologist and a fascinating philosopher of science, has time and again shown his boldness for tackling issues related to life and mind in the universe. It is no surprise that he was the first to try to describe, as seriously as possible, considering the subject matter, giga-anni-old civilizations, those which we would expect to exist on the basis of naive Copernicanism applied to the Lineweaver timescale in Eq. (2.3).

> **New cosmogony**: Very early cosmic civilizations ('the Players'; billions of years older than humanity) have advanced so much that their artefacts and their very existence are indistinguishable from 'natural' processes observed in the universe. Their information processing is distributed in the environment on so low a level that we perceive it as operations of the laws of physics. Their long-term plans include manipulation of these very laws in order to create new stages of cosmological evolution. Since the whole of the observable reality is, thus, partly artificial, there is no Fermi's paradox.

This is the most radical hypothesis which preserves the 'ordinary ontology' of our world (in contrast to the **Simulation hypothesis**): distant stars and galaxies do exist as tangible physical objects, at least roughly described by our astronomical textbooks and catalogues. The interpretation of this astronomical knowledge, however, can hardly be more different from the textbook one. Observe the distance Lem goes here—it is not just that *some extraordinary* astrophysical process is an artificial product of advanced sophonts (like Centaurus A in Sagan's *Contact*[81]). On the contrary, it is the *usual* processes which we consider entirely natural (orbiting of planets and moons, cooling and heating of interstellar gas, waves propagating in plasma, even possibly the reactions between nuclei and elementary particles) which constitute the thoughts and actions of the Players. The very difference between the natural and the artificial is thus utterly erased at some point in the course of astrobiological evolution. The answer to Fermi's question *Where are they?* is *everywhere*, literally.

A precursor of sort is contained in musings of the great cosmologist, Sir Fred Hoyle, who, as one of the authors of the steady-state cosmological model naturally wondered about the prospects of life and intelligence in an eternally existing and globally unchanging universe. If the universe has remained essentially the same since past temporal infinity, a strange question arises: why don't we perceive traces of *arbitrarily old* extraterrestrial civilizations? This is a particular version of **KardashevFP**, but much stronger than what we have considered so far (recall the discussion of horizons in Chapter 2): past temporal infinity allows—and even guarantees—for anything that is not specifically forbidden by laws of physics to occur. Since there is seemingly no law of physics prohibiting Kardashev Type 3 civilizations (and even higher types, if any), their absence is indeed puzzling. Why are there no galaxies in the shape of Platonic solids, for example?[82] Some discount is obtained by the fact that steady-state cosmology possesses an event horizon, which old galaxies eventually cross and disappear from view (so we can claim that extreme intentional influences, like dodecahedron-shaped galaxies, are safely behind the horizon). But the leeway here is much smaller than in the standard Big Bang cosmology, where evolutionary processes create a clear cut-off in the age distribution, as discussed in Chapter 2. Therefore, Hoyle speculated that, at some point in their development, civilizations become indistinguishable from their environment, which enables them to persist for arbitrarily long times allowed by the steady-state theory.[83]

The modern cosmologists Paul C. W. Davies and Frank Tipler have also independently noted this strange feature of cosmologies with past temporal infinity.[84] Simply speaking, such cosmologies do not mesh well with astrobiological and SETI considerations. If we wish them to, we need to perform a sort of radical, foundational turn, as Hoyle did (with gusto). Now, as discussed in Chapter 2, we know that we are not living in a universe with infinite past; quite to the contrary, the history of at least our part of the physical world started abruptly in the Big Bang about 13.8 Ga ago. But Hoylean considerations might still be relevant if we, enlightened by the shortness and explosive pace of human history, consider this age as being *effectively infinite*—so much larger than the timescales for evolution of civilizations (once biological prerequisites have been met) that we can still be sure that *whatever's possible according to the laws of physics would have occurred at least once*. This is the crucial signpost on the road taken by Lem.

Another obvious source is the *Star Maker*.[85] In the last quarter of the celebrated 1937 novel, we are offered an 'internal' perspective on the creation of universes. As the narrator accesses larger and larger spatial and temporal scales in Stapledon's strange world, the distinction between untamed nature and the 'artifice of eternity' (W. B. Yates) dissolves. 'The cosmos which he now created was that which contains the readers and the writer of this book' (p. 238). The creation here is naturalistic, but it is the human language which, indeed, fails in this limit of physical capacities: 'To speak thus of the universal creative spirit is almost childishly anthropomorphic' (p. 239). Some of the description is surprisingly fresh, considering its vintage:

> This most subtle medium the Star Maker now rough-hewed into the general form of a cosmos. Thus he fashioned a still indeterminate space-time, as yet quite ungeometrized; an amorphous physicality with no clear quality or direction, no intricacy of physical laws.

Stapledon perhaps did not entirely succeed in conveying such a grand vision, but it is a glorious failure, and the one still awaiting its proper inheritor; perhaps s/he will belong to the realm of physics, rather than—or in addition to—the one of poetry.[86]

In an essay on another of Stapledon's thought-provoking novels, *Last and First Men*, Lem outlines the central paradox of long-term prediction exactly in the context of **StrongFP**:[87]

> Predictions beyond 80 or 100 years inevitably fail. Beyond that range lies the impenetrable darkness of the future, and above it, a single definite sign indecipherable, but impinging on us all the more: the Silence of the Universe. The universe has not yielded to the radiance of civilizations; it does not scintillate with brilliant astro-technical works—although that is how it should be, if the law of psychozoic beings were an aspect of the exponential ortho-evolution of instrumentality in cosmic dimensions. This is absolutely certain, and it is worth keeping in mind.
>
> It is also worth considering the significance of the fact that the innovations Stapledon distributed over immeasurable time are already real, that decades have already completed Stapledon's Billion Year Plan. This means that everything that 40 years ago was imaginable only with the greatest exertion of fantasy will be realized eventually. What we have not realized, we have considered impossible; but in fact we no longer consider anything to be absolutely and finally pure fantasy.

How far can this non-fantasy be sustained, and what are the consequences for SETI? Lem's answer to this is given—through a fictional 'external perspective'—in the mind-boggling story/essay/paper 'The New Cosmogony', which was originally written in 1971.[88] In contrast to the timid observer in Stapledon's novel, Lem's narrator is a modern-day Nobel Prize winner, Prof. Alfred Testa,[89] a somewhat cynical physicist and philosopher who describes his 'trivial' solution of **StrongFP**. Why don't we perceive artefacts and engineering of super-civilizations billions of years older than ours? No, no, no, he says, we do!—the laws of physics themselves are the product of their engineering, of the Great Game, played on the largest possible spatial and temporal scales. The Players' pool of knowledge is so vast and their capacities so overwhelming, that only 'by their fruits ye shall know them'—and those are the (effective) laws of nature:[90]

> If one considers 'artificial' to be that which is shaped by an active Intelligence, then the entire Universe that surrounds us is already *artificial*.... Instrumental technologies are required only by a civilization still in embryonic stage, like Earth's. A billion-year-old civilization employs none. Its tools are what we call the Laws of Nature. Physics itself is the 'machine' of such civilizations!

He continues to explain the hierarchical levels of various 'physicses' and the strategies of reconstructing, using game-theory approaches, the evolutionary pathway of our own effective laws. But the main point is clear:[91]

> Theoretically, if the energy Earth's science invests in elementary particle research were to be multiplied $10^{19}$ times, that research as a discovering of the

> state of things would turn into a changing that state! Instead of examining the laws of Nature, we would be imperceptibly altering them.

Only seemingly paradoxically, Lem was an optimist in both epistemological and ethical terms as far as this particular story is concerned. Recounting an anecdote at the end of his speech, Prof. Testa says:[92]

> I should like to quote here the words of Professor Ernst Ahrens, my teacher. Many years ago, when, still a youth, I went to him with my first drafts containing the conception of the Game, to ask him his opinion, Ahrens said: 'A theory? A theory, yet? Maybe it is not a theory. Mankind is going to the stars, yes? Then, even if there is nothing to it, this thing, maybe what we have here is a blueprint, maybe it will all come to pass someday, just so!'

So, Lem concludes his splendid essay, we could not only hope to understand the 'true' palimpsest of reality created by the Players, but also hope to become a Player one day. Rejecting scientific realism need not necessarily mean the defeat of science—it could (in an admittedly contrived context) be the ultimate vindication of its power. To think that 'The New Cosmogony' predates—for many decades!—such relevant scientific ideas and concepts as distributed computing, cosmological computational bounds, the Internet of things, and the holographic principle, not to mention speculations about multiple vacuums of superstring theory or eternal inflation, is truly flabbergasting.[93]

Does it work as a resolution of Fermi's paradox, however? It might be an explanatory overkill, since it could in principle explain not only the empirical Great Silence (and any other form of Fermi's problem), but also *any other* particular astrobiological situation. At some point, that a posteriori explanatory power must become a disadvantage. Besides, **New cosmogony** might have a problem with the lack of our detectable peers, that is, civilizations which are not the inscrutable Players, but only somewhat older than us. This problem is minor in comparison to the bulk of **StrongFP**, and it might be explicable as well within the **New cosmogony**, but it requires some further work. Finally, in spite of Testa's fictional achievements, to make progress in quantitative elucidation of this hypothesis seems like a very tall order. At least for now.

## 4.8 A Solipsist Résumé

A rather wild journey, isn't it? It is difficult to objectively assess the value of solipsist hypotheses as solutions to Fermi' paradox, since we have grown so much accustomed to scientific realism that even very modest violations, like

**Directed panspermia**, sound outlandish at times. Most of these hypotheses are either untestable in principle, like the eponymous metaphysical doctrine, or testable only in the limit of very long temporal and spatial scales (like the **Zoo hypothesis** or the **Planetarium hypothesis**), so that they do not belong to the realm of science, conventionally understood. In other words, they miss the stepping stone upon which practically the entire scientific endeavour of today is founded. Their proponents are likely to retort that the issue is sufficiently distinct from other scientific problems to justify a greater divergence of epistemological attitudes—but this is hard to defend as long as one could pay a smaller price.

Some of them violate the non-exclusivity requirement as well; this is, for instance, obvious in the **Zoo, Interdict**, or **Planetarium** scenarios, since they presume a large-scale cultural uniformity. This is not the case, however, with the **Simulation hypothesis**, since the simulated reality is likely to be clearly designed and spatially and temporally limited. Depending on the unknowable goals of the simulation Programmers, it might be the case that we are truly alone in the simulated universe or that the detectability of other sophonts has been intentionally suppressed. **Directed panspermia** has additional problems—notably the absence of any further manifestations of our 'parent civilization', in spite of its immense age. If they became extinct in the meantime, what did happen with other seeded planets? The Copernican reasoning suggests that we should expect evolution to occur faster at some places than on Earth (and, of course, slower at other sites as well)—where are our interstellar siblings, then? Shouldn't we perceive the Galaxy *fuller of life* with the **Directed panspermia** then without it?

Observation selection effects are important ingredient in these hypotheses. **Directed panspermia** could, for instance, be linked with a curious puzzle posed by Ken Olum and which also helps illustrate the intriguing interplay between modern cosmology and astrobiology.[94] Starting from the assumption of an infinite universe (following from the inflationary paradigm), Olum conjectures that there are civilizations much larger than ours (which currently consists of about $10^{10}$ observers). Spatial extent and the amount of resources at the disposal of such large civilizations would lead, in principle, to a much larger number of observers (e.g. $10^{19}$ observers in a Kardashev Type 3 civilization in a galaxy similar to the Milky Way). Now, even if 99 per cent of all existing civilizations are small ones similar to our own, the anthropic reasoning offers an overwhelming probabilistic prediction that we live in a large civilization. Since this prediction is spectacularly unsuccessful on empirical grounds, with a probability of such failure being about $10^{-8}$, something is clearly wrong here. In our nomenclature, Olum's problem would correspond to **KardashevFP**. Olum presents a dozen or so hypothetical solutions to this alleged conflict of the anthropic reasoning with cosmology, one of them being the possibility that *we are indeed part of a large civilization without being aware of that fact*. The **Directed panspermia** hypothesis can be regarded as operationalization of that option—and so could be, with some semantic manoeuvring, the **Planetarium hypothesis**, the **Simulation hypothesis**, and even the **New cosmogony**. There are systematic deficiencies in Olum's conclusions[95] but, in any case, the very fact that some form of the principle of indifference and the counting of observers is used in this discussion shows how closely the theory of observation selection effects[96] is tied with the issues at the very heart of Fermi's paradox and, in particular, its solipsist sector.

Jumping ahead, a clearly non-exclusive solution obeying all philosophical desiderata has not been found thus far—therefore, we should not reject solipsist hypotheses out of hand. Even the most sober mathematical studies, such as the one by Newman and Sagan, were compelled to, somewhat resignedly, conclude that 'it is curious that the solution to the problem "Where are they?" depends powerfully on the politics and ethics of advanced societies.'[97] There is—and there will probably be for a long time to come—something deeply unsatisfactory about this sort of answer. It is especially disappointing to encounter it after a lot of mathematical analysis by the same authors, and keeping in mind by now more than half a century of sustained and often carefully planned and executed SETI efforts. Something is rotten in this state.

This is, in fact, the council of despair, and it is not surprising to notice that sophisticated religious thinkers like Alvin Plantinga at least honestly admit that it is naturalism that bothers them, not science and accompanying scientific realism.[98] This circumstance, as well as occasional (sub)cultural and even political appeal, explains why solipsist hypotheses are likely to reappear from time to time in the future, especially if the current erosion of secularist and modernist values of the Enlightenment continues.

While it is easy to mock solipsism, the temptation should be resisted. After all, the paranoia with which some people resist any idea of METI might conceivably grow in the future of humanity and result in the future (post) human civilization taking an isolationist and stealthy pathway. If we suppose that most or all advanced civilizations in the Milky Way take the same course, we could end up in a situation where contact will be shunned in spite of the possible physical presence of advanced sophonts in other inhabited planetary systems, including the Solar System. Such an extremely stealthy approach might have the same effect as some of the solipsist hypotheses; we might find ourselves in a **Zoo** not because we are inherently interesting, worthy of preservation or observation, but because generic pathways of advanced technological civilizations involve isolationism and stealth. To a degree, this conjecture exhibits continuity with some of the logistic hypotheses, to be discussed in Chapter 7.

Contradictory messages and mixed signals—like those surreal shadows in Resnais's movie—can be empirical clues for testing the solipsist hypotheses.[99] More research is necessary, however, especially on building specific quantitative models of at least some of those hypotheses. The difficulties in conducting such research are obvious: from mockery and institutional resistance to the lack of relevant background multidisciplinary work, especially in behavioural and social sciences. Post-positivist epistemology makes it somewhat easier to conceive such a research program, but the obstacles are still enormous. Writing a funding proposal for research on such 'mad' ideas still sounds like the proverbial *mission impossible*.

Another important thing to keep in mind—at least until we return to a comparison of hypotheses in Chapter 8—is that the solipsist category is the one containing extraterrestrial intelligent beings similar to those invented by human imagination and represented in the pop-cultural discourse. Together with some of the logistic solutions (Chapter 7), the sophonts implied by solipsist solutions are the ones we could, in principle, communicate with or (if communication is not possible, such as under **Directed panspermia** or **New cosmogony**) at least vaguely understand their motivation. Even Lem's Prof. Testa belongs to the near future of humanity—not do only he and his colleagues firmly belong to people like us and not some superior post-persons, but it is also quite improbable that cultural practices like the Nobel Prize speeches will persist for cosmological timescales in the future! And yet, there is a substantial sense in which the 'Testan research programme' could reveal key aspects of the existence of giga-anni-older sophonts. Perhaps we should go back and read more Lem!

In a similar vein, Benoit Lebon has suggested 'progenitive conception' as a possible future of humanity: the construction of an artificial system containing all our knowledge and experiences as a safeguard—and transferring it to another animal species if and when humanity goes extinct.[100] While some of Lebon's concerns are obviated by the emergence of the concept of postbiological evolution, the central idea remains intriguing. The environment might indeed contain much more cognition in general than is seen from afar and much more than it is naively thought. We can imagine the safeguarding system as a kind of 'Krell machine' from the legendary science-fiction movie *Forbidden Planet*: an astro-engineering feat not easily—or at all—detectable from afar.[101] We can also envision the transferring/embodying process to take a significant interval of time—for instance, if no convenient species of animals survive the ending of the sophonts' civilization, and the system has to wait for such a convenient species to evolve. On the other hand, once copied into the artefact, the civilization might indeed prefer such a virtual form of existence and not be embodied at all.[102] Alternatively, one can imagine a purely biotechnological form of transplanting consciousness into nature, as vividly suggested by Shane Carruth in his magnificent *Upstream Color*—a movie which can serve as an appropriate bookend to *L'Année dernière à Marienbad* in discussing this thematic circle.[103]

In this intuitive, but relevant, sense, the assumed sophonts are *like us*. In contrast, under the 'rare-Earth' or catastrophist categories, either there are no sophonts whatsoever or they are short-lived or dramatically different from us. While this is in part a simple consequence of anthropocentric selection effects—hypotheses for resolving Fermi's paradox being product of human minds and culture and thus more likely to involve drives and motivations 'like ours' than something radically different—it is worth considering that the explanatory project should not be *entirely driven* in this manner. I shall return to this point in Chapters 8 and 9.

CHAPTER 5

# Terra Nostra

## *'Rare Earth' and Related Solutions*

In the complex 1975 novel *Terra Nostra* ('Our Earth' in Latin), the great Mexican writer Carlos Fuentes boldly attempted to distil the entire historical experience of Latin America and, by extension, the whole of the West, into a single fantastic narrative. The story is centred on the key decades of the European colonization of the 'New World' during the reign of King Philip II of Spain (1556–98). The baroque plot shifts dramatically between the sixteenth and the twentieth century, with excursions into the more distant past and an alternative future, and is obsessive about seeking the causes and roots of seemingly narrow, particular, parochial, contemporary phenomena in a wider and richer tapestry of the underlying cognitive, scientific, technological, and artistic processes.

And it is not just the novel's title which relates to our present topic. At one point, a character, Valerio Camillo, presents a wondrous invention, his 'Theatre of Memory' (a suitably fantastic version of the project actually suggested by the very real Italian humanist namesake Giulio Delminio Camillo, *c.*1480–1544), which is capable of showing not only possible future histories, but also possible *alternative pasts*. Such an undertaking is not devoid of danger, as Valerio's friend, student Ludovico, points out:[1]

> 'What will they give me, the kings of this world, in exchange for this invention that would permit them to recall what could have been and was not?'
>
> 'Nothing, Maestro Valerio. For the only thing that interests them is what really is, and what will be.'

> Valerio Camillo's eyes glistened as never before, the only light in the suddenly darkened theater: 'And is it not important to them, either, to know what never will be?'
>
> 'Perhaps, since that is a different manner of knowing what will be.'
>
> 'You do not understand me, monsignore. The images of my theater bring together all the possibilities of the past, but they also represent all the opportunities of the future, for knowing what was not, we shall know what demands to be: what has not been, you have seen, is a latent event awaiting its moment to be, its second chance, the opportunity to live another life. History repeats itself only because we are unaware of the alternate possibility for each historic event: what that event could have been but was not.'

We are in a similar situation when trying to explain the causes of historical events, be they part of a political, cultural, cosmological. or planetological history. In a sense, we need to reverse our perspective when tackling those questions: in contrast to the usual, quasi-Baconian perspective which abstracts away particulars of ourselves as observers, our special position in time and space, and so on, we need to ask directly about the origin and impact of such peculiarities. As Fuentes's inventor tells the student: 'Here roles are reversed... You, the only spectator, will occupy the stage. The performance will take place in the auditorium.'

On the astrobiological stage we—the denizens of Earth's biosphere—are still the only performers *and* spectators. So, any explanatory hypothesis of a perceived general fact needs to be qualified on the particular requirements for our existence and our perceptions of that alleged general fact. But we (obviously) do not know the entire list of requirements and requisites for the emergence and evolution of the observers similar to us; if we knew that, much of the fields like astrobiology, evolutionary biology, cognitive sciences, and so on, would be simple and boring. Therefore, a shady area exists in which we are not yet able to separate lawful necessities of the evolution of matter from accidental, historical happenstance. Either one or both led to our present-day observations and our present-day empirical facts, but in what proportion and to what extent is not clear.

This shady area opens up possibilities for explaining a high-level general fact like the absence of extraterrestrial sophonts and their traces and manifestations from our past light cone. If such sophonts—and observers generally, including us—are more the result of happenstance than a lawful outcome of cosmic evolution, then even bare intuition tells us that our attempts to find a general (causal/lawful) explanatory mechanism are misguided. That is the road taken by the 'rare-Earth' theorists, considered in this chapter.

The same conclusion can be reached in yet another way. From the *general* point of view of intelligent observers, *any* intelligent observer will find a situation specified in **StrongFP** problematic. The problem, according to solipsist hypotheses, is in our perceptions of the world; neocatastrophic hypotheses (to be considered in Chapter 6), as well as logistic hypotheses (Chapter 7), locate the problem in our actions and reactions in the world. 'Rare-Earth' hypotheses, to be discussed here, are somewhat closer to the solipsist perspective, since they problematize not only perceptions, but more inclusive properties of observers, including the limitations their very existence imposes on the world itself. In the words of Fuentes's Chronicler, we aspire to achieve 'the impossible: a perfectly simultaneous narration'—simultaneous, that is, in the sense important for this book, in terms of understanding both our contingent properties and how the answers to our questions are constrained by the very same contingency. So, is the Fuentesian 'opportunity to live another life' open to the Galactic sophonts?

## 5.1 Down with Copernicanism!

One might naively think that rejection of Copernicanism is a steep price, unlikely to be paid by anybody except a few religious zealots stuck in the Middle Ages. Unfortunately, this is far from being the truth and, if anything, the anti-Copernican cartel has grown stronger in recent decades. I call it a cartel, since it gathers wildly heterogeneous groups, individuals, and ways of thinking, with the common denominator of either vested interests in anthropocentric institutions permeating our society, or ideological blindness for reality underpinning the successes of the scientific method, especially in the course of the last two centuries. An extremely wide anti-Copernican front encompass people ranging from opponents of animal rights and other defenders of anthropocentric legal orthodoxies to various conservative 'warriors on science' and their various allies, from the Discovery Institute, to anti-vaccination lobbies, to self-proclaimed 'progressive humanists' incapable of dealing with the rational facts of science on a psychological level (including indubitably enlightened people like Hannah Arendt or Michael Frayn[2]), to radical futurists believing we need ideological anthropocentrism to ensure the perceived desired future of humanity. Fighters against perceived 'scientism' and the alleged 'coldness' and 'inhumanity' of modern science, à la Francis Fukuyama or Mary Midgley,[3] hold hands *both* with anti-environmentalists who do not recognize Genesis 1:28–30 as the harmful Bronze Age nonsense/superstition it really is, and extreme new-age environmentalists worshipping Gaia as—no surprise there!—the centre of the universe. Concerned guardians of the 'humanistic canon', worried about the position of social sciences and humanities in university curricula (allegedly under attack from the evil forces of science and engineering), join forces in the anti-Copernican camp with assorted media and arts pundits portraying science and scientists in a uniform Dr Victor Frankenstein's mould. And to all these one should add legions of their less sophisticated counterparts in much of the developing world, often blending local superstitions into the antiscientific mix (e.g. the 'explaining', in large parts of Africa, of illnesses—including AIDS—as being caused by black magic) and preying on poor educational standards. In spite of much effort by various environmental groups, in the twenty-first century, a mass murderer of animals, including our closest mammalian and even primate relatives, is still celebrated as a 'capable hunter', while nobody would attach that label to, for instance, the Norwegian far-right terrorist Anders Behring Breivik, who was convicted for killing seventy-seven people, including sixty-nine children, in 2011. It would not be an overstatement to claim that anti-Copernicanism in one form or another dominates 99 per cent of public life and thought on this planet—which still serenely revolves around the Sun, an insignificant speck on the periphery of the Milky Way, with our Galaxy itself being only a smudge of light at the outskirts of the Local Supercluster.

Thus, the job of the Copernican revolution is still quite an actual, timely, and risky concern. While the Inquisition which condemned Galileo seems unlikely to receive any open support today, I submit that this is more due to their old-fashioned garments and politically incorrect language than any true dissonance of ideas. After all, the underlying concern stays the same: worry about the perceived 'well-being of humanity' and its institutions being threatened by 'cold' and 'soulless' science and its discoveries, never minding the truth. The rulers of today may not style themselves 'the kings of this world' any more, but they are still utterly disinterested in the Theatre of Memory. The focus of the odium has shifted from astronomy in Galileo's time to evolutionary biology, computer science, and environmental science today, but the underlying reality remains the same: below a thin skin of modernity often threatens a surprisingly medieval anthropocentrism. As summarized by Edsger Dijkstra, 'Science is hated because its mastery requires too much hard work, and, by the same token, its practitioners, the scientists, are hated because of their power they derive from it.'[4]

Even in science itself, the Copernican revolution often looks like an unfinished business, and indeed many scientists aid and abet the tide of anti-Copernicanism in various ways: by condoning various anthropocentric social and political mores, especially in animal ethics and environmental science; by reintroducing teleological elements into science even where it is clearly unnecessary; by accepting some of the postmodern nonsense about the social construction of physical reality; by seeking 'deep' reasons beneath obvious coincidences; by subscribing to conspiracy theories outside one's own discipline; by condoning the abuse of science by politicians and clergymen whenever it suits their own ideological prejudices; and so on—and, most pertinently for our purposes here, by postulating various 'rare-Earth', anti-Copernican hypotheses. This is not to say that such hypotheses cannot be the best explanations of the empirical phenomena; but I maintain that we need to be fully honest and upfront about their wider context and ramifications.[5]

The way in which Fermi's paradox hinges on our acceptance of Copernicanism was explained in Section 3.4. What remains to be done is to demonstrate how abandoning—even for the sake of speculative discussion—this philosophical assumption generates plausible explanatory hypotheses.

The ground was set by the celebrated book *Rare Earth*, which was written by Peter Ward and Donald Brownlee and whose appearance in 2000 heralded the birth of the new astrobiological paradigm.[6] The authors expounded the view that, while simple microbial life is probably ubiquitous throughout the Galaxy, complex biospheres, like the terrestrial one, are very rare due to the exceptional combination of many distinct requirements. The following ingredients for a *rare-Earth hypothesis* are well known to even a casual student of astrobiology:

- **a circumstellar habitable zone**: a habitable planet needs to be in a very narrow interval of distances from the parent star
- **'rare Moon'**: having a large moon to stabilize the planetary axis is crucial for long-term climate stability
- **'rare Jupiter'**: having a giant planet ('Jupiter') at the right distance to deflect much of the incoming cometary and asteroidal material enables a sufficiently low level of impact catastrophes
- **'rare elements'**: Radioactive *r*-elements (especially U and Th) need to be present in the planetary interior in a sufficient amount to enable plate tectonics and functioning of the carbon–silicate cycle
- **'rare Cambrian-explosion analogies'**: the evolution of complex metazoans requires exceptional physical, chemical, and geological conditions for episodes of sudden diversification and expansion of life

Many further items can be added to the list. Each of these requirements is unlikely and they are (or at least seem to be) causally independent, so that their combination is bound to be incredibly rare and probably unique in the Milky Way. The probability of conjunction of many independent events is just the product of individual probabilities; if individual probabilities are small—or if at least one of them is *very small*—the conjunction must be astronomically improbable. In addition, Ward and Brownlee break new grounds with revealing the importance of hitherto ignored or downplayed factors like the importance of plate tectonics, inertial interchange events, or 'Snowball Earth' episodes of global glaciation for the development of complex life. In many ways, the rare-Earth hypothesis has become somewhat of a default position in astrobiological circles,[7] and—since it predicts the absence of targets for SETI—a mainstay of SETI scepticism. Thus, its challenge to Copernicanism has been largely accepted (although, as argued below, there are lower prices to be paid on the market of ideas) as sound in the mainstream astrobiology. Particular rare-Earth hypotheses (insofar as we may treat them as separate) are difficult to assess, as we lack first-hand knowledge of other Earth-like planets, but some of the difficulties have been discussed in the literature thus far.[8]

To see how this could help us with Fermi's paradox, consider a chronological chain of the following evolutionary events:

**Planet formation → Abiogenesis → Evolution of complex metazoans → Noogenesis → Emergence of technological civilization → Emergence of astroengineering technology**

This is written only symbolically, without any claims of being exhaustive or complete. Some of these events have already occurred on Earth, while the last one lies in our future, unless—per processes to be discussed in Chapter 6—prevented by some defeating catastrophe. The last one is a placeholder for 'any form of visitation, messaging, or leaving of traces detectable over interstellar distances', that is, a placeholder for whatever causes the unpalatable conclusion in **StrongFP**. Let us neglect this last event and concentrate on the chain of processes leading to it. If any particular event in this chain is *inherently* extremely unlikely to be happening anywhere beyond what we positively know happened on our planet, we need not worry about the paradox at all. And perhaps there are links in this chain which are unknown to us at present but are inherently extremely unlikely.

The qualification 'inherently' is of central importance here: we need to distinguish between the inherent unlikelihood of an event taking place or a process being completed anywhere and at any epoch, and historical 'effective' improbability which is due to an external interference, be it Divine or quasi-Divine intervention (which was considered in Chapter 4 within the pool of solipsist hypotheses), or meddlesome supernovae/GRBs (to be considered in Chapter 6, within the pool of neocatastrophic hypotheses) causing an adverse outcome in almost all cases. Insofar as we are capable of separating between the two—which is a reflection of the age-old problem in the philosophy of science, namely that of separating 'law-like' probabilistic statements from those representing just contingent, historical happenstance—we obtain a new lever to suppress the last event in the chain, the emergence of astroengineering technology, with all its paradoxical consequences when put into the realistic spatio-temporal context of the Milky Way. By the same token, if we are unable to completely separate them, we have a shady situation in which the proposals for abandoning Copernicanism might smoothly join with those rejecting either realism or gradualism as well.

To make matters more specific, let me consider one particular requirement for the rare-Earth hypothesis, namely the requirement that a habitable planet must have a massive moon in order to stabilize its rotational axis and prevent chaotic changes in obliquity. The latter would act as causes of possible climate chaos.[9] Ward and Brownlee argue that changes similar to those characterizing the rotational history of Mars (variations in obliquity of more than 50°) would be adversarial to the emergence of complex life forms, at least on land, although the authors allow for the possibility that marine life could be affected less and so proceed to evolve complex metazoans. And yet, the giant impact, the 'Big Splash' (or 'Big Splat') creating our Moon, was seemingly a highly improbable event.[10] Therefore, such stabilization of the rotational tilt via a massive satellite would, in the general Galactic context, be exceedingly rare—so it is yet another 'rare-Earth' factor violating Copernicanism with the respect to Earth's habitability.[11] However, it clearly does not influence all crucial steps in the sequence above. After all, the very admission that simple and possibly even complex life forms could survive in oceans in spite of dramatic climate changes on land speaks to that effect. So, the space of possibilities leading to a habitable planet seems to be much wider than we—in our poor imagination—tend to believe.

All this could be given another, more empirically minded slant. Abandoning Copernicanism in this particular context does not merely means that Earth is *un*typical in the set of all habitable locales in the universe; as previously discussed, for a particular choice of metrics, this will always be true. There are always those characteristics of our planet which make it truly unique: for instance, we can define a quantity $\sigma_{2408}$(planet) = the fractional surface area above

2408 metres above the sea/reference level at the present time.[12] Abandoning Copernicanism certainly does not mean that Earth is somehow 'more unique' as to the value of $\sigma_{2408}$. It is highly unlikely that there is a Planet X anywhere within our cosmological horizon such that $\sigma_{2408}(X)$ differs from $\sigma_{2408}(\text{Earth})$ for less than one part in a billion. This is true irrespectively of any philosophical or methodological commitments. However, $\sigma_{2408}$ is as irrelevant for astrobiology as it is arbitrary. Abandoning Copernicanism means, instead, that Earth is untypical in one or more parameters necessary for being actually inhabited by intelligent beings. Examples of non-arbitrary and relevant parameters could be the oxygen content of the atmosphere, the stability of various orbital and rotational motions, the stability of the climate, and so on.

Is this really so, however? We feel strongly that (say) the oxygen fraction $f_{O2}(\text{planet})$ is relevant, in contrast to $\sigma_{2408}(\text{planet})$, but how do we proceed to prove it exactly? Remember, we are discussing the habitability for intelligent beings *in the most general available context*. If we argue that the oxygen fraction played an all-important role in the evolutionary history of *our* biosphere, which resulted in the emergence of intelligent beings, we are subject to the simplest observation-selection effect whenever we try to generalize such a conclusion. We cannot hope to obtain a sufficiently general conclusion without trying—and failing—to envision other pathways to intelligence, including those hypothetical paths which progress in an anoxic environment. Since intelligent beings, in contrast to simpler organisms, can use technology to adapt to a hostile environment, we need a very strong refutation of the existential quantifier in order to suppress any traces and manifestations of intelligence at all past epochs and thus resolve **StrongFP**. Any vision, artistic or technological, of a lunar colony testifies that it is entirely conceivable to search for intelligent beings in places which are non-habitable in their own right.

Therefore, not just any rare-Earth hypothesis will do the explanatory task. At least in some respects, rare-Earth hypotheses for resolving Fermi's paradox must go a bit further *beyond* the original rare-Earth hypothesis of Ward and Brownlee. The rare-Earth hypothesis argues that the emergence of Earth-like complex biospheres is unlikely and that therefore such complex biospheres are rare at all times. We might differ in the estimated level of 'rareness', but let us, just for the sake of argument, assume that a unique complex biosphere per galaxy similar to the Milky Way is the required measure.[13] In the present epoch, our Earth is such a complex biosphere. But there were epochs before the formation of Earth and the Solar System, which are rather young features, cosmologically speaking. So, have there been complex biospheres in the Milky Way before Earth?[14] If there were—even a few of them—we still need *some* additional explanation for why they have not spawned advanced technological civilizations which could be observable way beyond their vicinity in space and much farther in time.

As discussed in Chapter 1, there is a fundamental asymmetry between conditions for the emergence of life and intelligence, and conditions in the region where advanced technological civilizations could spread and live. The former are much more restrictive, which is why it is easy to conceive of advanced technological civilizations utilizing available resources in almost any Galactic locale, no matter how inhospitable to life it might generically be. This is almost trivial: even the primitive human civilization of today utilizes resources like oil from the bottom of the sea, and humans certainly did not evolve and could not have evolved (with our current morphology) in the ocean.

If intelligent beings and civilizations emerging within those *rare* complex biospheres fail to spread throughout their home galaxies (and, ultimately, the universe beyond), perhaps we need some specific explanation for that. The explanation cannot now be based upon any form of influence of other sophonts, since we have assumed that the Galaxy and the wider universe are essentially empty. And, by any reasonable scenario, expansion into an empty universe has to be easier—and *faster*—than the expansion into an already partly inhabited universe. So, why don't we observe Kardashev's Type 3 civilizations in other galaxies? Should they not be more feasible?[15]

For all these reasons, when I speak about rare-Earth hypotheses here, I consider this wider range of issues, going beyond the ideas astrobiologists usually put under that title. The entire *class* of hypotheses has, as a common denominator, the abandoning of Copernicanism in at least one of its many aspects. This, of course, does not mean that all the hypotheses are of equal merit. The situation can be envisioned through an analogy with the theories of abiogenesis: depending on the resolution of the crucial dilemma of the origin of life, (almost) all explanatory hypotheses can be divided into 'metabolism-first' or 'replication-first' classes. Both classes contain hypotheses which are more or less likely to be the true description of the origin of life; our basic choice of class, however, does not impact their intrinsic likelihood. Even if we are firmly in the 'metabolism-first' camp, this will not make our subjective probability of two hypotheses, A and B, say, which postulate that replication came first, equal to zero. Our evidence (real or alleged) that metabolism came first might depress the probabilities of A and B in the standard Bayesian manner—but it cannot and should not make them zero. For this rather prosaic reason, even those astrobiologists who reject the rare-Earth hypothesis of Ward and Brownlee[16] should do well to consider the rare-Earth explanations for Fermi's paradox. The major lesson that a rare-Earth hypothesis needs to be somewhat augmented in order to cope with the possibility of a 'rare *early spreading of intelligence*' should be kept in mind.

## 5.2 Modern Rare-Earth Hypotheses

If one, for reasons which may have nothing to do with Fermi's paradox, concludes that we are the only technological civilization in the Galaxy, the problem is seemingly resolved, although we still need to make the notion more precise. Here, Hanson's concept of the *Great Filter* comes in very handy. The Great Filter is in part synonymous with whatever process is the explanation of Fermi's paradox: whatever is out there and acts to reduce the great number of sites where intelligent life might arise to the tiny number of intelligent species with advanced civilizations actually observed (currently just one: human). This probability threshold, which could lie behind us (in our past) or in front of us (in our future), might work either as a barrier to the evolution of intelligent life, or as a high probability of destruction or self-destruction. Thus, the Great Filter covers the solutions discussed in this chapter and in Chapter 6; it is much less clear whether it could be formulated in a way which would encompass the solipsist solutions of Chapter 4, or the logistic ones to be considered in Chapter 7.

For the present purpose, it is particularly convenient to link the issue with the critical steps which occur in the history of life and intelligence. Hanson suggests the following: [17]

> Consider our best-guess evolutionary path to an explosion which leads to visible colonization of most of the visible universe:
>
> 1. The right star system (including organics)
> 2. Reproductive something (e.g. RNA)
> 3. Simple (prokaryotic) single-cell life
> 4. Complex (archaeatic & eukaryotic) single-cell life
> 5. Sexual reproduction
> 6. Multi-cell life
> 7. Tool-using animals with big brains
> 8. Where we are now
> 9. Colonization explosion
>
> (This list of steps is not intended to be complete.) The Great Silence implies that one or more of these steps are *very* improbable; there is a 'Great Filter' along

the path between simple dead stuff and explosive life. The vast vast majority of stuff that starts along this path never makes it.

While the exact list may differ from author to author, the general concept seems well defined and clear. Hence, the following is a generic possible answer to Fermi's paradox:

> **Early Great Filter**: At least one of the early crucial steps in biological evolution is very hard, typically requiring time much longer than the age of the universe given in Eq. (2.1). The fact that this *hard step* (or steps) actually occurred on Earth is explained through observation-selection effects, since our existence as observers would not have been possible otherwise. Since the hard step is inherently astronomically improbable, no other biospheres in the Galaxy has passed it yet—so we are truly alone and there is no Fermi's paradox.

This is the default explanation of those who choose to accept the negative horn of Fermi's dilemma ('we are alone').[18] Authors favouring this type of explanation are often labelled 'SETI sceptics', although this is misleading for several reasons, since there is no a priori reason to regard uniqueness as the default position. In this view, Earth is truly unique—in the Milky Way context, if not in the wider universe—and SETI searches are futile and hopeless from the start. Occasionally, rather simplistic ethical conclusions are drawn as well: that we need to re-appreciate our value in the universe, that humanity is more important than hitherto assumed, and so on.[19]

In one important way, this is just a 'passing-the-buck' strategy, since it immediately prompts questions such as why, exactly, are abiogenesis/complexity/etc. so astronomically improbable? This is even more forceful considering that proponents usually fail to identify which of these early steps of the ladder leading to observership is actually that *very hard* one. In this important respect, **Early Great Filter** can be considered not wrong, but incomplete. (A resolute defender of the rare-Earth hypothesis might argue that other explanatory approaches motivate further hard questions as well. For instance, all sociological explanations like the **Zoo hypothesis** or the **Interdict hypothesis** prompt questions about cultural and political convergence. So, this is certainly not a fatal objection—but it is a strike against accepting **Early Great Filter**, and the uniqueness hypotheses in general, as baseline or default explanations.)

In the case of abiogenesis, it seems that this strategy has already failed—which is why, in the case of the rare-Earth hypothesis, the hypothesis's founders themselves emphasized that simple life forms are probably ubiquitous in the

Galaxy. While the full story is beautifully complex and has been told several times already, the advances in the theory of abiogenesis have by now gone far beyond what has for long been called the 'lucky-accident' way of thinking about the origin of life.[20] If the probability of abiogenesis—even under favourable physical and chemical preconditions—is astronomically small, say $10^{-100}$, but one still professes that it was a completely natural event, than a curious situation arises in which an opponent can argue that the supernatural origin of life is clearly a more plausible hypothesis.[21] Namely, even a fervent atheist and naturalist could not rationally claim that her probability of being wrong on this metaphysical issue is indeed smaller than $10^{-100}$, knowing what we know about the fallibility of human cognition. According to the predominant rules of inference, we would have been forced to accept the creationist position, if no other hypothesis were present.[22] Fortunately, the situation is quite different: huge efforts in both theoretical and experimental domains have led to great progress in the study of abiogenesis, and we have much better physical and chemical insights into the processes which led to early life. Contemporary theories, such as those of chemical synthesis and autocatalysis, have achieved such a great progress that we do not need to invoke astronomical odds at all, but instead we understand abiogenesis as a modular process, which has a high likelihood whenever relaxed physical and chemical preconditions are satisfied.[23]

Failing the challenge of extremely rare abiogenesis, proponents of **Early Great Filter** and the rare-Earth hypothesis in general now put most credence in the emergence of multicellular complexity as *the* hard step in the ladder leading to detectable technological civilizations (Hanson's Assumption #4). Even this is undermined by works such as those on cellular automata and other pathways towards high complexity, as have been elaborated in recent decades: for example, phase transition and other forms of spontaneous symmetry breaking have been demonstrated to lead to high-complexity dynamical behaviours.[24] Great advances in the development of molecular clocks, cladistics, and other methods of reconstructing the history of very early life have enabled a much better reconstruction of this historical transition on Earth. Although the question is still open, we do not have much reason to believe that the complexification of early life is an astronomically improbable step.

Note, however, that the complementary 'late Great Filter' simply does not work as an explanation of Fermi's paradox. At best, it can be a placeholder for some of the hypotheses to be considered in Chapter 6. For instance, if one insists that building of an advanced interstellar civilization is impossible because all intelligent species blow themselves up in a nuclear holocaust soon after discovering nuclear energy, then this reduces to one of the catastrophic solutions. In any case, additional explanatory mechanisms are needed if we wish to account for such late obstacles. This indicates another issue with solutions that are similar to the rare-Earth hypothesis: they work in those areas where the causal structure of events is not well established. Just as in the case of 'rare Jupiter', we are actually not certain what the causal role of Jupiter in the specific case of Earth's habitability is; in the more general issue of the number and role of hard steps in the evolutionary chain, we are forced to reason without sufficient knowledge of the (proximate) causes and effects of those evolutionary transitions. Therefore, any conclusions we reach are necessarily provisional: while it is *plausible* that such-and-such conditions reduce the chances for the emergence of advanced technological civilizations, we cannot really be certain that evolution—both biological and cultural—will not find a way around them.

Finally, we might have a truly cosmological version of the rare-Earth hypothesis, such as that suggested by Paul Wesson:[25]

> **Horizon to the rescue:** The probability of replication of abiogenesis or any other critical step is so superastronomically small that there is only one, and at most a few, technological species within our cosmological horizon. Since the distances involved are orders of magnitude larger than those within our Galaxy, we cannot hope to establish the assumptions leading to any form of Fermi's paradox. Although we might not be alone in the universe in toto—and, indeed, there might be infinitely many extraterrestrial sophonts—we are effectively alone as far as detectability is concerned.

The idea here is that the density of civilizations is so low that only a few of them, if any, are located within our (even hypothetical) reach. In such a context, the question of the lack of signals or manifestations is dissolved—but, unfortunately, this is just begging the question, since such an extremely low density of inhabited sites—less than 1 $\text{Gpc}^{-3}$, say—is not only un-Copernican, but clearly requires some additional explanatory mechanism. This may be biological contingency, rarity of Cambrian-explosion analogues, rarity of Jupiter analogues, or any number of others invoked by the proponents of the rare-Earth hypothesis, but it is clearly necessary.

So, **Horizon to the rescue** is once again a 'passing-the-buck' hypothesis—we still need to explain why the emergence of sophonts is so fantastically, cosmologically rare. Its virtue, however, is that, being so extreme, it is easily falsifiable: it would be enough, for example, that an extragalactic Ĝ infrared survey[26] discovers a signal from a Kardashev's Type 3 civilization, that conventional SETI succeeds, or that we discover traces of independent abiogenesis on Mars; and there are many other ways to undermine the idea that realistic habitats for life are intrinsically cosmologically rare. Recall the important conclusion of Chapter 1 that, while a single SETI success would not in itself automatically resolve Fermi's paradox in its stronger versions, this does not imply that it would not have an important role to play in answering the puzzle. In particular, even if the only success were the discovery of a persistent and undecipherable 'Wow!' signal (as in *His Master's Voice*), this would enable the rejection of a whole slate of hypotheses postulating very rare abiogenesis and/or noogenesis, headed by **Horizon to the rescue**. Therefore, while I shall return to the issue of falsifiability in Chapter 8, at present we need another mechanism of limiting the spread and detectability of Galactic sophonts.

## 5.3 Gaia or Bust?

A novel strategy located broadly within the rare-Earth-hypothesis paradigm has been suggested by Aditya Chopra and Charles Lineweaver: this approach relies on the complexity of the relationship between life and its physical environment.[27] We regard our terrestrial biosphere as optimal—'Gaia-like'—without bothering much to ask questions about particular timings of individual steps in its early evolution. Chopra and Lineweaver dare to ask such difficult questions and find rather disturbing answers. While these ideas partially overlap with **Early Great Filter**, they are still sufficiently precise to get a separate heading.

> **Gaian window**: Complex biospheres are rare because only a few Earth-like planets can develop a stable biotic feedback in which life forms stabilize planetary conditions. In particular, greenhouse gases and planetary albedo are regulated by life in a particular and often narrow 'temporal window' in which it is feasible to do so, which is created by stellar, atmospheric, and tectonic evolution. In the terrestrial case, this occurred between 1.0 and 1.5 Ga after the formation of our planet. If the analogue opportunity is missed elsewhere, the ubiquitous *abiotic* feedbacks are likely to lead to the extinction of the biosphere—or at least to its subsistence in the form of a specialized extremophile community unlikely to evolve towards the complexity necessary for cognition and technology. So, there are particular *bottlenecks* which need to be passed before observers emerge, and some of these bottlenecks might be quite narrow. Therefore, there are

few places where sophonts could evolve, and their emergence is strongly suppressed throughout the Milky Way—so there is no Fermi's paradox.

This is one of the most ingenious variations on the general theme of the rare-Earth hypothesis. Its major advantage is that it uses those pieces of data we have established from studying Earth's past and present—in contrast to those speculating about what could have gone wrong. It also expands our conceptual horizon by introducing and promoting philosophically sound thinking about habitability as a well-defined concept divorced from each particular instantiation—what Chopra and Lineweaver call 'recipe-based' habitability/bottlenecks.[28] The relationship to ecology is an important merit of this hypothesis: the understanding that it is only in the ecological, holistic context that we can understand what it means for a complex biosphere to evolve. A similar idea of intermittent habitability has been recently put forward by Fergus Simpson.[29] Further inquiry into the physical reasons for the intermittency will lead us to some of neocatastrophic scenarios to be investigated in Chapter 6.

But why Gaia? Is not that asking for *too much, too soon*? The original Lovelock–Margulis Gaia hypothesis has been criticized, among other things, for presuming a Leibnitzian context of 'the very best of all worlds', with homeostases and biotic feedbacks regulating too much of our environment.[30] The criticism itself is not devoid of anthropocentrism, since critics lack a standard of comparison: too much in comparison to what, exactly? Astrobiology can offer help in this respect. The discovery of other habitable planets will certainly give us the required yardstick. As an antidote to our lurking anthropocentrism, one could do well to consider the concept of a *superhabitable planet*, as proposed by René Heller and John Armstrong: planetary habitats *friendlier* to life than our humble terrestrial Gaia.[31] Another option is given by 'empty habitats': perfectly habitable planets where abiogenesis never happened (and panspermia remained inefficient).[32] In view of the large number of planets and potential habitats, it seems only reasonable and prudent to conclude that, instead of binary yes/no habitability, we need a continuous—or finely grained—*spectrum of habitability*, something which tallies well with the astrobiological landscape metaphor.

And where exactly is Gaia—of the sort required for evolving technologically advanced sophonts—on this spectrum? Biotic regulation of the planetary environment is certainly important—but is it *that* important, as Chopra and Lineweaver suggest? The adequacy of the **Gaian window** as an explanatory hypothesis crucially hinges on the assumption that evolution cannot find other ways to advanced cognition except through a terrestrial-like Gaia. This could be only the failure of our imagination (which Chopra and Lineweaver quite honestly admit). Further theoretical research in the size of the relevant parameter space seems necessary before we determine how narrow the bottlenecks actually are—and whether there are other, discontinuous bottlenecks leading to similar outcomes by other trajectories.[33] One such alternative trajectory could be the emergence of swarm intelligence, known for a long time from science-fiction discourse but recently elaborated by David Schwartzman and George Middendorf.[34]

An important take away from this is that the **Gaian window** does not work without many *disruptive* feedbacks and processes favouring extinction over persistent survival of the biosphere. In other words, the universe in both space and time must be more dangerous and more risky than we have hitherto assumed. Extinction is the major game in astrobiology overall—and only our particular luck (coupled with the obvious observation-selection effect) has led us to downplay this fact. Sufficiently strong extinctions are usually linked with *catastrophic* processes.

Puzzlingly enough, the latter need to be somewhat fine-tuned as well. Notably, we might wish to retain a 'pump of evolution', as proposed by the physicist John Cramer in his column in the *Analog* magazine in 1986.[35] He acknowledges random catastrophic events—notably cometary/asteroidal impacts, as in the case of the extinction of dinosaurs and many other taxa at the Cretaceous–Paleogene boundary—as strongly impacting (macro)evolution, while noting that not all effects of catastrophes are detrimental. On the contrary, some killing and displacement of species might be highly desirable in order to achieve greater complexity and there is again a Goldilocks effect in play:

> If this crisis pump cycles too fast the catastrophes occur too often and are too wasteful of life, too likely to extinguish promising species, and too close for the species to fully adapt to the reshuffled ecological niches before the next cycle, and so evolution proceeds more slowly. If the pump cycles too slowly the rate of crisis-stimulated evolution will also be slowed, and species may become so firmly ensconced in their niches that they will die rather than change when displaced... So there must be an ideal crisis/catastrophe rate which speeds evolution along at an optimal speed by blasting species out of their niches and weeding nature's garden at just the right time to promote improvement.

This idea has become one of the planks of rare-Earth hypotheses and is discussed by Ward and Brownlee in their Chapter 8, but without quoting Cramer.[36] We shall return to this key point when discussing the neocatastrophic family of solutions.

## 5.4 An Adaptationist Solution?

So, we reach the adaptationist version of the rare-Earth hypothesis, which in this case could truly be dubbed a 'rare-mind' hypothesis. It was hinted at by the palaeontologist David Raup in 1992, but was developed in more detail a decade later in the novel *Permanence* by the Canadian author Karl Schroeder:[37]

> PERMANENCE: Conscious toolmaking and civilization building are ephemeral adaptive traits, like any other in the living world. Adaptive traits are bound to disappear once the environment changes sufficiently for any selective advantage which existed previously to disappear. In the long run, the intelligence is bound to disappear, as its selective advantage is temporally limited by ever-changing physical and ecological conditions. The outcome of cultural evolution in the limit of very long timescales is a reversion to the direct, non-technological adaptation and consequent loss of spacefaring and astroengineering capacities—or extinction. Since such non-technological adaptation obviates the need for any detectable traces and artefacts, there is no Fermi's paradox.

This intriguing hypothesis uses the prevailing adaptationist mode of explanation in evolutionary biology to argue that the phase of detectability in the sense which connects to SETI is an ephemeral, passing stage in the overall astrobiological evolution. As Schroeder's protagonist argues:[38]

> The truth is that we are intelligent animals, but animals just the same, subject to the inescapable laws of our evolution. Our first theories about alien intelligence were *providential*: we believed with Teilhard de Chardin, that consciousness is a basic characteristic of complex thinking entities....
>
> What we found instead was that even though a species might remain starfaring for millions of years, consciousness does not seem to be required for toolmaking. In fact, consciousness appears to be a phase. No species we have studied has retained what we could call self-awareness for its entire history. Certainly none has evolved into some state *above* consciousness.

The key insight here is that such state of affairs need not limit us as much as we hitherto thought. The necessity of consciousness is an illusion. Consciousness may be unrequired for toolmaking; in the words of Julian Jaynes, it may not be necessary for thinking, either![39] The impact on our SETI prospects is dramatic; but that is just the very beginning.

The hypothesis Schroeder derives is quite complex, belonging to perhaps the top three most complex discussed in this book, and I shall therefore devote a bit of space to it, in order to demonstrate how fruitful thinking about Fermi's paradox can be over a broad sweep of research areas. Another reason is that it offers more than just an explanation of Fermi's paradox: it is an entire new—depressing or revolting, according to one's sensibilities!—way of looking not only at the problem of intelligence in the universe, but also at the future of humanity and human-made values as well:[40]

> We are left with a *selectionist* theory of sentience: consciousness and space-faring toolmaking ability arise by chance from countless combinations of traits that in the vast majority of species fail to produce results. Our studies have turned up thousands of species that 'might have been' like ourselves. One, for instance, has all our traits, except that it lacks a tolerance for remaining stationary for long. Its people roam across the plains of their world, incapable of creating tools larger than they can carry.
>
> Countless other species are similarly close, but also miss the mark, some for want of a single trait.

This is the crux of the problem (it is only a problem for the astrobiologist protagonist of the novel; for us, who want to resolve Fermi's paradox, it is a solution): our estimates and expectations of key cognitive phenomena like intelligence or consciousness—which are, above all, biological phenomena—are wrong. Intelligence is significant in the overall history of life only insofar as it offers an evolutionary advantage, a meaningful response to the selective pressure of the fluctuating environment. Consciousness might not offer any advantages—in the long term—at all. Even worse, it could be maladaptive and even fatal.

This approach, derived from biology, is known as *adaptationism*; its major proponents are distinguished biologists such as Richard Dawkins and John Maynard-Smith, as well as contemporary philosophers such as Daniel Dennett and Eliot Sobber. An adaptation is a trait that has been selected for by natural selection. The term 'adaptationist hypothesis' can be conventionally defined as 'a statement that asserts that a given trait in a population is an adaptation'. In other words, natural selection is the major (if not the sole) cause of the presence and persistence of traits in a given population. Sober's definition of adaptationism, given in his influential textbook, goes as follows:[41]

> Adaptationism: Most phenotypic traits in most populations can be explained by a model in which selection is described and nonselective processes are ignored.

Examples of adaptationist explanations abound. The camouflage colours of birds and insects; the Inuit face; the horns of *Ontophagus acuminatus*; and myriads of other observed properties of living beings are interpreted as giving their carriers an advantage in the endless mill of natural selection. Their genes are more likely to propagate along the thousands of generations of natural history. The most extreme version of adaptationism is sometimes called gene-centrism and is expounded by Richard Dawkins, being neatly encapsulated in the title of his best-selling book *The Selfish Gene*: genes are using (in a sufficiently impersonal sense of the word) organisms to propagate their own copies as efficiently as possible in time.[42] It may be of historical interest that adaptationism is usually traced back to Alfred Russel Wallace, one of the two great biological revolutionaries, who was also, as previously mentioned, one of the forefathers of modern astrobiology. Since his remarkably prescient *Man's Place in the Universe*, there has always been an adaptationist strain within astrobiological thought.[43]

Thus, intelligence/sentience is an adaptive trait, like any other. Adaptive traits are bound to disappear once the environment changes sufficiently for any selective advantage which existed previously to disappear. In the long run, intelligence is bound to vanish, as its selective advantage is temporally limited by ever-changing physical and ecological conditions. (Of course, such traits could be *exapted*—appropriated for some other utility—and thus preserved even when the original utility vanishes.[44]) Lacking better insights into the nature of sentience, this is as good a hypothesis as any. How it can be operationalized in the context of SETI is shown by Schroeder via his example of a fictitious advanced civilization whose traces human interstellar archaeologists find at some point (obviously, sufficiently in the future, from our present point of view):[45]

> They wanted their civilization to last forever—that's the one thing we do know about them. They built for the ages in everything they did. The evidence is that they did last a very long time—maybe eighty million years. But early on, they discovered a disquieting truth we are only just learning ourselves. It is this: Sentience and toolmaking abilities are powerful ways for a species to move into a new ecological niche. But in the long run, sentient, toolmaking beings are never the fittest species for a given niche... It doesn't matter how smart you are, or how well you plan: Over the longest of the long term, millions of years, species that have evolved to be comfortable in a particular environment will always win out. And by definition, a species that's well fitted to a given environment is one that doesn't need tools to survive in it.
>
> Look at crocodiles. Humans might move into their environment—underwater in swamps. We might devise all kinds of sophisticated devices to help us live there, or artificially keep the swamp drained. But do you really think that, over thousands or millions of years, there won't be political uprisings? System failures? Religious wars? Mad bombers? The instant something perturbs the social

> systems that's needed to support the technology, the crocodiles will take over again, because all they have to do to survive is swim and eat.

In 1992, Raup suggested that animals on other planets may have evolved, by natural selection, the ability to communicate by radio waves (and, by analogy, at least some of the other traits we usually think about as possible only within a civilization of intelligent beings). This would not be random noise, which is characteristic for astrophysical processes, but neither would it be structured, meaningful information, which would be characteristic for intentional messages of human or extraterrestrial intelligence—it would be something in-between. We know that animal communication on Earth is often tremendously complex and information rich; in some better-studied cases, its most general meaning can be established.[46] This does not mean that we can decode the communication between blue whales, the communication between bees in a hive, or even the chemical communication between bacteria in the same manner as we decoded the Mycenaean Linear B or even as British cryptanalysts in Bletchley Park decoded German Enigma ciphers. So, according to Raup's idea, we could encounter a new form of potential SETI signals, which could persist for millions of years, due to the slow pace of evolutionary change, in sharp contrast to the ephemeral nature of any particular communication *technology*. Since this is not really what SETI is after—although, of course, the discovery of such life forms would be tremendously interesting in its own right!—such unconscious signals should be treated as confusing noise; however, such noise would be persistent and very difficult to remove.[47]

Generic phases of evolution, according to **Permanence**, can be schematically represented, as in Figure 5.1. Direct adaptation as the outcome of natural selection (and other evolutionary processes, if present) leads, on long timescales and opportunistically, to the emergence of technological civilizations; in turn, such civilizations can create short-term technological adaptations for a far wider range of Galactic habitats. Notably, one can think about various, wildly different planets colonized by advanced sophonts in their period of ascendancy. However, technological adaptation is quite complex and henceforth unstable, reverting—in Schroeder's scenario—to direct adaptation, which is now specific to each particular habitat (i.e. non-local from the point of view of the originating habitat).

As much as it is appealing as an explanation, there are many difficulties with **Permanence**. Its insistence on adaptationism at all times is a form of inductivist fallacy. As in earlier times, inductivists argued that it is natural to assume a

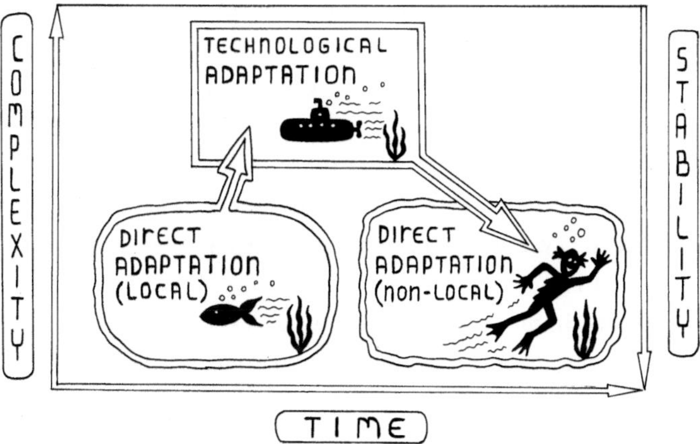


**Figure 5.1** A schematic presentation of the three phases of adaptation to any particular Galactic habitat.

*Source*: Courtesy of Slobodan Popović Bagi

meta-rule of inference along the lines of 'the future will resemble the past', so there persists a creeping prejudice that the present and future modes of evolution need to be the same as those of the past. This works in conjunction with the old-fashioned idolatry of adaptation: the almost reflex and non-thinking assumption that any evolution has to be adaptationist.[48] In spite of such fashionable views like evolutionary psychology/behavioural ecology/sociobiology, there is no reason to believe that all complex living systems anywhere in the universe evolve according to the rules of functionalist natural selection. Instead, hypothetic sophonts could have evolved, for instance, in a Lamarckian, orthogenetic, or saltationist manner. In fact, it is exactly in the domain of astrobiology that the most general role of individual evolutionary mechanisms (natural selection, genetic drift, lateral gene transfer, etc.) will be tested.[49]

Besides, even if *all* giga-year-old civilizations are now extinct, what about their astroengineering traces and manifestations? After all, 99 per cent of *human* civilizations in the historical record are extinct as well—if not for their traces, their manifestations, and particularly their *monuments*, they would be *as unknown to us at present as any number of extraterrestrial societies!* Schroeder uses the example of a long-extinct advanced civilization ('Dis builders') as the placeholder for his exposition of the adaptationist idea, but is it really the case that future humans will have to go excavate in person in order to uncover the civilization's secrets? Why, in the words of Weston, as quoted in Chapter 1, has the evidence been so long in coming and why it is so indirect?

Note also that, in order to function as envisioned, **Permanence** must also incorporate a measure of the 'passing-the-buck' approach previously discussed. True noogenesis—that is, the emergence of species like humans or the Dis builders as described by Schroeder—has to be rare for the explanation to work; not so rare as in **Horizon to the rescue**, but still rare in comparison to the naive Copernican baseline picture.[50] If the Dis builders lasted about 80 Ma, and if we allow that the average age of civilizations is half that, while controlling/surveilling at least 10 per cent of the Galaxy, the rate of emergence of new intelligent communities in the entire Galaxy must be less than about 0.1 per million years. While a more precise model should take into account factors such as pre-emption (the impossibility of new civilizations to evolve in already controlled volume), panspermia, and others, this is, in any case, an incredibly low rate—especially when contrasted with the wealth of potentially habitable planets in our stellar system.

In the space of explanatory hypotheses we are navigating in this book, **Permanence** is directly and irreconcilably opposed to **Transcendence** (see Section 6.7) and all its versions. In fact, the dichotomy is so pronounced that there is a very close scenario of dysgenic degeneration which, as shrewdly noted by the Russian astrophysicist Vladimir Lipunov, is diametrically and utterly anti-Tsiolkovskian. Unused facilities, in the course of evolution, either are exapted for other purposes or devolve; why couldn't this process operate much more quickly in the domain of mind and brain, which anyway possess high plasticity? Hence we have the following hypothesis:[51]

> **Thoughtfood exhaustion:** Intelligence/consciousness/toolmaking ability functions only as long as there is significant new content to process. If the exponential growth of knowledge and understanding, which humanity has been experiencing during at least the last 500 years, continues for the foreseeable future, it might push against the limits set by the world itself. It is quite possible that the world—physical, mental, and social—is of finite complexity and that it is possible for an advanced civilization to reach the state of 'knowing everything (worth knowing)' in a period of time that is short by evolutionary and astronomical standards. Co-evolutionary feedback between biology and culture could further accelerate the downward path. In contrast to Schroeder's scenario of slow dysgenic decline and devolution, isn't it possible that mind reacts to the true lack of new inputs in a manner similar to body reacting with the true lack of nutrients: by rather quick extinction?

I leave this as *food for thought* (for now).

## 5.5 Unphysical Ceteris Paribus, and Other Problems

Clearly, any general argument against the rare-Earth hypothesis would be problematic for the explanatory hypotheses based on it. Although it is not possible to review all the problematic aspects of the rare-Earth-based hypotheses here, it is enough to note some of the difficulties, and point the interested reader towards the growing literature on this topic.[52]

For instance, the famous argument about Jupiter being the optimal 'shield' of Earth from cometary bombardment has been brought into question by multiple studies conducted by Jonti Horner and the late Barrie W. Jones, who used numerical simulations to show that the off-handed conclusion that Jupiter acts as a shield against bombardment of inner Solar System planets is unsupported. Moreover, they conclude 'that such planets often actually increase the impact flux greatly over that which would be expected were a giant planet not present.'[53] If the results of Horner and Jones withstand the test of time and further research, it is hard to imagine a more detrimental result for the entire rare-Earth paradigm.

This is just the start of problems with the original rare-Earth hypothesis, however. In supposing how the state of affairs could be different, rare-Earth theorists assume simple, linear change, not taking into account self-organizing nature of the relevant physical systems. The example of Jupiter is again instructive, since asking about the fate of Earth in the absence of Jupiter is self-contradictory: Earth is a part of the complex system which includes Jupiter as a major component, so there are no guarantees that Earth would have existed at all if Jupiter were not present. Even if it existed, we would have to account for many other differences between that counterfactual situation and the actual one, so the question of to what degree one is justified in calling such a body 'Earth' would be very pertinent.

Thus, we unexpectedly enter the murky metaphysical waters of identity and persistence. We know for certain that there are no round squares or married bachelors—those entities are self-contradictory. What about non-habitable Earths? (Here, of course, I do not consider Hadean Earth in the first 500 million years or so after its formation; instead, what we need to consider is *an alternative non-habitable present-day Earth*. More precisely, we need to consider a planetary body at the same spatio-temporal coordinates and with other relevant parameters identical—except for a single requirement for the rare-Earth hypothesis.) In terms of contemporary analytical metaphysics, is there a possible world in which Earth is not habitable? We are accustomed to thinking about Earth as habitable and will even go so far as to use Earth as the prototype of any and all habitable planets, as Ward and Brownlee consistently do in building the framework for the rare-Earth hypothesis and as NASA does, for example, in discussing habitability in the TRAPPIST-1 system. But is that truly a de re statement, namely a statement describing some real, physical relationship between things in the world?[54] And if it is a de re statement about a particular individual—'our Earth' or '*the* Earth'—how are we to extend this to discuss de re modal versions of statements, like 'Earth could not be habitable without Jupiter/a large Moon/etc.'?

It is hard to overemphasize how difficult it is to discuss the modal de re statement without invoking or implying 'magic-wand' (i.e. unphysical) alterations of context. In the particular case, we may compare the actual Earth's habitability with Earth's habitability with Jupiter *being removed from its present location by waving a magic wand*—which is essentially what rare-Earth-hypothesis theorists do. Clearly, this does not look as appealing in this formulation, not the least for the unfortunate lack of magic wands, does it? The 'magic-wand' procedure is what I have dubbed the unphysical ceteris paribus fallacy.[55] In 'possible-world' semantics, the rare-Earth-hypothesis claim is equivalent to the claim that there is a possible world in which Earth exists without Jupiter (and is less habitable). It seems to follow that the one and the same individual planet—*the* Earth—exists in some merely possible world, as well as in the actual world, so that there is an identity between Earth and some individual planet in another possible world. Even if the original rare-Earth-hypothesis claim has nothing to do with possible worlds, instead talking about presumably actual Earth-like planets in our Galaxy, the argument involves a commitment to the metaphysical view that some individuals exist in more than one possible world, and thus to what is known as 'identity across possible worlds,' or *transworld identity*.[56]

Transworld identity is a highly non-trivial matter—and we cannot delve deeper into it in this book, for obvious reasons. The very fact that the validity of counterfactuals to the rare-Earth hypothesis depends on the particular account of transworld identity should give us pause. Did they really have *this* in mind? Could statements of obscure metaphysics meaningfully influence our degree of confidence in a sound scientific hypothesis in planetary science? The unfortunate tendency to believe in magic wands especially applies to those willing to accept the rare-Earth hypothesis as truth in advance, usually for extrascientific reasons (e.g. Genesis 1:28–30). Numerical studies aiming to at least roughly delineate part of the parameter space dedicated to Earth-like habitability, such as those of the Potsdam group, or Forgan and Rice,[57] are less affected by this. In general, however, the validity and applicability of novel astrobiological concepts are much less obvious and require significantly more epistemological and methodological work than hitherto assumed.

## 5.6 A Rare-Earthist Résumé

It might be the case that we are indeed alone in the Milky Way—but the full context and ramifications of such a particular form of astrobiological landscape need to be fully understood. This group of hypotheses offers more or less plausible explanations of **StrongFP** in terms of the 'rare-Earth' paradigm—and the arrow of explanation certainly does not point in the opposite direction, as sometimes is confusedly thought. The rare-Earth hypothesis is the foundation (arguably, not as firm as one would desire), and the explanatory account for Fermi's paradox is a building erected on top of these foundations. On one hand, it is a good model of interdisciplinary synthesis within modern-day astrobiology, and an example of how SETI studies could, in spite of all their extrascientific baggage, be smoothly joined with mainstream astrobiology. On the other hand, weaknesses in the rare-Earth-hypothesis framework—like, for example, the one opened by the work of Horner and Jones—therefore necessarily reflect on the prospects of these explanatory hypotheses.

Their common property is that some fine-tuning is desirable in order to complete the explanatory task. If the filter in **Early Great Filter** is sufficiently early, it requires probably the least amount of fine-tuning. Unfortunately, it passes the buck so far into the realms of biochemistry and molecular palaeontology that many people remain unaware that any further explanation is necessary. On the plus side of the ledger, all uniqueness hypotheses completely obviate the difficult issue of the non-exclusivity principle: it cannot be applied because, according to these views, the prerequisite for its application, namely the set of Galactic sophonts, does not exist. Some of the options, notably **Gaian window** and **Permanence**, hold the promise of exciting new research directions. On the minus side, testing rare-Earth hypotheses is bound to be extraordinarily difficult and impractical, as with testing any general negative existential claim. A naturalistic lucky accident is distinct from a miracle, as Iris Fry persuasively discussed, but both tend to stymie the advance of science.

As we have already noted, there is an overlap between different categories in nuances, if not in a substance. Some of the rare-Earth requirements actually rely on catastrophic mechanisms—if Jupiter were not there to protect us, allegedly, a comet might exterminate all complex life forms on Earth. Therefore, Earth analogues elsewhere are likely to pass through many cycles of destruction without evolving intelligent beings. But cataclysms such as cometary impacts certainly require us to reject gradualism as well, don't they? Wouldn't cometary and asteroidal impacts be enough of an explanation for the Great Silence in their own right? It is to these issues that we must now turn.

CHAPTER 6

# At the Mountains of Madness

## *Neocatastrophic Solutions*

History of human civilizations can be read as a form of punctuated equilibrium between periods of slow, gradual processes and more or less sudden catastrophic outbursts.[1] What about non-human ones? Insofar as we are forced to rely on fiction, we had better go along with those examples least burdened by an anthropocentric bent. In one of the most brilliant such instances, Howard Phillips Lovecraft's short 1936 novel *At the Mountains of Madness*, Prof. Dyer, the narrator, tries to infer the wakes and tides of the seemingly extinct civilization of the Old Ones on the basis of their material culture, traces, and artefacts.[2] The project of this keen and observant scientist—one of the best, although unappreciated portrayals of the scientific approach in the popular fiction—has many similarities with different aspects of the SETI studies of the last half-century. At one point, Dyer directly considers the items dealing with *habitability*:[3]

> The abyss, it seems had shelving shores of dry land at certain places, but the Old Ones built their new city under water—no doubt because of its greater certainty of uniform warmth. The depth of the hidden sea appears to have been very great, so that the earth's internal heat could ensure its habitability for an indefinite period.

Perhaps nowadays we might find this and other similar passages unremarkable—but this only testifies how much our outlook has been transformed by the ongoing astrobiological revolution. Lovecraft's novel was written more than three-quarters of a century ago. Here we see, many decades before the astrobiological revolution of the mid-1990s, and much before the ecological awareness of the 1960s and 1970s, the idea that the survival or collapse and extinction of even quite sophisticated and powerful communities crucially depends on the state of habitability of their planetary environment. When that state is dramatically altered, in either a gradual or a catastrophic fashion, the very existence of intelligent communities and their capacities for influencing their environment (including the capacity for contact with other sophonts) are brought into question.

We have realized recently that, in contrast to the apparent peacefulness of our physical environment on the timescale of human life, such changes can be dramatic indeed. Events such as the impact of Comet Shoemaker–Levy 9 onto Jupiter, or superluminous supernovae such as SN 2006gy, have played key roles in this sea change. The neocatastrophic class of hypotheses for resolving Fermi's paradox reflects this new awareness. The novelty is occasionally hidden behind complex concepts—after the last two chapters, however, it certainly should not discourage us!

## 6.1 Down with Gradualism!

The category of hypotheses described here is the most heterogeneous one, containing both some of the oldest speculations on the topic of the history and fate of advanced intelligent species, and some of the newest ones. Before we review the main contenders, it is important to emphasize that the prefix 'neo' is used almost reflexively with this mode of thinking, for historical reasons. The defeat of the 'classical', nineteenth-century catastrophism of figures such as Cuvier, d'Orbigny, de Beaumont, Agassiz, or Sedgwick in the grand battle with the gradualism of Charles Lyell and his pupils (including Darwin) imposed a lasting stigma on views which were perceived as belonging to this tradition of thought. This has clearly impeded the development of the geosciences.[4] In addition, the association of catastrophism with the pseudoscientific (although often thought-provoking!) views of Immanuel Velikovsky in 1950s has brought an additional layer of suspicion upon the label itself.[5] Thus, the resurgence of catastrophism after 1980 and the discovery of Alvarez and collaborators that an asteroidal/cometary impact was the physical cause of the extinction of ammonites, dinosaurs, and other species at the Cretaceous–Paleogene boundary

65 Ma ago is often referred to as 'neocatastrophism'.[6] The new, post-1980 trend has accelerated so far that, as argued in Chapter 3, it has become something of a new normal. The history of life during the Phanerozoic aeon (the last 541 Ma) is now commonly seen to be punctuated by a spectrum of mass extinction events, apparently caused by severe disruption of the physical environment.[7] As a modern-day geoscientist summarizes,[8]

> Because catastrophism, strictly speaking, contrasts only with the substantive doctrine of gradualism, it is not surprising to see modern scientists embracing its position. There is nothing contradictory in adhering to uniformity of law plus uniformity of process (actualism), while also preferring catastrophist to gradualist explanations of geological phenomena. The term 'catastrophic' only applies to the intensity and duration of a particular geological process. It does not necessarily have anything to do with whether or not such a process is manifest today (actualism) or even with the well-known methodological claim that simpler explanations are to be preferred to more complex ones.

So, as the Cold War cultural pessimism retreated, neocatastrophic hypotheses obtained a strong boost from the resurgence of catastrophism in Earth and planetary science, as well as in astrobiology. This occurred hand-in-hand with the emergence of a new, general paradigm according to which the Solar System is an open system, strongly interacting with its Galactic environment. As already discussed, this neocatastrophist tendency has recently become almost default in a wide range of fields, from research on abiogenesis, to aspects of macroevolution, to the debates on the evolution of humanity, to future studies, but all its ramifications have not yet been elucidated in any detail.[9] The major feature of these solutions is, of course, the abandoning of the classic gradualist dogma that 'the present is the key to the past', and the acknowledgement that sudden, punctuated changes present a major ingredient in shaping both Earth's and the Milky Way's astrobiological history. In terms of the central metaphor I use here, the astrobiological landscape is rough and rugged, with many steep cliffs and deep gorges.

Obviously, not all hypotheses in this category involve particularly destructive events. At least one of the items reviewed in this chapter has universally been considered the ultimate boon for many centuries or even millennia in almost all human cultures (see Section 6.7). Some ideologically minded people might feel the same in connection with the **Introvert Big Brother**-type scenarios (see Section 6.5). But the common feature is a large, dramatic change—a sort of phase transition in the relevant parameter space. Suddenly, everything changes—if not literally in the blink of an eye, then certainly on an evolutionary brief timescale. The system passes from one (quasi)stable configuration to another (quasi)stable configuration on timescales which are short in comparison with the average evolutionary timescales of such systems. The new state need not necessarily be extinction, but extinction is the simplest case and an attractor in the parameter space, so I will start with a few apocalyptic hypotheses.

## 6.2 Natural Hazards I: Random Delays

It is self-evident (in a Cartesian manner!) that our universe is habitable. This is based upon the self-evident fact that our immediate surrounding—say, our planet or the Solar System—is habitable. As discussed in Chapter 5, it is the Copernican tenet (or prejudice, if you incline towards the rare-Earth views) that this is a 'typical' or an 'average' case. A further, this time gradualist, prejudice is that this habitability we observe today has stayed roughly the same throughout our cosmological past. If we reject these two sorts of prejudices, we obtain a transitional hypothesis which joins the 'rare-Earth' ideas with neocatastrophism.

> **The Gigayear of Living Dangerously**: The Galaxy at large is much less habitable than we infer from our historical experience, due to intermittent and random catastrophic perturbations. This applies to both the spatial and the temporal extensions of our location. Therefore, the evolutionary chain leading to intelligent beings is regularly interrupted, and no technologically advanced species arise except by a freak exception, so there is no Fermi's paradox.

In other words, random natural hazards make the Galaxy much more hostile to life and intelligence than we perceive on the basis of our environment and our past history.[10] This is a hypothesis presenting clear continuity with the rare-Earth ideas discussed in Chapter 5. Obviously, as previously discussed, we should not be surprised by the fact that we perceive our environment to be hospitable to life. This is a historically contingent fact, however, not some necessity built in the laws of nature at the Big Bang (or on a higher, multiverse level).

Quite to the contrary, we are well aware of many catastrophic threats the terrestrial and any other biosphere in the Galaxy is bound to face in the course of time. Examples of relevant catastrophic processes include:

- cometary/asteroidal bombardment[11]
- supervolcanism[12]

- nearby supernovae, magnetars, and/or GRBs (provided that their 'lethal zone' is small)[13]
- global glaciations ('Snowball Earth' analogues)[14]
- runaway *natural* greenhouse effects[15]
- 'wild cards': something unforeseen, but deadly and *of natural origin*[16]

(Qualifications serve here just to delineate this hypothesis with respect to others considered in this chapter. For instance, the 'lethal zone' of cosmic explosions needs to be small *in comparison to the characteristic length scale* of the GHZ in order to avoid *correlated* Galactic catastrophes, which are discussed in Section 6.3. Catastrophes caused by the actions of intelligent agents are considered in Sections 6.4–6.7.) All these processes—except the wild cards, of course—have been observed in the past in the Solar System and its vicinity, and their destructive capacities are uncontroversial. What is controversial, though, and where the possibility for accounting for Fermi's paradox opens is how typical the *local* frequency and severity of those processes are. Both the frequency and severity of those catastrophic processes might, in principle, be much higher than we infer from the history of our Earth and the Solar System. (Note that, in accordance with the discussion in Chapter 5, we should not be surprised that we find ourselves in a rare quiet corner of the tumultuous Galaxy: this would be just a consequence of the expected anthropic selection effect.)

If catastrophic processes are indeed more intense on a typical habitable planet in the GHZ than on Earth, these natural hazards are likelier to break the evolutionary chain leading to the emergence of intelligent observers, so we should not wonder why we do not perceive manifestations and traces of older Galactic communities. For instance, one well-studied case is the system of the famous nearby Sun-like star Tau Ceti; this system contains both planets and a massive debris disc which is analogous to the Solar System Kuiper belt. Modelling of Tau Ceti's dust disc indicates, however, that the mass of the colliding bodies up to 10 kilometres in size may total around 1.2 $M_\oplus$, compared with the only 0.1 $M_\oplus$ estimated for the current Solar System's Edgeworth–Kuiper belt.[17] The Tau Ceti system is a very complex planetary system, and it is likely to harbour one or more terrestrial planets. It is only reasonable to conjecture that any such terrestrial planet of this extrasolar system is subjected to much more severe impact stress than Earth has been during the course of its geological and biological history.[18]

If we use the Tau Ceti system as a placeholder for those locales in which random catastrophic stress is much stronger than the one testified on by the Solar System history, the question becomes whether Tau Ceti or our Sun is the prototype of a star harbouring life. While strong impacts like those presumably occurring on Tau Ceti planets are unlikely to completely sterilize them (although some such might occur from time to time), they are likely to be sufficiently destructive to reset the evolutionary clock several critical steps back. And if the timescales for critical steps are large, the stage of technological civilization might not be reached within the window of habitability created by the Main Sequence evolution of their parent stars. Obviously, this hypothesis can be empirically tested by observing more detail in more planetary systems. So far, it seems that *Tau Ceti*, rather than our Sun, is an odd exception instead of a rule.

## 6.3 Natural Hazards II: Synchronized Delays

If random resetting of astrobiological clocks is insufficient to suppress the contact cross section as much as the Great Silence requires, a natural step is to ask whether that effect can be achieved with non-random—or *correlated*—resettings. Correlated effects are usually more efficient than random ones; consider how much more difficult it is to fight *organized* crime as compared to spontaneous, unorganized criminal behaviour. Thus, we face the next prototype hypothesis in the neocatastrophic category:[19]

> **ASTROBIOLOGICAL PHASE TRANSITION**: There are global (i.e. Galactic) regulation mechanisms, resetting and re-synchronizing biological evolution all over the GHZ. Those global regulation mechanisms secularly evolve in such a way that the Galaxy becomes more hospitable to complex life and observers in the course of cosmic time. The net effect is that the emergence of early Galactic sophonts is unlikely (in contrast to late sophonts, like us) and there is no Fermi's paradox.

Obviously, this hypothesis is grounded in the notion of a *global regulation mechanism*. Such a mechanism could occasionally reset astrobiological 'clocks' all over the GHZ and, in a sense, re-synchronize them. This is a prototype *disequilibrium* hypothesis: there is no paradox, since the relevant timescale is the time elapsed since the last 'reset' of astrobiological clocks and this can be substantially smaller than the age of the Milky Way or Lineweaver's scale in Eq. (2.3). So, is there any sign of such a mechanism?

In a pioneering paper published in 1999, the Fermilab astrophysicist James Annis suggested that GRBs, whose cosmological and extremely energetic nature had by then been increasingly well understood, serve as such a global regulation mechanism.[20] The idea of Annis was especially beautiful, since it argued that the emerging solution to one 'big puzzle' (the origin of GRBs) could lead us to solving another, perhaps even bigger one (Fermi's paradox). A subsequent important study by Scalo and Wheeler confirmed the suspicion that GRBs are likely to have adverse consequences over a large part of, if not the entire, GHZ.[21] In particular, their results suggested that GRBs can be directly lethal up to a distance of 14 kpc for eukaryotes, and up to 1.4 kpc for prokaryotes. These length scales should be compared with ~11 kpc for the outer rim of GHZ, as given, for instance, by Lineweaver and his collaborators.[22]

(Much earlier, in 1981, Jim Clarke had speculated that occasional explosions in the centre of our Galaxy could serve as such a regulation mechanism. The idea is that the Milky Way could intermittently behave in a manner similar to Seyfert galaxies, whose nuclear regions undergo dramatic explosive processes accompanied by massive releases of energy in the form of both electromagnetic and cosmic-ray radiation. With some additional assumptions, such nuclear outbursts could sterilize all budding biospheres on terrestrial planets—thus forcing biological evolution there to start over from scratch. Since the transit time of this sterilizing radiation would be equal—or at least comparable—to the light-crossing timescale, the sterilizations would have been tightly correlated over the entire history of the Galaxy. While the physical causative mechanism does not work, as we now know, the idea that a particular event could correlate astrobiological evolution over the entire home galaxy has remained the essence of **Astrobiological phase transition.**[23])

Subsequent studies have shown that the situation is much more complex. In brief, there are three types of effects of Galactic GRBs on biospheres: (1) the direct effects of ionizing photons; (2) the direct effects of cosmic rays accelerated in the GRB; and (3) the long-term effects of GRBs on atmospheric chemistry and climate. In addition, if a GRB is *very* close, there could be, as in the case of close supernovae, direct deposition of matter; since the subset of potential biospheres located that close to a GRB is certainly very small, I shall neglect this possibility here. As far as the first class of GRB effects is concerned, all such effects would be strongly attenuated by planetary atmospheres and, to an even higher degree, oceanic waters. While the UV contingent of the incoming photons, transmitted through an atmosphere similar to Earth's, would undoubtedly cause ecological damage and even the disruption of food chains throughout the biosphere affected, it seems quite implausible that a strong astrobiological 'reset' could be achieved in this way. A better candidate would be the second class of GRB effects, since high-energy cosmic rays and their secondary products, especially atmospheric muons, penetrate water and even rocks to several kilometres of depth.[24] However, the cosmic-ray fluence created by GRBs remains highly uncertain to this day, as well as the level of anisotropy of their acceleration. Biospheres encountering tightly collimated jets of cosmic rays are likely to experience a biotic cataclysm, but we need both more detailed models of the ecological effects and better astrophysical understanding in order to establish how often that happens per typical $L^*$ galaxy.

The importance of the third class of GRB effects seems indisputable—laboratory experiments, atmospheric nuclear tests, and all ozone-layer research confirm that strong ionizing fluxes leads to the formation of nitrogen oxides (generically named $NO_x$), which efficiently destroy ozone layer and lead to the prolonged exposure of planetary surface to stellar UV flux, with detrimental consequences for many living species.[25] Other significant effects might include extreme acid rains and increased rate of cloud formation, leading to strong climatic cooling and possible ice-age triggering.[26] That all these have catastrophic effects for biospheres is indisputable; what is disputable, however, is how efficient is the destruction wrought by such explosions and how often they affect a typical habitable planet in the Galaxy. The last question is equivalent to asking how distant an explosion needs to be to affect an Earth-like planet (and how anisotropic is the affected spatial region).

Seemingly, this question could be resolved by looking to Earth's past in a manner similar to that used for estimating the impact risk on the basis of Earth's impact history. We encounter insurmountable difficulties in the case of cosmic explosions, however. It was exactly desperation stemming from the *lack* of traces of causal mechanism which led Otto Schindewolf to propose originally that a close supernova explosion caused the end-Permian mass extinction.[27] For quite a long time, the idea has been downplayed in the biological and geoscience communities,[28] to be revived only very recently after the magnitude and astrophysical significance of cosmic explosions was ascertained. As a recent paper emphasizes,[29]

> There is a very good chance (but no certainty) that at least one lethal GRB took place during the last 5 gigayears close enough to Earth as to significantly damage life. There is a 50% chance that such a lethal GRB took place during the last $500 \times 10^6$ years, causing one of the major mass extinction events.

Adrian Melott and his collaborators have promoted the hypothesis that the end-Ordovician mass extinction (one of the 'Big Five' such episodes) was caused by a GRB.[30] While this has attracted much attention and provoked further work so far, it is still quite early to claim that the link truly exists. Of course, in the

Precambrian superaeon, even that tenuous empirical evidence on the basis of which we could judge events and their causes vanishes. Since the Precambrian contains roughly six-sevenths of the history of life on Earth, it is unavoidable that most of even potentially discriminating evidence is lost; this is enhanced further by the core assumption of the **Astrobiological phase transition** hypothesis that potentially lethal events were more frequent in the past. Even in the case of much later events, positively identifying traces of a cosmic explosion on Earth would be a devilishly difficult task. After all, there is considerable debate whether a supernova exploded close to Earth within *thousands* of years;[31] while this is likely to remain an active field of research for years to come, it is hard to envision that reconstructing ancient cosmic explosions and their local biotic impact will ever become an easy task.

There is still a large amount of work to be done on elaborating the precise effects of cosmic explosions on a typical biosphere in the Galaxy. However, it seems unavoidable that at least some of them are occasionally exposed to an extreme environmental stress due to such explosions in their astronomical vicinity. This constitutes one of the terms in the total risk function, which determines the outcome in **The Gigayear of Living Dangerously**; the present solution assumes that this term is in fact dominant and that its evolution forces the density of biospheres with complex life forms throughout the Galaxy—ultimately forcing the density of intelligent observers as well. As the frequency of catastrophes decreases with cosmic time, the window of opportunity for complex life forms and intelligent observers to arise becomes longer. Eventually, this leads to the emergence of those biospheres which evolve intelligent observers and their technological civilizations capable of protecting themselves from future 'resets'.

In this manner, **Astrobiological phase transition** leads to the situation schematically envisioned in Figure 6.1, where we are within the temporal window of a 'phase transition'—from an essentially dead place, the Galaxy will be filled with intelligent life on a timescale similar to $t_{FH}$. So, the major problem is whether it is possible to ensure that the transition is sufficiently sharp. It might be the case that it requires *some* fine-tuning to account for the observed Great Silence in the Milky Way.[32] Whether this is better or worse than the amount of fine-tuning necessary in other explanatory hypotheses for **StrongFP** remains to be seen.

In general, the transition time needs to be of the order of the Fermi–Hart timescale. For short values of $t_{FH}$ (those obtained with self-reproducing probes), it is hard to obtain a sufficiently sharp transition. If the lethality zone is smaller than the entirety of the GHZ, this would entail synchronization of resets, which

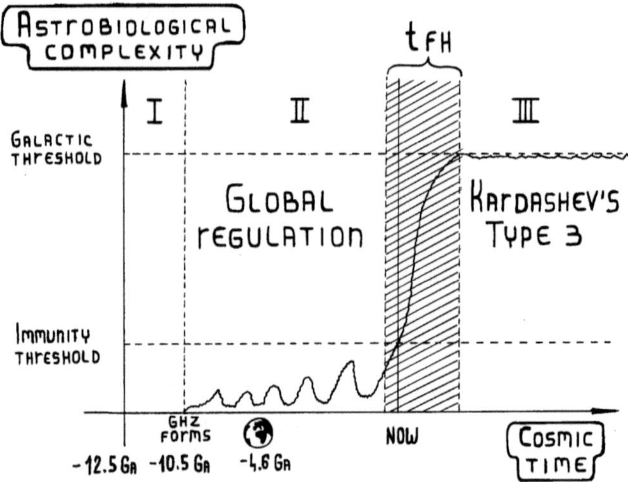


**Figure 6.1** Very simplified scheme of the phase-transition hypotheses: an appropriately defined astrobiological complexity will tend to increase with time, but the increase will not become monotonous until a particular epoch is reached; $t_{FH}$, Fermi–Hart timescale.

*Source*: Adapted from Ćirković M. M. and Vukotić, B. 2008, 'Astrobiological Phase Transition: Towards Resolution of Fermi's Paradox', *Orig. Life Evol. Biospheres* **38**, 535–47. Courtesy of Slobodan Popović Bagi

is hardly realistic from an astrophysical point of view. In this limit, the degree of non-exclusivity is diminished, since we still expect *some* civilizations to reach the level of an advanced technological civilization or even Kardashev's Types 2.x and 3 civilizations; and the dreaded question *why not in our part of the Galaxy/universe?* reasserts itself.

On the other hand, if Newman and Sagan were right in claiming that $t_{FH}$ is on the order of $10^9$ years, things are much easier for the proponent of the phase-transition hypothesis.[33] Such a situation does not require precise synchronization, and the current astrobiological snapshot of the Milky Way would reveal a much larger distribution of the ages of intelligent species. Of course, the price one needs to pay here is having to explain the apparent *absence* of von Neumann probes; this is possible to do (see Section 6.6), but adds another layer of assumptions. Further work on getting a better numerical hold on $t_{FH}$—which is essential for progress to be made in resolving Fermi's paradox in any case!—will enable a better assessment of whether the transition needs to be unrealistically sharp.

All in all, **Astrobiological phase transition** is one of the most provocative explanatory hypotheses—it offers an astrobiologist or a SETI student much to do. GRB physics, the spatial distribution of progenitors in the Galaxy (and in other galaxies possessing habitable zones), the ecological impact of high-energy photons and cosmic-ray particles, possible forcing mechanisms other than GRBs, biospheric signposts of this kind of mass extinction event, paleo-environmental effects detectable in the fossil record on Earth, even the ways that advanced technological civilizations might mitigate such catastrophes—all these topics are a challenge for theoretical studies, empirical studies, or both. So, this hypothesis meshes very well with many active currents of astrobiological research, as these focus on the interplay between astrophysics and habitability.

And it does offer a number of, at least potentially, falsifiable predictions. Among those are the spatial and temporal correlations between civilizations in the Milky Way, as well as the absence of any astroengineering signatures in outer galaxies.[34] I shall return to the discussion of interstellar archaeology, as an emerging subfield of SETI, in Chapter 8. Although some of these predictions are rather long shots, overall, the prospect of submitting **Astrobiological phase transition** to conventional scientific testing seems better than for most rivals discussed in this book.

## 6.4 Intentional Hazards I: Self-Destruction

The capacity for self-destruction is an obvious trait of any technological civilization after it passes some threshold level. In the case of humanity, the threshold was passed in 1945, with the development of nuclear weapons. This leads us to the most classical of all catastrophic hypotheses.

> **Stop Worrying and Love the Bomb**: Almost all Galactic civilizations self-destruct on a short timescale upon the discovery of nuclear weapons. This prevents sophonts of greater age and wisdom, capable for leaving traces and manifestation detectable over interstellar distances, from emerging in the first place. Hence, there is no Fermi's paradox.

Clearly, this has been the hallmark hypothesis of the Cold War era and it is no surprise that several disenchanted SETI pioneers embraced it.[35] It seems that its popularity was in part due to the extreme uncertainty accompanying the threat of global nuclear war between superpowers. Accordingly, it became much less fashionable after the end of the Cold War—but if anything, ephemeral cultural changes in our recent history should not really modify prior probability for this dramatic scenario.

Problems with the exclusive and anthropocentric nature of such a hypothesis—considering the fact that social and political developments on habitable planets throughout the Galaxy are quite unlikely to be correlated—are rather obvious.

Before we assign an extremely low grade to **Stop Worrying and Love the Bomb**, please note that this hypothesis may, unfortunately, receive some help from quite a distinguished quarter, namely game theory. If, as it has been suggested, any situation with many political and military actors of different sizes will tend to evolve into a situation with almost all actors falling into one of the two large camps, this will strengthen the case for nuclear holocaust as a more general phenomenon.

Unfortunately, self-destruction options have multiplied in the meantime, since the spectrum of potentially destructive technologies in human history has recently broadened, especially when those forthcoming technologies which are not yet operational are taken into account. This now includes the intentional or accidental misuse of biotechnology, and is likely to soon include the misuse of nanotechnology, robotics, artificial intelligence, and geoengineering.[36]

> **Self-Destruction, Advanced Version:** Almost all Galactic civilizations self-destruct on a short timescale upon the discovery of post-nuclear technologies capable of mass destruction. Therefore, sophonts significantly older than humanity are extremely unlikely. Since self-destruction occurs prior to any significant astroengineering activities, there are no traces or manifestations detectable over interstellar distances and there is no Fermi's paradox.

The spectrum of potentially self-destructive technologies is wide indeed. On the most realistic edge of it, we encounter the possibility of the intentional or accidental misuse of *biotechnology*, for example, through the creation and release of new pathogens, against which sophonts' biological immune systems would be helpless. Further on in the near future, we might face the danger from the misuse of *nanotechnology* in various specific ways. Then, there are more speculative possibilities with potentially even greater adverse impact, such as the misuse of *artificial intelligence*, the use of *geoengineering* to alleviate global climate change, or the creation of an *asteroid-defence* system intended to alleviate the threat of impacting small bodies. In addition to all these advanced technologies which could be misused, with cataclysmic consequences, one should always add a 'wild-card' category, since the future of technology is notoriously difficult to predict, and human experience shows that almost any technology could be misused for destructive purposes.

If most of the technological societies in the Galaxy self-destruct through any of these means before reaching the level of an advanced technological civilization, this would be an explanation for **StrongFP**. Quite clearly, the same qualms about exclusivity apply as before. However, the extension of the spectrum of self-destructive options does bring a qualitative difference, since the effects are clearly cumulative: humanity needs to survive *all* those separate threats in order to become what I have denoted as an ATC. So, if a fraction $x_1$ of Galactic civilizations succumbs to nuclear holocaust, and a fraction $x_2$ succumbs to an artificial pathogen, and so on, we need to exempt the whole $\sum_i x_i$ fraction from our considerations of Fermi's paradox, with *i* running over the entire list of hazards. All these existential risks, generalized from the usual considerations pertaining to the future of humanity to the entire set of intelligent species, act to reduce the contact cross section and therefore at least partially resolve Fermi's paradox.[37] One exception to this might be the risk from advanced artificial intelligence, which need not result in decreasing the contact cross section and may, in fact, increase it in some scenarios (see **TRANSCENDENCE** in Section 6.7). This is another example of the blurred boundaries between different solutions to Fermi's problem.

Parenthetically, one might ask to what extent *external* wars—that is, those involving two or more different extraterrestrial civilizations—could contribute to resolving Fermi's paradox. It seems quite reasonable to conclude that, by itself, the reduction of the total cross section of all Galactic civilization in this manner would be small, as long as such wars were zero-sum games. If there were a winner in any of such hypothetical conflicts, its cross section would increase as the loser's one decreases; we can imagine an old-fashioned space-opera-like interstellar conquest, with the planets and other resource bases of one civilization coming under the rule of the other. Only if such interstellar wars are profoundly in the domain of *negative-sum games* (where total losses vastly overweigh the gains of any side) can we expect a significant reduction in the total cross section. The catch here, however, is that civilizations capable of waging interstellar wars are, of course, certain to know and simulate, to flabbergasting detail, all of this—and will refrain from entering any negative-sum games. The exception to this rule might be wars waged by autonomous proxies (which I shall discuss in Section 6.6) or those cases in which the convergence of goals leads to an equilibrium in which expansionist sophonts are militarily suppressed in a quick and efficient way (hypotheses to be considered in Chapter 7).

Finally, in this day and age, one might also ask whether advanced sophonts might actually intentionally and willingly self-destruct to avoid a *still worse* fate. While most ethical systems reject the idea that there is anything worse than total annihilation, it is not entirely inconceivable, especially after the traumatic experiences of the twentieth century. In both deontological and consequentialist moral theories, death in either the individual or collective sense might, under some assumptions, be preferable to an eternity of suffering, slavery, torture, and degradation.[38] This leads us to the next set of possibilities, involving analogues of the greatest evils begotten by human history.

## 6.5 Intentional Hazards II: Self-Limitation

Suppose that a sufficiently large fraction of Galactic civilizations is capable of overcoming both natural hazards and the urge for self-destruction and thus advances to technological maturity. Are they all possible SETI targets—and sources of the paradoxical conclusions with respect to Fermi's puzzle? Not necessarily, since there are at least two problems which are likely to remain worrisome even after such a transition: self-limitation, and the threat from deadly probes, to be considered in this section and in Section 6.6, respectively.

Self-limitation means that a civilization decreases its detection cross section through some intentional process.[39] Self-limitation can take many possible forms. One of them is the **Hermit Hypothesis** discussed in Chapter 1. 'Self-limitation' can be regarded as an umbrella term for many different scenarios. However, one of them is particularly worrisome: the establishment of permanent or near-permanent totalitarian control over all individual intelligent agents in the civilization, coupled with a set of goals leading to a small cross section for detection.

> **Introvert Big Brother:** If all Galactic civilizations, instead of self-destruction, slip into permanent totalitarianism, this will dramatically decrease the contact cross section, making them essentially undetectable. In addition, this circumstance will increase the civilizations' susceptibility to other hazards, like those falling under **The Gigayear of Living Dangerously** category. Since permanently totalitarian civilizations are likely to be spatially small (in Galactic terms) and technologically stagnating, they are unlikely to leave traces and manifestations detectable over interstellar distances, so there is no Fermi's paradox.

Since we obviously know nothing about the distribution of historical trajectories of different intelligent species, this hypothesis is extremely speculative—no less so than those of Chapter 4. However, given the overall gravity of its subject matter, as well as its extreme relevance to the future of humanity, **Introvert Big Brother** should not be neglected. The Orwellian state is quite disinterested in the external universe; even if it were willing to communicate, its inherently paranoid nature would have made any opportunity for contact orders of magnitude more difficult.[40] Remember Comrade O'Brien's solipsist geocentrism in *1984*:[41]

> 'What are the stars?' said O'Brien indifferently. 'They are bits of fire a few kilometres away. We could reach them if we wanted to. Or we could blot them out. The earth is the centre of the universe. The sun and the stars go round it.'

Historians have demonstrated how classical totalitarian systems like Nazism and communism undermined science and technology for ideological benefit.[42] Robert Zubrin has argued that 'bad memes' could even destroy a large Galactic civilization which would otherwise be immune to all other natural and artificial threats; this is **Introvert Big Brother** taken to the extreme.[43] Lem's novel *Eden* paints another bleak picture of extraterrestrial totalitarianism.[44]

How could such a state of affairs emerge? Obviously, there are many possible ways of establishing a totalitarian state, but one is becoming more and more actual with time: in order to *avoid self-destruction* or other global catastrophic risks, the infrastructure for such a state could be set up, with broad societal acquiescence.[45] Such an infrastructure would include global and detailed surveillance, advanced methods of data processing, genetic screening, and so on. All such measures—and other more intrusive ones, not considered today in this relatively benign, nearly totalitarianism-free moment in human history—may have entirely legitimate justification within a liberal governance; however, once in place, they might be subverted for totalitarian purposes much more easily than in the case of setting up totalitarianism ab initio. We have witnessed that even the most liberal and enlightened human societies can take illiberal measures, with broad acquiescence of the population, if sufficiently threatened. The relevant insight is that such a development will likely lead to the decrease of contact cross section, thus enabling the hiding of older intelligent communities, and the paradoxical conclusions of Fermi's problem.

A particularly troubling feature of this type of hypothesis is that a *single type* of totalitarian state could arise as a consequence of many *different* sorts of crises (including preventing global catastrophic/existential risks), but there are few ways—if any—it could be dismantled if it is technologically sufficiently advanced. In other words, it could be regarded as an *attractor* in the space of the possible historical trajectories of civilizations. If that is indeed so, the disturbing **Introvert Big Brother** must play at least some role in the ultimate resolution of Fermi's paradox, since it is hard to imagine that at least some sophonts in any sufficiently big sample manage to avoid falling into the totalitarian trap.

The hypothesis also faces serious problems, however. *At least* as much as **Stop Worrying and Love the Bomb**, it suffers from violation of non-exclusivity, since in contrast to, for example, **Astrobiological phase transition**, both warfare and totalitarianism on different Earth-like planets in the Galaxy are independent factors and are extremely unlikely to be correlated.[46] However, while self-destruction is one endpoint of the historical trajectory of a civilization, totalitarianism does not mean extinction. No matter how advanced, a totalitarian government cannot have the same finality as extinction; even a very improbable event like the overthrow of a technologically advanced totalitarian government, can happen in the fullness of time. While this is good news for the enslaved populations of such a regime, it is rather bad news for **Introvert Big Brother** as an explanatory hypothesis. It has to explain the suppression of the detection cross section over much longer times than **Stop Worrying and Love the Bomb**. This automatically decreases our credence in it.

Again, totalitarianism might actually work in conjunction with other destructive process to suppress detectability. It is reasonable to expect that totalitarian regimes are more vulnerable to some kinds of natural or artificial disasters than more open societies are—even our limited human experience suggests it. (Consider, for example, the appalling inefficiency of Soviet response to the Chernobyl nuclear plant accident in 1986.) So, a mixture of the **Introvert Big Brother** and **The Gigayear of Living Dangerously** can be a potent suppressor of detectability over vast regions of space and time. There will be more on those synergistic solutions in Chapter 8.

A further line of criticism of **Introvert Big Brother** and similar scenarios suggests that, in fact, without a more extensive sociological and political theory, we cannot be sure that technologically advanced totalitarian regimes will not be extrovert and colonization oriented. At least, without any further specification, we cannot be *more sure* of that than we are in their long-term stability. The Orwellian vision encompassed the entire surface of the globe, but we can imagine totalitarian dictatorships willing to go much further. Philip K. Dick, in the most classic of all alternate history novels, *The Man in the High Castle*, has victorious Nazi Germany vaguely attempting to colonize Mars.[47] Another canonical literary dystopia, Yevgeny Zamyatin's novel *We*, even revolves around a protagonist who is constructing a spaceship for a totalitarian state.[48] Perhaps a totalitarian regime in the Scutum-Crux spiral arm might wish to arrange local stars into an image—including details of the mandibles and pseudopodia—of the Beloved Leader?[49] After all, most monuments of extinct human civilizations not only were created by oppressive social systems, but also were created for *purposes intrinsically linked to the oppressiveness* of such systems (pyramids, cathedrals, mausoleums, etc.). If, contrary to most of our intuitions and recent trends in our history, the current expansion of individual freedoms and human rights is just a fluke or a lull in the storm, and the future belongs to totalitarian ways of governing and social regimentation, could it be the case that human macro-engineering projects or artefacts will become *likelier*? While it is difficult to conclude either way in the absence of further insights into the social dynamics of totalitarianism, this very uncertainty ought to count as a weakness of **Introvert Big Brother** and similar hypotheses.

An Orwellian superstate governing through 'Ignorance Is Strength' is not the only form of self-limitation, though. Another is the (in)famous concept of 'return to nature': the either voluntary or involuntary relinquishment of technology, reversing the technological progress which on Earth has lasted for many centuries before suddenly accelerating with the Industrial Revolution. The latter is usually seen, according to ideologues of a return to a simpler, pastoral lifestyle, as the chief culprit for all of humanity's problems. Of course, pastoral civilizations are, by definition, **Hermits**, since they have no means of communicating over interstellar distances, let alone leaving traces in the form of astro-engineering. And, in contrast to totalitarianism, where stability is the ultimate goal of a system to be actively pursued, pastoral civilizations are unstable against both natural catastrophes and, ahem, *progress*.[50] Preventing all members, everywhere, from engaging in science and technology tinkering would require efficient totalitarian enforcement—which is likely to be impossible in the absence of technology. So, without going into the merits and demerits of 'return-to-nature' pastoralism, we can say that, although some intelligent species may opt for it, there can be few more exclusivist and hence unlikely options around.

The opposite of pastoralism, namely rapacious industrialization and exploitation of natural resources, presents a more serious catastrophic threat. This is the subject matter of another of the 'almost-default' hypotheses discussed by Brin, Hanson, and others, and one which becomes more and more of practical interest for intelligent beings here, on Earth:[51]

> **Resource Exhaustion**: Interstellar expansion by technological civilizations tends to consume the material resources of any planetary system on an exponentially short timescale. Thus, interstellar colonization creates a bubble of systems with exhausted resources, and eventually leads to the collapse of large civilizations. Within these exhausted bubbles, natural processes lead to slow, gradual replenishment—which, in turn, leads to the whole cycle repeating itself. We have emerged within one such bubble, at the 'down' part of the cycle (which is

no coincidence, since it could be argued that no young sophonts could evolve in the 'up' part of the cycle), so there is no surprise that we do not perceive large Galactic civilizations around us. It is yet another observation-selection effect.

This explanatory hypothesis is related to neocatastrophic, as well as solipsist, and logistic options (to be considered in Chapter 7). Obviously, it presupposes sudden, catastrophic processes violating gradualism; but, at the same time, it subtly assumes that the 'obvious' absence of any traces of other Galactic sophonts from the Solar System is an illusion. If the Solar System had been colonized even in the very distant past, say 2–3 Ga ago, some traces of that event should, in principle, be found, most probably in the asteroid belt or possibly on the Moon.[52] This should in no way be confused with the 'ancient astronauts' speculations discussed in Chapter 4: in sharp contrast, it is exactly the *lack of apparent traces* which makes **Resource Exhaustion** a functioning hypothesis. After so much erosion and local changes on the billion-year scale, there is hardly any chance of 'archaeological' evidence to be found on Earth; however, the question of the evidence in the asteroid belt (and other low-erosion environments) could be regarded as open, especially if it is of a purely geological nature, namely the traces of ancient mining activities.

There is no reason to doubt—and all reasons to suspect—that **Resource Exhaustion** could cripple civilizations, irrespective of their age or other parameters, *up to some critical technological level.* As long as we stick to the laws of physics as the only ultimate constraint and do not concern ourselves with the extreme timescales of physical eschatology (i.e. we consider only our astrophysical environment up to $10^{15}$ years in the future, the epoch in which galaxies remain as well-defined bound entities), **Resource Exhaustion** should not happen at all! Namely, the elemental composition of any particular material resource could, in principle, be replicated using nuclear transmutation, starting from any other chemical element, including hydrogen. After all, nature has created all isotopes heavier than helium (apart from a minuscule primordial abundance of $^{7}$Li) in exactly that manner, mainly by thermonuclear fusion in stars, with additional fission of unstable nuclides and occasional spallation reactions with high-energy cosmic rays. There is no fundamental obstacle to recreating all these reactions in a lab and even to perform them in bulk for industrial purposes, provided sufficient energy is available. The real question is not one of fundamental physics but is rather one of an economic nature: will bulk nuclear transmutation to create a particular resource X ever become *cheap enough* for a sufficiently advanced civilization to outweigh the costs of bringing the same amount of X from elsewhere in the universe?

If the answer is yes, advanced technological civilizations need not fear resource exhaustion, as long as they have any matter to work with and a sufficient energy supply. As previously argued, energy is plentiful both in the era of shining stars (when it can be captured via Dyson spheres and similar contraptions) and even later, in utilizing the gravitational collapse. On longer-still future timescales, exotic energy sources, like the bulk annihilation of CDM particles and antiparticles, or even Hawking evaporation of black holes, could be put to industrial use.[53] Baseline matter also need not be too big a problem if the civilization is compact—that is, far from being a Kardashev's Type 3; we observe a huge amount of unused matter in our astrophysical environment anyway. So, **Resource Exhaustion** will not really resolve our difficulties in this case.

If, for at least some vital resources, cheap transmutation always remains unfeasible, civilizations will *really* have an economic impetus to expand (which is the default position in many superficial treatments of Fermi's paradox). In this case, the expansion might form a roughly spherical wavefront which will gradually turn into a sphere of mostly exhausted planetary systems, surrounded by a thin shell of 'normal' activity. If this were to occur with the full knowledge of other sophonts who were doing it in other parts of the Galaxy (knowledge which would be easy to get with miniaturized robotic interstellar probes), we could have a particularly nasty form of the Hansonian 'burning the cosmic commons' scenario.[54] Considering the fact that there are some weak analogies in humanity's history for how quickly this form of escalation can ruin otherwise prosperous cultures (e.g. Rapa Nui), we should take the lesson seriously.[55]

So, **Resource Exhaustion** can, with some auxiliary assumptions, explain at least weaker forms of Fermi's paradox. On the other hand, the very dependence on these assumptions of a sociological and cultural nature—and the fact that it utilizes disequilibrium conditions (although conditions that are very slow—on timescales of the billions of years necessary for the natural replenishment of resources—to relax)—count against it in comparison with those hypotheses which offer at least a prospect of explaining the equilibrium. To one of those, and a particularly disturbing one, we now turn.

## 6.6 Intentional Hazards III: Deadly Probes and Unstable Equilibria

The well-known physicist and SETI sceptic Frank Tipler has repeatedly promoted his view that a Fermi's-paradox solution based on self-replicating probes offers an unbeatable argument against the existence of advanced extraterrestrial civilizations at this particular epoch of cosmic time.[56] While his argument is strong—and the SETI proponents might wish to devote more time and attention to analysing it in quantitative detail than has been historically done—it is not really unbeatable, due at least to one disturbing alternative. It might be the case that the very existence of older Galactic civilizations and their self-replicating probes is the reason for our observing the 'Great Silence'.

> **Deadly probes:** The Galactic ecosystem is dominated by self-replicating (von Neumann) machines either originally programmed to destroy intelligent species or having subsequently mutated to that effect. Released by an early civilization, they have either destroyed or simply outlived their creators and, on a short timescale ($t_{FH}$), spread throughout the Galaxy. We have been overlooked by these deadly probes so far—but this may not continue to be the case. Since deadly probes are relatively small and stealthy in the astronomical context, they cannot be detected directly—until it is too late—and they have been destroying intelligent species and presumably their habitats before the targets could acquire significant cosmic technology of their own. Thus, there is no Fermi's paradox—and our turn is still to come...

We are facing a reductio ad absurdum of the Tiplerian scenario, which in itself purports to be a reductio ad absurdum of the naive, optimistic SETI picture! The **Deadly probes** scenario presumes that self-replicating von Neumann probes are not peaceful explorers or economically minded colonizers,[57] but intentionally or accidentally created destructive weapons. They might occur due to

- malevolent creators (which in that case had to be the first or one of the first technological civilizations in the Galaxy, close to the Lineweaver limit), or
- the intentional subversion of legitimate self-replicating probes, or
- a random dysfunction ('mutation') in a particular self-replicating probe which has passed to its 'offspring'.

Despite the massive popularity of this hypothesis in fiction, little academic research has been done on the topic so far, although the situation has somewhat improved in very recent years.[58] In his seminal review, Brin considers it one of only two hypotheses which maintain wholesale agreement with both observation and non-exclusivity.[59] Since this hypothesis is extremely relevant and is routinely neglected in the discussions since Brin's study, I shall analyse it qualitatively in some more detail.

In both the case of the intentional creation/subversion of von Neumann probes, on the one hand, and the case of their accidental mutation, on the other, it seems that the originators of the probes either have vanished or are in hiding, while the Galaxy is a completely different (and more hostile) ecological system than is usually assumed. Depending on the (unknown) mode of operation of the destructive von Neumann probes, they might be homing in on sources of coherent radio emissions (indicating young civilizations to be eliminated), automatically sweeping the Galaxy in search of adversaries, or using some advanced astrobiological markers to detect biospheres likely to evolve intelligent beings. The effectiveness of the **Deadly probes** hypothesis does not crucially depend on this mode of detection, although future numerical modelling should take these differences into account.

If there really have ever been extraterrestrial civilizations in the Galaxy, and our observations are valid so far (meaning that we reject both the solipsism and the 'rare-Earthism' of Chapters 4 and 5), they either have or have not constructed self-reproducing interstellar probes. Under such assumptions, we have two independent things to explain: **StrongFP** *and* our existence on Earth at present (= failure of deadly probes to exterminate us so far). There are essentially four groups of scenarios for this:

1. The technology required for von Neumann probes is unfeasible.
2. The technology required for von Neumann probes is feasible, but is intentionally and successfully suppressed (always and everywhere) as too dangerous.
3. The technology required for von Neumann probes is feasible, and von Neumann probes have been released and destroyed their creators, subsequently dying off.
4. The technology required for von Neumann probes is feasible, and von Neumann probes have been released, have destroyed their creators, and are on their way to Earth.

(Since this is a book about Fermi's paradox and its specific explanations, we do not need to concern ourselves here with scenarios in which von Neumann probes are released, but their creators survive and thrive—that would be begging the question 'where are they?' that we started with.[60]) The first option does not look likely from the current human technological perspective, and its rejection was used quite efficiently by Tipler and others to argue against the premise that there are extraterrestrial civilizations in the first place. It seems that the near future (say, the next hundred years) of human technological development could, barring some global catastrophe, lead to the required levels of both technology and economy required for the successful construction of our von Neumann probes. While the obstacles do not look too formidable conceptually, we still need to be cautious, since some part of the picture is possibly missing. The prevailing idea today is that molecular nanotechnology will enable the construction of self-replicating probes by essentially using the natural mechanism for the molecular self-reproduction of living beings. However, this is not necessarily so; John von Neumann obtained his celebrated result decades before there was any serious thought about nanotechnology. If it turns out that, for some reason (such as insufficient stability when faced with the duration and environmental stresses of an interstellar journey), nanotechnological von Neumann probes are unfeasible, it is possible to envision a scenario in which a civilization possesses the technology to build a large von Neumann interstellar probe, but declines to do so for economic reasons (because such a probe would need, to give an intentionally hyperbolic example, a billion tons of very rare element like gallium). This would also count under Scenario 1.[61] All the variants of Scenario 1 neither look appealing nor are satisfactory in general.

Scenario 2 seems acceptable as a solution to **WeakFP**, as long as we are ready to meet the same limitations we discussed in Chapter 4 with regard to the uniformity of motivations and goals. For instance, we have seen that the successful implementation of the **Zoo Hypothesis** would seem to imply the singleton organization of society; clearly, the same is the case with the desired suppression of the von Neumann probes technology. This is another case in which non-exclusivity plays the pivotal role; if there are no hidden attractors somewhere in the parameter space describing the structure of advanced technological civilizations, it is hard to reconcile this option with the non-exclusivity requirement. Solutions in which self-replication has finite number of cycles, such as those proposed by the German systems biologist Axel Kowald, belong to this category.[62] And we still need something else entirely for **StrongFP**.

Scenario 3 is not really a solution, since we only substitute one mystery ('What has removed old advanced technological civilizations?') for another ('What has removed the deadly probes?'). Even worse, while civilizations are complex (eco)systems, subject to many destructive and limiting processes—which this chapter is devoted to anyway!—an interstellar population of deadly probes seems to be a simple and extremely robust ecosystem.[63] It is possible that even vastly technologically superior civilizations would be essentially incapable of removing the deadly probes 'infection' from the Galactic system once it had begun: the story here could be eerily similar to the human fight against common pathogens like the flu virus. Although humans are immensely more powerful than viruses in physical or cognitive terms, the fight against them is a stand-off—at best—since they mutate quickly enough to counter any particular strategy humans devise in real time. There is no reason conceivable at present why such a stand-off could not be a good model for a galaxy containing multiple advanced technological civilizations *and* infected by the deadly probes. However, one might argue that this situation of interstellar warfare and defence against a quickly replicating and persistent enemy could actually be quite conspicuous for newcomers like ourselves, thus in fact *aggravating* Fermi's paradox!

Therefore, we are left with Scenario 4—which seems to be what Brin and others had in mind. And it is truly disturbing, for several reasons. First of all, the analogy with pathogens shows that self-replicating threats are persistent and cumulative: once the deadly probes 'infect' a galaxy, it would be very difficult to get rid of them. Since, in the Copernican view, they have had *at least* the Lineweaver timescale to emerge, this would mean that, unless one can show that they are impossible or incredibly improbable, their probability has to be significant *even prior to invoking any form of Fermi's paradox.* Since such a demonstration of impossibility/extreme improbability is not forthcoming—if anything, recent developments in the fields such as molecular nanotechnology and the theory of constructors suggest that the opposite is the case[64]—the hypothesis that von Neumann machines were released in the Milky Way at some point in the past should be taken seriously. Furthermore, self-replicating technology seems to be extremely dangerous even without assuming malign intent or mutation: if left uncontrolled, self-replicating probes would represent a grave danger to the Galaxy's material resources, which could be consumed on fairly short timescales, thus creating ecological devastation.[65] Finally, in contrast to an aggressive interstellar civilization with colonizing ambitions—which was the hypothetical of the original Fermi's paradox—deadly probes might have a much smaller detection cross section. They might be extremely stealthy and, while the construction of copies in a given planetary system is probably detectable in principle, it would still be much harder to detect than an aggressive colonizing and expanding activity of Kardashev's Type 2.x civilizations.

If self-replicating probes are stealthy, this would imply, in the first approximation, that they are either small or take very long times to assemble their copies; the opposite conjecture, that they are either large or replicate quickly, would lead to processes detectable to even primitive technological societies (like us) when occurring in their home planetary system, or in the nearby planetary systems. The slow cycle of self-reproduction would lead to a minuscule dissipation of energy and a very low degree of any activity, which could be recognized through the emission of electromagnetic radiation, the emission of particles, or dynamical effects such as non-natural deviations of cometary/asteroidal bodies from Keplerian motion. Such artefacts could sit unrecognized in the Solar System asteroid belt (not to mention the Kuiper belt and the much less surveyed areas beyond it) and remain undetectable as anything but rocky or metallic asteroids, cometary nuclei, or other naturally occurring objects. Some of the SETA studies suggested over the last half-century have searched for exactly such artefacts,[66] but if the designers of these probes are advanced enough and bent on remaining undetected, there is little that a civilization in its pre-ATC or pre-Kardashev's Type 2 stage can do.

Therefore, it is possible that the Milky Way is home to a strange ecosystem which exists in parallel with astrobiological sites and consists of one or more species of self-replicators intentionally designed for interstellar travel. Some of the 'species' (or strains) of self-replicators would be performing their original task, either peaceful or violent, while others would have 'mutated' and would pursue other agendas. Some could be stealthy, while others could be visible or not care much about detectability at all. Such a von Neumann ecosystem, although originating by design and not through evolution, would be subjected to at least some of the evolutionary mechanisms known from terrestrial biology, and so would have its own particular evolution.[67] Since the amount of resources available for self-replication is finite and rather sparsely distributed, although huge in total, a form of fitness increase and natural selection is likely to arise. Although, unfortunately, the relevant processes of the evolution of non-natural self-replicators have not been studied so far,[68] some vague general conclusion can be drawn. Such evolution, in particular, would be subject to the same processes leading to the emergence of evolutionary stable strategies on the part of its actors.The reason why this whole bunch of options should perhaps be taken more seriously than has been the case so far lies in the simple fact, well known from our experience with *biological* self-replicators: if self-replicators are small enough and even moderately efficient in self-reproducing, it is essentially impossible to *ever* get rid of them. All human campaigns against various insects or—an even better model, if nanotechnological von Neumann probes are feasible—microorganisms have met with very limited success so far. Those species which humans have managed, after a prolonged and difficult struggle, to place under control, like the smallpox virus or the causative agent of diphtheria, *Corynebacterium diphtheriae*, are only those which self-replicate quite inefficiently.[69] In most cases, as with mosquitos, potato beetles, or most pathogenic bacteria, the best humans can achieve, in spite of our superior intelligence, is a stable stalemate. If the self-reproduction process is still more efficient, allowing for a wider range of mutations, as with the flu virus, even the 'stalemate' equilibrium often becomes unstable, to extreme human cost and detriment. It is reasonable to assume that, given enough time and resources—certainly available in the Galaxy—intentionally constructed self-replicators could be at least as efficient as naturally evolved ones. So, if the Galaxy were to be *infected* with von Neumann probes just once, it might remain infected, at least in some of its many dark corners, *forever*.[70]

Obviously, an advanced technological civilization will have much better insight into the merits and demerits of constructing von Neumann probes than we have today. Such insight might lead to the prohibition of such construction, and this prohibition then would have to be universally enforced; apart from societal development models resulting effectively in a singleton, it is hard to see how such enforcement could be possible at all.[71] Even at superior ethical levels, the presence of independent actors greatly empowered by the technological progress leading to the advanced technological civilization level would mean an a priori unlikely convergence of motives. This is even more so when one takes into account the possible *advantages* conferred by sufficiently widely spread von Neumann probes on their creators (be it an entire civilization, a fraction of it, or even just a single individual).

Thus, we are led to the following rough conclusion: advanced technological civilizations are effectively singletons tightly regulating von Neumann technology, there is at least one species of rogue von Neumann probes loose in the Galaxy, or von Neumann probes created by advanced technological civilizations are generically stealthy. 'Or' is inclusive in all these cases, and exceptions created by recent (as measured by the Fermi–Hart timescale, adjusted for the reproduction time of self-replicators) events are possible. This is not very optimistic, and it is not surprising that Brin, after a much more rudimentary analysis, came to the following conclusion in 1983, proclaiming that[72]

> the frightening thing about 'Deadly Probes' is that it is consistent with all of the facts and philosophical principles described in the first part of this article. There is no need to struggle to suppress the elements of the Drake equation in order to explain the Great Silence... It need only happen once for the results of this scenario to become the equilibrium condition in the Galaxy.

Therefore, what is *really* surprising here is that there has been so little study of the **Deadly probes** scenario, post-1983.

A future theoretical analysis should indicate how big a set of parameters determines the taxonomy of possible self-replicators. Numerical simulations could be used to establish how much of the parameter space describing von Neumann probes is consistent with their *apparent* absence from our past light cone. Finally, improved observations ought to confirm that absence or provide additional constraints from the absence of colonization activities in nearby planetary systems. All in all, the idea of self-replicating interstellar probes is extremely important and it is one of the hallmarks of the actual neglect and underdevelopment of the SETI theory that the relevant discourse is left to science-fiction artists (no disrespect towards the latter, of course); for example, see Figure 6.2).

In contrast to the stable equilibrium of a deadly probes-dominated Galaxy, we can imagine several unstable equilibrium scenarios of intentional suppression

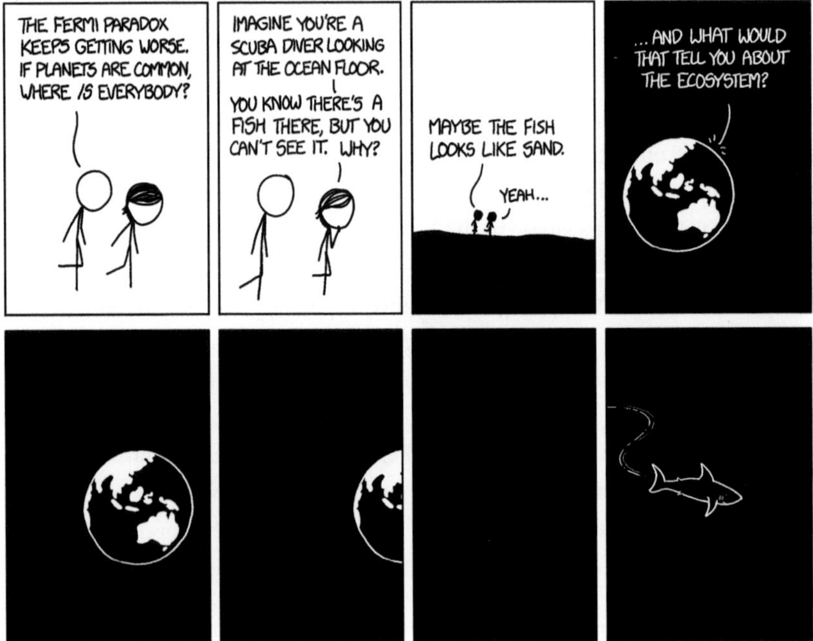


**Figure 6.2** *Xkcd* comic #1377, entitled 'Fish', originally published June 4, 2014. The two most popular stick figures in the xkcd universe, Megan and Cueball, discuss Fermi's paradox. Cueball notices that the glut of newly discovered extrasolar planets actually aggravates the paradox—in sharp contrast to the pundits of the SETI establishment—while Megan suggests an approach which could be interpreted as THE PARANOID STYLE IN GALACTIC POLITICS meets DEADLY PROBES.

*Source*: Courtesy of Randall Munroe and http://xkcd.com/

of contact/detection cross section. By their very nature, such hypotheses are in tension with the non-exclusivity principle, but they might be still worth considering, at least to highlight by contrast how serious the **DEADLY PROBES** menace is:[73]

> **INTERSTELLAR CONTAINMENT:** A form of 'Galactic Club' of advanced technological civilizations communicates and coordinates in order to implement the common strategy of containment. They undertake measures to prevent each and any of them, as well as newcomers, from increasing their contact/detection cross section above some threshold values. In particular, they prevent any aggressive, large-scale colonization policies, resulting in the same observed dearth of interstellar empires.

While one might find a clear motivation for implementing such a collective policy in avoiding the wholesale ecological destruction and exponential 'burning of the cosmic commons' of Hanson, much more problematic is finding the means to ensure uniformity of policy and compliance. If there is no faster-than-light signalling, any coordination has to occur on timescales of an order ranging from $10^0$ to $10^4$ years. This would imply tremendous social stability and goal convergence over a wide range of evolutionary origins and trajectories. One could argue that such a 'Galactic Club' itself represents de facto a Kardashev Type 3 civilization—which would itself be somewhat paradoxical.

Even if there is a common rationale for preserving a large part of the Galaxy as a 'wilderness preserve' and suppressing the expansionist outliers, there seems to be none whatsoever for refraining from astroengineering feats and 'miracles'. In particular, the very existence of the Club will obviate the need for secrecy, so any social motivations not to engage in astroengineering will be nullified. In contrast, the Club will have a strong incentive to introduce itself and its goals to any newcomer on the Galactic scene. One could go even further and suggest that this solution actually aggravates Fermi's paradox, since part of the **INTERSTELLAR CONTAINMENT** would be a motivation to erect a Galactic equivalent to traffic signs visible even to primitive civilizations of Kardashev's Type <1. In this, it represents a link to logistic solutions that will be considered in Chapter 7.

There is a further bit of doubt about the capacity to successfully intervene over interstellar distances and enforce the Club's convergent goals. Finally, the effects of such intervention and 'pacification' of outlying expansionists on the Club members themselves raise many doubts about feasibility. It is possible that very advanced social engineering will one day enable such deep intervention as to entirely erase those impulses rooted in the evolutionary past of any intelligent sophonts, such as territoriality or aggression, but it is in no way certain. Lacking such technologies, the maintenance of large forces capable of performing police actions over interstellar distances could be a destabilizing factor undermining the equilibrium.

## 6.7 Transcendence, Transcension, and Related Scenarios

As previously mentioned, the category of hypotheses violating gradualism needs not be violently destructive; it simply means that we should expect *large* changes in an astronomically brief amount of time. While the hypotheses presented in Sections 6.2–6.6 are certainly violent and destructive, we now need to consider an entirely different option, pushing the limits of the modern, non-gradualist understanding of evolution in the opposite direction.

> **Transcendence**: Advanced technological civilizations have neither destroyed themselves nor spread through the Galaxy, but have transformed themselves into 'something else', not recognizable as a civilization and certainly not viable as a SETI target. Since that development has annihilated—or at least dramatically reduced—the contact cross section, there is no Fermi's paradox.

This is the perfect antithesis of the **Hermit Hypothesis**, discussed in Chapter 1. While the **Hermit Hypothesis** argues that, essentially, everything stays the more or less the same and comparable to what we have on Earth right now, **transcendence** postulates that, indeed, everything changes—and dramatically so. That both can be solutions to the same problem testifies to the depth and complexity of **StrongFP**. The famous Italian writer Giuseppe Tomasi di Lampedusa (1896–1957) put that same paradox into his major novel about social and historical change, *The Leopard*, where a character memorably utters: 'For things to remain the same, everything must change.'[74]

Another literary source—even more popular outside the *literati* circles!—gave an alternative name to this possibility. *Eucatastrophe*, according to none else than John R. R. Tolkien, is a sudden and dramatic *positive* turn of events, usually with religious overtones.[75] **Transcendence** could be regarded as an ultimate form of eucatastrophe, which could advance humanity or any other Galactic civilization 'to the next level'. Some authors, clearly exhibiting a positive attitude towards such a dramatic event, compare it with the Cambrian explosion and other major evolutionary steps.[76]

Historically, this has been the first solution to **StrongFP**, offered by Konstantin Tsiolkovsky who posed the paradox in the first place. Tsiolkovsky, under the influence of his teacher Nikolay Fedorov and other Russian cosmists, concluded that the only reason we do not perceive manifestations of much older civilizations is their evolving into a form of 'superreason' with near-godly powers and, presumably, inconceivable interests.[77] The ideas of Tsiolkovsky have some similarities with both the **Zoo hypothesis** and the **New Cosmogony** discussed in Chapter 4 (with a sprinkling of **Special Creation** added for good measure). However, the bottom line of his view was that this mystical ascension is a quasi-regular destiny of all intelligent beings, leaving the physical universe much as it is in its naturalistic and non-intentional state.

Roughly speaking, transcendence has been regarded as a mystical concept for a long time, before it became a (quasi)scientific concept. Russian cosmists represented a kind of transitional period; a similar form of transition can be found in some speculations of J. B. S. Haldane and Teilhard de Chardin and in Stapledon's *Star Maker*. Today, a similar idea is often formulated in terms of 'technological singularity', the concept envisioned by Stanislaw Ulam and I. J. Good and popularized in the 1990s by the mathematician and author Vernor Vinge.[78] Of course, it is usually linked to the future of *humanity* but, as I have argued in Chapter 1, all such hypotheses could and *must* be mirrored in the astrobiological landscape. The vagueness and flexibility of the idea goes a long way towards explaining the apparent popularity of the concept in the science-fiction genre.[79]

It is a remarkable recent development that this concept has received something approaching full legitimacy in scientific and philosophical circles.[80] Before that, transcendent scenarios abounded in fiction for quite some time, mostly under Stapledon's influence. A prototypical example can be found in Arthur C. Clarke's *Childhood's End*.[81] The technologically superior, interstellar-travelling species 'the Overlords' is just a servant of the really transcendent power, the 'Overmind'—a vast, non-material, cosmic intelligence into which (a part of) humankind is invited to join. Although no specifics are given, it is suggested that the Overmind encompasses many Galactic species—therefore giving a sort of solution to Fermi's paradox, although it is rather peripheral to the novel's main concerns—and that it is essentially mystical. (All sorts of psychic powers, including clairvoyance, spiritism, telepathy, telekinesis, and even *psychic* astro-engineering are discussed and invoked.) Since the Overlords, in spite of all their scientific and technological sophistication, are unable to join the Overmind themselves, a significant part of the problem from our point of view remains unresolved. This highlights the general issue with **Transcendence**: it must be pretty non-exclusive to work at all as a solution to Fermi's problem.[82]

In contrast, John Smart's 'transcension' represents a modern, scientific, computer-science-informed version of this explanatory approach.[83] Smart starts from evolutionary developmental biology to draw a big analogy between contingent and convergent elements in biological and cultural evolution. Therefore, we have a contemporary variation on the age-old theme:

> **Transcension hypothesis**: Evolutionary development guides all advanced technological civilizations towards a compact, optimized 'inner space' (STEM—space, time, energy, matter), presumably linked to black holes and their harnessing for superior computation. Exponential progress in information creation, processing speed, and connectivity at the leading edge of complex systems will entail each medium being replaced by more efficient STEM substrates. Almost all activity within such a compact configuration will occur at the level of information flows. Therefore, the detection cross section of advanced societies is necessarily small. Outliers are suppressed by a complex combination of external and internal (e.g. ethical/legal) constraints. Since only very early, pre-transcension civilizations are easily detectable through SETI searches, there is no Fermi's paradox.

This is a modern-day, naturalistic, and rational version of Teilhard's Omega-point theory encompassing the idea of grand eucatastrophe. It has a notable relation with the **Zoo hypothesis** as Smart himself argues, since the road to transcension will mean leaving primitive societies alone to find the same attractor (or to destroy themselves or to succumb to external threats). We shall encounter the same idea of very-high-level natural selection acting to filter out unsuccessful civilizations, in Section 7.5. The value of the **Transcension hypothesis** as a solution to Fermi's question critically depends on the assumed attractor nature of the transcended state within the parameter space of cultural evolution. Its advantage over other versions of the **Transcendence** idea is that it offers a number of potentially testable predictions, notably the decreased flux of outgoing, detectable signals from all artificial sources, and some predictions related to the outcome of optical/infrared SETI surveys.

**Transcendence/Transcension** is a unique item in this whole taxonomy of hypotheses in that it is not only hard but rather *intrinsically impossible* to say whether it blunts or sharpens the issue. No matter whether the event of transcendence is mystical or technological (if such a distinction can be strictly made for very advanced technology[84]), no matter whether it is a single momentous event or a series of protracted processes, it is easy to conceive of scenarios going both ways: decreasing the detection cross section, as well as increasing it. Clarke's Overmind engages in astroengineering on huge scales, thus increasing the detection cross section and aggravating Fermi's paradox. On the other hand, Smart's transcended civilizations are likely to be compact, ultra-efficient, and hardly detectable over interstellar distances. Those properties might be combined with a degree of safety and/or extra-high-efficiency concerns to make them even less visible from afar.[85] So, there is a wide range of scenarios driving detectability in either direction.

There is more than our current ignorance on the possibilities and capacities of transcendence: there is a difficulty *intrinsic* to the very concept. Since its central event depends on achieving a superior level of various cognitive phenomena, for example, superior intelligence and superior volition, we are again in the position of Darwin's dog speculating on Newton's mind, or Lem's ants feeding on the brain of a dead philosopher. That is an even worse relative position than us versus the hypothetical advanced sophonts of Lineweaver's timescale—which creates a problem in hypotheses such as the **Zoo hypothesis** or the **Interdict hypothesis**. Even if we could *understand* some of the actions of those sophonts (e.g. building a Dyson swarm, remaining hidden from sight, and migrating elsewhere in the Galaxy are behaviours which do have a *meaning* to us), we are certain to encounter difficulties in trying to *predict* them. The entire reasoning is much more forceful in the case of the **Transcendence** hypothesis in its general form.[86] Since these are the very same actions which create (or remove) Fermi's paradox, we seem to be in a blind alley here.

## 6.8 A Neocatastrophist Résumé

Denying gradualism leaves us with many different options in what amounts to true embarrassment of riches. The neocatastrophist portfolio contains serious and even frightening explanatory hypotheses, bearing deep relevance for the future of humanity. From a strictly scientific point of view, this has both merits and demerits: while wider interest is likely to bring more bright people and funds for research, it is also likely to introduce systematic errors, prejudices, and anthropocentric biases. The story of **Stop Worrying and Love the Bomb** is paradigmatic here: it was taken seriously during the Cold War and it has ceased to attract much interest since then (which is quite worrying).

Among the non-exclusive hypotheses, the **Astrobiological phase transition** model has a temporary and very parochial advantage in comparison to **Deadly probes**, since we understand the possible dynamics of global regulation mechanisms. This might change soon, however, since at least some theoretical insight into various quantitative models of **Deadly probes** seems feasible at present. In a sense, this is more an artefact of the human history of science that we know more about the astrophysical environment than about the space of possibilities opened up by self-replication technology.

Moreover, global catastrophic events affecting large parts of the GHZ will tend to reset many local astrobiological clocks nearly simultaneously, thus significantly decreasing the probability of existence of extremely old civilizations, in accordance with Annis's scenario. In both these hypotheses, however, it is possible that pockets of *old* (in effective, astrobiological terms) habitable sites remain, either through the purely stochastic nature of lethal regulation mechanisms, or through dysfunction of destructive von Neumann probes. In other words, the reset is not 100 per cent efficient. In this manner, we shall connect this to a logistic hypothesis, **Persistence**.

The predictions of these two prototypical hypotheses, and their ramifications for ongoing SETI projects, cannot differ more dramatically. While the **Deadly probes** scenario is particularly bleak and offers no significant prospect for SETI—or the future of humanity, by the way, the science-fiction sagas of Benford and Saberhagen notwithstanding—punctuation of the astrobiological evolution of the Milky Way with instances of an **Astrobiological phase transition** will, somewhat counterintuitively, have the net effect of strengthening the rationale for our present-day SETI efforts. As the secular evolution of the regulation mechanisms leads to an increase in *average* astrobiological complexity (as in Figure 6.1), we might expect more and more civilizations to enter the 'contact window' and join efforts in expansion towards Kardashev's Type 2 and 3 status. It is exactly at this stage when they are most amenable to detection and possible—though always uncertain—communication, as envisioned in the classical SETI portfolio that a sizeable fraction of them are *now*.

Intuitively, it seems clear that any form of catastrophic events affecting planetary biospheres in the Milky Way will reduce the hypothetical extraterrestrial civilizations' ages and thus reduce the tension inherent in **StrongFP**. In terms of the Drake equation, 'neocatastrophic' denotes everything which makes $L$ shorter; **Transcendence/Transcension** could do it, as well as any number of self-destruction visions. If such events are spatially and temporally uncorrelated—as in **Stop Worrying and Love the Bomb** or **Introvert Big Brother**—such an explanation is obviously low on the non-exclusivity scale. In contrast, hypotheses with correlated events—such as **Deadly probes** or **Astrobiological phase transition**—fare much better here. In some cases, it is still impossible to estimate how tightly correlated some of the postulated events might be; this applies in particular to the transcendence-type scenarios, where the extent and the nature of **Transcendence** remains a mystery.[87]

So, troubling questions remain. Are there cosmic civilizations which are, in the infamous financial phrase, *too big to fail*?[88] If not, are their failures entirely due to catastrophic events, or there is an additional slow mode of failure which has prevented them from arising in the Galactic history so far? If the answer is positive, does this slow failure threaten humanity as well? And to what extent are factors other than age crucial for the growth and expansion of technological civilizations? These and other questions cannot be meaningfully discussed in isolation, outside of the big picture—we shall return to them after we consider the final class of explanatory hypotheses for Fermi's paradox.

CHAPTER 7

# The Cities of the Red Night

*Logistic Solutions*

A small number of hypotheses have been proposed which do not fall easily into any of the broad categories previously described in this book. Although the total variation of approaches to Fermi's problem is already stupendous, it is remarkable how a small number of ideas escape the general philosophical categories discussed in Chapters 4–6. In a nutshell, the remaining option for hypotheses is that *although other Galactic sophonts exist and are in principle detectable, the economic cost of maintaining detectability over astrophysical timescales is too high, for one reason or another.* The emphasis of such hypotheses usually lies on the perennial puzzle already considered in Chapter 1: how wide is the spectrum of responses of different sophonts to the same evolutionary challenges? Are there invisible walls in the vast space of historical trajectories of advanced intelligences?

William S. Burroughs's 1981 masterpiece, *The Cities of the Red Night* describes, with somewhat brutal irony, a prediluvian human civilization containing both utopian and dystopian elements and whose breakdown contained the seeds of all future conflicts of history. The old master of the Beat movement (and much of modern pop culture) weaves his grand theme thus:[1]

> The Cities of the Red Night were six in number: Tamaghis, Ba'dan, Yass-Waddah, Waghdas, Naufana, and Ghadis. These cities were located in an area roughly corresponding to the Gobi Desert a hundred thousand years ago.

> At that time the desert was dotted with large oases and traversed by a river which emptied into the Caspian Sea.
>
> The largest of these oases contained a lake ten miles long and five miles across, on the shores of which the university town of Waghdas was founded. Pilgrims came from all over the inhabited world to study in the academies of Waghdas, where the arts and sciences reached peaks of attainment that have never been equaled. Much of this ancient knowledge is now lost.

Each city has a slightly different character. Tamaghis is city of 'contending partisans' where 'everything is as true as you think it is and everything you can get away with is permitted.' Ba'dan is a city of commerce where 'everything is true and everything is permitted.' Yass-Waddah is a 'female stronghold' where 'everything is true and nothing is permitted except to the permitters.' Waghdas is a university city where 'complete permission derives from complete understanding.' Naufana and Ghadis are both 'cities of illusion where nothing is true and *therefore* everything is permitted.' Burroughs notes that 'the traveler must start in Tamaghis and make his way through the other cities in the order named. This pilgrimage may take many lifetimes.'

This characterization—entirely different, but still on the same evolutionary trajectory (influenced by catastrophic events such as the eponymous red nights, which are speculated to be the consequence of a meteorite impact)—does seem a fine metaphor for the topic of this chapter. But, wait! The story Burroughs weaves in this excerpt is just one of the three very different strands of the novel. In another, located in the contemporary world, the (anti)hero, the private detective Clem Snide, gets a job to obtain nothing less than the original of the book *The Cities of the Red Night*:[2]

> 'Why do you want the originals? Collector's vanity?'
>
> 'Changes, Mr. Snide, can only be effected by alterations in the original. The only thing not prerecorded in a prerecorded universe are the prerecordings themselves. The copies can only repeat themselves word for word. A virus is a copy.'

So, the central dilemma of stability versus evolvability (e.g. the capacity for alteration and change in a sufficiently general sense) appears as a powerful motivator in fiction as well as in science. Not only the varying outcomes of this eternal tug of war, but also the possibility of variability itself, in the context of Galactic sophonts, comprise the subject of this chapter.

## 7.1 Down with '-Ism'!

There is no doctrine formulating and promoting a lack of economic prerequisites, no adequate '-ism' to reject here. Instead, we have the seemingly commonsensical idea that various 'microassumptions' we have made in applying the non-exclusivity requirement to the real Galactic scene are, in fact, invalid: that interstellar travel is impossible, that other sophonts live in unexpected places or have all evolved in some unrecognizable direction, that there are global sustainability limits, and so on. The lack of a coherent doctrine—even in rejection—should not lead us into downplaying these possibilities. While this might be the most heterogeneous category of solutions, this has nothing to with their plausibility, at least not a priori.

The main ingredient here is the issue which can, by analogy with one of the crucial concepts in mathematical physics, be dubbed the *ergodicity* (or the lack of it) in the development of Galactic civilizations. Recall that the ergodic hypothesis states—in a cartoonish oversimplification—that the ensemble (or spatial) and temporal averages of any function describing a dynamical process should be the same 'almost everywhere and everywhen.' Rougher still, this common-sense conclusion enables us to study ageing in humans without observing individuals over their lifetimes—we take representative samples from various age ensembles instead. The same procedure even more forcefully applies to studies of stellar evolution: stellar astrophysicists most certainly cannot 'wait and see' in order to observe changes in important properties in stars. And yet, the enviable precision of our modern theory of stellar evolution comes from investigating huge samples of various stars in different stages of their evolution and comparing their observed properties with theoretical prediction, *assuming* that the ergodic hypothesis is valid.

There are many non-ergodic systems, however, a nice example being amorphous solids with a disorganized structure, like glass: over very long timescales, glass behaves as a liquid, but no amount of individual samples of glass ('ensemble average') will reveal that in a laboratory. Even the oldest stained glass windows in medieval cathedrals will not—contrary to a popular myth—exhibit glass flowing; the relevant timescale is simply far longer than the duration of human civilization.[3] However, this does not impact the major point of non-ergodicity: we could consider spin glasses (or disordered magnets) instead. Their dynamical evolution visits only a tiny subset of their possible states, and hence exhibit history in a deep sense. Another interesting example of non-ergodicity is the economic growth exhibited by free markets.[4] So, the relevant question for

Fermi's paradox can be formulated as, are extraterrestrial civilizations more like stars or more like spin glasses and free markets? Do they pass through a well-defined set of stages governed by some general, albeit unknown, principles, on the one hand, and local initial conditions, on the other? Is there any form of *regularity* with which they enter the stage of being detectable over interstellar distances?

Even more pertinently, do they become *more* ergodic with the passage of time? Namely, we might expect that the *initial* conditions in which other sophonts evolved are very heterogeneous—even if we limit ourselves to life based on chemistry, life based on carbon chemistry, or even life based on the same basic compounds, proteins, and nucleic acids as found in terrestrial life. Wildly varying extrasolar planetary habitats and contingent, opportunistic evolutionary histories are likely to produce wildly different outcomes, even if they all belong to the same chunk of the astrobiological landscape labelled 'sophonts'. Different morphologies and other biological baggage will certainly make the properties of civilizations emerging in this way very heterogeneous—that much seems obvious nowadays (although it was not appreciated enough in the early days of SETI, nor has it, for obvious practical reasons, been taken to heart in pop-cultural representations of extraterrestrial intelligence).

In contrast, there has been much informal talk (mostly on the future-oriented Internet forums, boards, and mailing lists) to the effect that, at the later stages of civilizational development, we should expect higher level of convergence. In other words, we should expect older and more advanced civilizations to become more similar than they were in their early days. This might be regarded as a particularly broad and long-term generalization of the 'end-of-history' thesis.[5] The suggested rationale is that the universe we all inhabit is, after all, unique and common for all and contains the same kind of resources; and that expansion into this common habitat will be more and more influenced by the general properties on large scales and less and less influenced by the particular properties of its starting point. Like gas injected into a container, the final state of the system will be determined by the 'global' properties of the container, such as shape, volume, and temperature, and not by the exact place in which, or even the exact manner in which, the injection was performed. Challenges of interstellar colonization will, according to this view, be so large and common for all sophonts that, in the end, it will be irrelevant whether the sophonts in question are marine- or land-based organisms, whether they have arthropod, vertebrate, or some yet unimagined morphologies, or whether their senses are based on visible light, radio waves, or X-rays. This view is in broad agreement with the hypothesis that most advanced civilizations are *postbiological*, an idea which, although hardly new, has gained traction recently; I shall return to this important concept in Chapter 9. But even biologically based sophonts should not experience too many difficulties in adapting to a form of Banksian 'Culture'.[6]

Still, this view of 'late convergentism' among different Galactic civilizations is controversial. One can just as easily find counterarguments. Not only does it go contrary to what we perceive in the biological world in both ontogenetic (babies are much more alike than adults) and phylogenetic (microorganisms inhabiting Precambrian Earth were much more similar than Earth's inhabitants are today) senses; it is unclear, to say at least, whether it is applicable to human culture and history either. Opening new frontiers (e.g. during the European Age of Discovery in the fifteenth and sixteenth centuries) has mostly resulted in *increased* diversity across the board. One of the major motivations for exploring and colonizing the universe since the times of Tsiolkovsky has always been the idea that the vast new frontier of our cosmic environment will offer prospects of new diversity, rather than new uniformity.[7] Finally, large spatial and temporal scales (provided the speed of light remains an absolute limit to information transfer) will of necessity undermine any attempt to impose uniformity, even if it is regarded as desirable. So this intriguing issue remains open—and offers a challenge for the further engagement of SETI with the social sciences.

Formally speaking, the hypothesis that interstellar travel is impossible in principle, for all intelligent species, should belong to this category—what is 'more logistic' than that? However, it will not be discussed in any detail, because not only does it not offer any response to stronger versions of Fermi's paradox (i.e. ones involving astroengineering and detectability), but also it could either be deemed a false hypothesis or be subsumed under other explanatory hypotheses. If the reason for the alleged impossibility is physical, the hypothesis can be regarded as false, since natural objects in interstellar flight—such as comets in hyperbolic orbits; free-floating planets; or, indeed, human space probes, starting with Pioneer 10—have been observed. That such form of interstellar travel is inefficient is irrelevant; we should remember Boltzmann's dictum:[8]

> It may be objected that the above is nothing more than a series of imperfectly proved hypotheses. But granting its improbability, it suffices that this explanation is not impossible. For then I have shown that the problem is not insoluble, and nature will have found a better solution than mine.

If the reason is social or cultural, for instance, due to self-destruction or stopping technological progress before interstellar flight could be realized, then it should be subsumed under **Stop Worrying and Love the Bomb, Self-Destruction, Advanced Version** (Section 6.4), or **Introvert Big Brother** (Section 6.5). In addition, one could list a host of important specific results on the *feasibility* of interstellar flight, with these results having been obtained within even such a technologically young civilization as ours[9]—and these will only add to our confidence in rejecting the claim that interstellar travel is impossible. There are other complexities and subtleties, however, which include realistic forms of interstellar travel to be considered in this chapter.

## Logistics, Cultural Attractors, and City State versus Empire State

Here we may pause for a moment to zoom out to a wider issue: what might be the non-instrumental goals of an advanced technological civilization? While this certainly is too formidable question for this book—and this epoch!—to address seriously, a cartoonish simplification, such as that shown in Figure 7.1, might still be useful food for thought. We can think about these issues in terms of *quantity*; for example:

- the spatial extent of the civilization
- the number of individuals in it
- its total economic output
- its Kardashev type

or in terms of *quality*; for example:

- the precision of its computations and simulations
- the 'depth' of its scientific understanding
- its ethical values (such as the autonomy of actors)
- its place on the Barrow scale describing microscopic control over matter[10]
- the extent it strives for 'Olympian perfection'

Of course, these axes are not truly independent in reality but we can still use the setup as a sort of 'zeroth-order' approximation, which is justifiable by our ignorance, to derive useful conclusions. The level of our present-day human civilization with respect to these issues is obviously quite low in terms of both quantity and quality, as it is a newcomer on the cosmic scene, so it is located in the area near the coordinate origin.

Within this setup, it seems reasonable to assume that, out of many different evolutionary pathways which advanced technological civilizations could take when striving for what I have labelled a 'naive paradise', namely maximization of

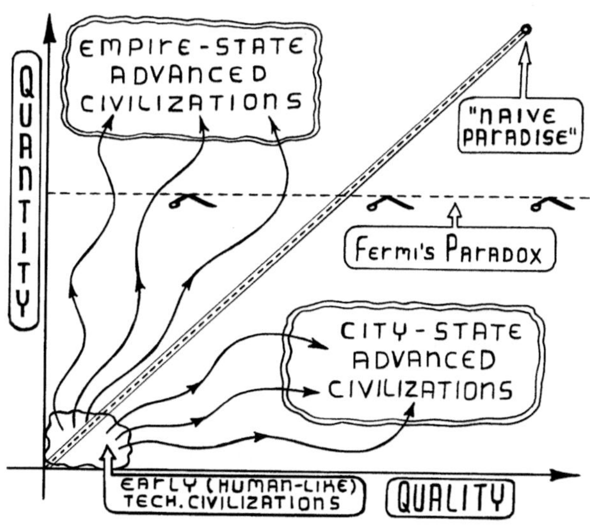


**Figure 7.1** The essential logistic dilemma: quantity vs quality in space colonization. Some possible evolutionary trajectories starting from the initial primitive state are shown as arrowed lines. Fermi's paradox acts a sort of cut-off, precluding the existence of at least some kinds of advanced civilizations favouring much interstellar expansion. Observe how it does not adequately account for StrongFP in general, since even city-state-type advanced civilizations can produce artefacts detectable over interstellar distances or send probes to our neighbourhood.

*Source*: Courtesy of Slobodan Popović Bagi

*both* quantity and quality, we may expect two attractors to form in accordance with the astronomical distribution of resources. One would correspond to a conventional expanding civilization of the conventional space colonization discourse: the empire-state model. This would correspond to civilizations which were Kardashev Type 2.x and higher. The opposite model—the city state—would encompass compact, highly efficient civilizations utilizing only the resources of a single planetary system and maintaining cultural unity in the face of the light-speed barrier. An example of the latter would be any civilization described by the **TRANCENSION** hypothesis (Section 6.7).

Fermi's paradox enables us to reject at least some of the empire-state evolutionary trajectories. Recall the definition of **KardashevFP** in Chapter 1. Even for Kardashev types lower than 3, it is only reasonable to conclude that the lack of detectable manifestations and traces tend to undermine the viability of evolutionary trajectories close to the empire-state model. Constraints on the city-state model for advanced technological civilizations are much weaker, especially since such civilizations could maintain stealth much more easily than empire states could over astronomically long periods of time.

## 7.2 Wrong Tree?

Are we possibly barking up the wrong tree in the whole SETI mess? It seems rather easy—puzzlingly easy, in fact—to muddy the waters of serious analysis of SETI-related issues. One question repeatedly comes across as confusing in the popular account: why do SETI projects routinely insist on searching just for emissions from the vicinity of Sun-like stars? Orthodox SETI thinkers tend to shrug this away either by vaguely appealing to Copernicanism or even by outright mockery, and proclaim that there are no viable alternatives.[11] However, even if Copernicanism is indeed one of the main criteria for SETI target selection, it certainly could not be the only one. (We will leave aside for the moment the problem of the proper sets—or reference classes—for the application of Copernicanism, as discussed in Chapter 3.) And the star of origin is not really relevant if we search for older, more advanced Galactic civilizations which are likely to be present in multiple planetary/stellar systems. Hence, here are a number of hypotheses suggesting that we have, indeed, barked up the wrong tree in the Galactic forest:

> **RED EMPIRE:** Extraterrestrial civilizations either emerge in or move into planetary systems of red-dwarf stars (spectral classes late K and M), the most numerous and longest-living type of normal star in the Galaxy. The extreme longevity and huge number of habitats available in this kind of systems ensure stable conditions for development and planning on extremely long timescales. In contrast, extraterrestrials tend to avoid more luminous Main Sequence stars like our Sun.[12] Therefore, they will avoid the Solar Systems and our planet, so Fermi's paradox is resolved.

According to modern-day numerical models, an M-type red dwarf of 0.1 $M_\odot$, with a present-day (effectively zero-age) surface temperature of about 2200 K and a luminosity of about $4 \times 10–^{4}$ $L_\odot$, spends an incredible $6.28 \times 10^{12}$ years (!) burning hydrogen into helium, as a Main Sequence star.[13] Obviously, it's a good place for long-term planning! Although such a red dwarf does change its luminosity during that fantastic lifespan, by about three orders of magnitude overall, the bulk of the change comes *only* in the last ~140 *billion* years, as it approaches the helium white dwarf final state. For the first ~6.1 trillion years, the mean variation in output is within a single order of magnitude and anyway occurs so mind-bogglingly slowly that *any* possible technological civilization in the area will be able to adapt to it easily. Of course, there might be other problems with the habitability of planets around red dwarfs, notably due to synchronous rotation and occasional strong flares; this is a much-discussed topic in contemporary astrobiology.[14] However, for our present purposes, it is enough to note that those problems are irrelevant once either (i) a technological civilization somehow evolves in a red-dwarf planetary system or (ii) a technological civilization *moves* into a red-dwarf planetary system from another place. The technology necessary for interstellar flight is surely more advanced than the one required for protection from flares or even for 'terraforming'.[15] Once a civilization is established in a red-dwarf planetary system, it has ensured a relatively stable haven for survival on truly cosmological/physical eschatological timescales.

Clearly, only weaker forms of Fermi's paradox can be resolved in this manner. In spite of all the 'cautious voices' from the SETI community, the most important ground fact in any discussion related to **StrongFP** has been and still is the absence of astroengineering manifestations and traces (and the absence of Kardashev's Type 3 civilizations). There is no a priori reason why red-dwarf-based civilizations would be less likely to construct a Dyson sphere or burn antimatter than those living around Sun-like stars.

Not surprisingly, some of these ideas have been prefigured in a loose form within the discourse of science fiction. Karl Schroeder not only formulated the above-mentioned **PERMANENCE** answer to Fermi's question (Section 5.4), but also envisaged an entire Galaxy-wide ecosystem based on brown dwarfs (and the halo population in general) and taking advantage of a low-temperature environment.[16] Hence, we have the following variation on this theme:

> **BROWN EMPIRE**: This is the same situation as described in **RED EMPIRE**, but based on brown dwarfs and free-floating planets ('rogue planets') instead of Main Sequence stars. Brown dwarfs may have planets and other kinds of circumstellar material which could be used for engineering purposes, just like normal stars, while keeping everything still colder (hence, more efficient). Since brown dwarfs are presumably still more numerous than Main Sequence stars, there would not be a lack of habitats to expand into, if desired, and multiple reasons to do with efficiency, culture, or ethics might lead to the avoidance of Sun-like stars. The consequences for Fermi's paradox are thus the same as for **RED EMPIRE.**

The extremely red L-type dwarf PSO J318.5338-22.8603 discovered in 2013 in the Beta Pictoris moving group of stars can be seen as a prototype of such targets. With a mass of about 6.5 Jupiter masses and a temperature of 1160 K, this free-floating brown dwarf/giant planet is a very slowly cooling infrared source—and there are indications that there are large quantities of dust and other material around it.[17] There is no reason it could not have large satellites, ones of the size of Earth, if not larger.

There is an important structural difference in comparison to the **Red Empire** case here: if our mainstream understanding of habitability is correct, a technological civilization needs to *move* into the realm of brown dwarfs, rather than evolve right there. This presumes the development of interstellar flight, as well as long-term logistic capacities. So, **Brown empire** would be more appealing to postbiological civilizations. Since brown dwarfs can be safely supposed to be quite energy efficient, the cold environment around these failed stars would be convenient for highly efficient computation and other relevant activities. Consequently, **Brown empire** civilizations would be hard to detect with our observational capacities since, as argued in Chapter 1, detectability is inversely proportional to energy efficiency, in at least a portion of a typical technological civilization's historical pathway.

Both **Red Empire** and **Brown empire** suffer from the same basic problem: they account only for weak versions of Fermi's paradox and—obviously, since they are described in this chapter!—are not very strong on non-exclusivity either. If we limit ourselves to **ProtoFP** (as, unfortunately, some critics have been all too willing to do[18]), these hypotheses fare better; in reality, **StrongFP** remains a problem, since we also cannot reasonably expect that the nature of the stellar habitat influences the long-range signalling or the astroengineering projects.

## 7.3 Persistence

To what extent does the Fermi–Hart timescale in Eq. (1.1) truly represent the timescale for touring the Galaxy? As any tourist knows, one can visit Paris without setting foot into the Luxembourg Garden, or be in Istanbul in a very real sense but missing a delightful cup of coffee at the top of *Galata Kulesi* in the Karaköy district. Simply, the number of sub-destinations (or, if you wish, locations at each particular destination) is too big from the practical point of view. If you have decided to spend time in the Louvre instead of in *Le Jardin du Luxembourg*, that can hardly be called an *irrational* decision or something which cries out for deeper explanation. Your assignment of priorities according to what moral philosophers and decision theorists would call your utility function is what makes all the difference.

Such could be the case with interstellar exploration/expansion as well. The resolution of weaker versions of Fermi's paradox crucially depends on the unknown *dynamics* of interstellar exploration and colonization. Since this is unknown, it is entirely legitimate to make more or less plausible assumptions in constructing various explanatory hypotheses. For instance, Geoffrey Landis,

Osame Kinouchi, Branislav Vukotić, and other researchers have investigated interstellar colonization models which, under particular assumptions, leave large bubbles of empty space surrounded by colonized regions.[19] An analogous phenomenon in the context of condensed-matter physics is known as *persistence* and appears in a wide range of phenomena, notably percolations.

> **Persistence**: Interstellar colonization proceeds in such a way that large regions of the Galaxy remain untouched by it for a long time. We happen to be inside one of these regions, 'bubbles' of wild, untechnologized universe. While the colonized region may spread across the entire Milky Way disc, dozens of kiloparsecs in each direction, it is highly *porous*, leaving large regions of 'wilderness' in the disc. Our Solar System is located in one of these bubbles where evolution of new intelligent species is possible. This explains the absence of extraterrestrials from our planetary system and our stellar neighbourhood. If old and large Galactic civilizations are discrete and efficient, this state of affairs explains away Fermi's paradox.

While **Persistence** is an interesting and thought-provoking idea, it is quite improbable that it can entirely resolve Fermi's paradox. An obvious weakness of the hypothesis is that it still implies cultural uniformity regarding the dynamical parameters of colonization, and this violates the non-exclusivity requirement. For example, in the generalized invasion percolation model of Galera and coworkers, all uncertainties related to the process of colonization are bundled together in a single colonization parameter, which is clearly unrealistic. In order to avoid that particular problem, we need to postulate restrictions on the unknown dynamics/economics of interstellar travel/colonization. In addition, we would expect either to detect extraterrestrial signals coming from outside of the local non-colonized bubble, or to detect manifestations of giga-anni-older technological societies, even in the absence of the direct presence of extraterrestrials in the Solar System.

Furthermore, there is a 'passing-the-buck' element here as well. We need to take for granted that colonization does proceed in a manner similar to, say, the way coffee percolates, or like similar phenomena in statistical physics. But why should it be so? For there are underlying physical reasons in the latter case, dealing with the maximization of entropy and scaling properties of the medium. Such physical principles impose a seeming order on the random motion of particles in systems like coffee grains in the water solution. For the former, the analogy must break down, since interstellar colonization is necessarily an *intentional endeavour*. There might be some very high-level laws or regularities of social dynamics which dictate such-and-such structure of interstellar colonization but, at the present level of our knowledge, this is pure speculation. If anything, we tend to believe the opposite: that colonization will be guided by detailed astronomical knowledge on the properties of target systems. And such knowledge is anything but random. One expects colonization to follow the distribution of resources, including habitable real estate. And we have no reason to believe that the distribution of resources is such that leaves large empty bubbles. In particular, our Solar System and its vicinity do not seem to be resource poor at present.

A similar approach has been favoured in Bjork's numerical simulations, although the timescales obtained in his model are quite short in comparison with Eq. (2.4), even with his explicit rejection of self-reproducing probes, thus being more in line with the older calculations of Hart, Jones, and Newman and Sagan.[20] Bjork concludes, rather too optimistically, that Fermi's paradox could be resolved through the statement that 'we have not yet been contacted by any extraterrestrial civilizations simple because they have not yet had the time to find us.' In view of the Lineweaver timescale, this is clearly wrong as long as we do not postulate some *additional reason* for the delay in starting the Galactic exploration. Recent detailed numerical simulations conclude that[21]

> while interior voids exist at lower values of [the colonization probability parameter] *c* initially, most large interior voids become colonized after long periods regardless of the cardinal value chosen, leaving behind only relatively small voids. In an examination of several . . . models with a large range of parameters, the largest interior void encountered was roughly 30 light years in diameter. Since humans have been broadcasting radio since the early 20th century and actively listening to radio signals from space since 1960 . . . it is highly unlikely that the Earth is located in a void large enough to remain undiscovered to the present day.

Finally, we should emphasize that **PERSISTENCE** helps only with weaker versions of Fermi's paradox. In the terms of Chapter 1, it explains **ProtoFP** and **WeakFP** well. One can claim that it explains **KardashevFP** as well since, as long as the large bubbles of non-technologized space persist in a galaxy, we can talk only of Type 2.x civilizations and not true Type 3 ones. On the other hand, it is doubtful that the existence of a local bubble resolves **StrongFP**, since there is, in principle, no reason why we should not detect either communications or astroengineering feats of the surrounding large civilizations. Some *exclusive* clause, such as 'they are discrete', 'they use only neutrinos for communication', or something similar must be added as an assumption, which obviously weakens the appeal of **PERSISTENCE** as a solution to the puzzle.

One interesting feature of **PERSISTENCE** is that it blends well with another analogy which condensed-matter physics lends us, namely phase transitions. It has been noted that systems exhibiting persistence tend to undergo dramatic phase transition(s) at some point.[22] This is intuitively reasonable: whatever it is that prevents the tendency towards equilibrium to be realized in a finite time will break down at some point, resulting in sudden, tremendous change which sweeps like a wildfire through the system. In this manner, there is an underlying relationship between this and some of the neocatastrophic solutions discussed in Chapter 7, notably the **Astrobiological phase transition**.

Note that **Persistence** is one of the more SETI-friendly or contact-optimistic solutions surveyed in this book. Not would only viable SETI targets exist, but there would be nothing, *in principle*, to prevent us from detecting old and large Galactic civilizations beyond the bounds of our local empty bubble; it would just be an issue of the sensitivity of our technology and our persistent (pun intended!) SETI efforts. Either for subtle culturological reasons or for pure contingent facts of history, large civilizations have not (yet) created truly large, obviously detectable astroengineering feats, but what they have done might enter into our horizon of detectability with each new improvement of our instruments and techniques. Extragalactic SETI observations could, in principle, discern the pattern of non-uniform, quasirandom changes in a galaxy being colonized on our past light cone. Also, there is nothing to prevent a peer civilization (a young one, similar to our own) to evolve somewhere within our local bubble, presenting us with an excellent SETI target. This falsifiable feature, which **Persistence** partially shares with other logistic solutions to Fermi's paradox, is an important methodological advantage, to be explored more fully in Chapter 8.

## 7.4 Migrations: To the Galactic Rim and Beyond

As previously mentioned, a large fraction of the red and brown dwarfs discovered in the vicinity of the Solar System are in fact old, metal-poor Population II objects belonging to (or originating in) the Galactic halo. Clearly, most of these objects are located at large distances from the Solar System and even from the entire GHZ. A next step—or a variation on the theme—for advanced Galactic civilizations could be to physically relocate to the region in which these objects are more common. Indeed, if the mountain won't come to Muhammad, then Muhammad must go to the mountain!

Hence, we come to the subject of migrations. An approach originally due to the late Robert J. Bradbury offers an alternative solution based on the assumption that most or all advanced technological societies will tend to optimize their resource utilization to an extreme degree.[23] It could be shown that such optimization will ultimately be limited by the temperature of the interstellar space—and that temperature decreases with increased galactocentric distance in the Milky Way. The Landauer–Brillouin's limit, known from the physics of computation, suggests that the maximal amount of information ($I_{max}$; in bits) which could be processed by any classical computation device using energy $E$ (in ergs) on the working temperature $T$ (in K) is given as[24]

$$I_{max} = \frac{E}{k_B T \ln 2} = 1.05 \times 10^{16} \frac{E}{T}, \tag{7.1}$$

where $k_B$ is the Boltzmann constant, which relates kinetic energy to temperature. How far down can the temperature $T$ go? Obviously, the ideal case is that of the CMB temperature at present epoch:[25]

$$T_{CMB} = 2.72548 \pm 0.00057\,\mathrm{K}. \tag{7.2}$$

It decreases with cosmological expansion, but extremely slowly. This is the temperature of the 'universal heat reservoir' filling the entire space. Any object, if no energy is supplied to it whatsoever, will eventually reach thermal equilibrium with the universe at this temperature (although, depending on the mass, shape, and specific heat capacity of the object, it might take a very long time). There are in the universe, however, many heat sources which act to prevent equilibration with CMB. In the Solar System, the source is obviously our Sun, although gaseous giants like Jupiter have small internal energy sources (due to residual gravitational contraction). As we go farther away from the Sun, objects generally become colder and colder. For example, the coldest known natural object so far is Neptune's giant moon Triton, for which both terrestrial instruments and those on the Voyager 2 space probe gave a surface temperature of $38.0^{+2.0}_{-1.0}$ K.[26] This might seem quite low by human standards but computation efficiency at this temperature, in terms of the standard set by Brillouin's inequality, is still grossly inefficient, as this temperature is more than an order of magnitude above the CMB limit. One cannot go much lower by simply going farther away from the Sun, since the Galactic radiation field (created mostly by O-, B-, and A-type stars) will prevent further cooling. The mean temperature of a solid body in radiative equilibrium with the radiation permeating the disc of the Milky Way is in the 10–30 K range, depending on its exact position.[27] If one wishes to have more efficient—by an order of magnitude—computation approaching the Landauer–Brillouin limit, one needs to go further out, towards the edge of the Galactic stellar disc. Hence we have the following hypothesis:

> **Living on the rim**: Advanced technological civilizations are located at the outskirts of the Galaxy, where temperatures are conductive to extremely efficient computing. Since efficient computing is an attractor in the space of evolutionary pathways, and the limitations which the laws of physics impose on computing efficiency are universally valid, long-term planning sophonts, at one time or another, will have migrated to the Galactic rim, far from star-formation regions, supernovae, and other energetic astrophysical events, in order to process information most efficiently. This will occur no matter where they have originated. They have little interest in the inner parts of the Galaxy (although they are likely to survey it via non-intrusive remote sensing probes and advanced astronomical techniques).

This hypothesis is a reflection on the general trend towards investigating *post-biological* options for evolutionary trajectories of intelligent species.[28] This is in part related to—and indeed does not mutually exclude—hypotheses involving some form of the transcendence discussed in Section 6.7. Recall that, for example, under the **Transcension hypothesis** of Smart and others, advanced sophonts are focused on the 'inner space', requiring high computing efficiency. In the long term, gains achieved by moving into a colder environment will certainly compensate for the expenses of the migration itself; this is especially so if the civilization consists of uploaded minds (and their cities, similar to those fictionally portrayed in Egan's *Diaspora*[29]) or even more compact substructures within a superintelligent artificial intelligence. We may imagine that the migration itself will involve rather minimal physical transfer and consequently small expenses: only a single 'universal constructor' coupled with a receiver needs to be sent physically. When it arrives, the relevant information could be beamed at the destination using gamma-ray lasers, neutrinos, or some other high-density communication channel.[30] Such a migration will in itself be a feat of astroengineering—but (and that represents an epistemological advantage for **Living on the rim**) not of the kind which will be readily detectable from interstellar distances.

This solution modestly violates the non-exclusivity requirement, depending on how universally valid the assumption of resource optimization as the major motivator of advanced sophonts is.[31] It has been claimed in classical SETI literature that interstellar migrations will be forced by the natural course of stellar evolution.[32] However, even this 'attenuated' expansionism—delayed by $\sim 10^9$

years or more—is actually *unnecessary*, since the naturally occurring thermonuclear fusion in stars is an extremely inefficient energy source, converting less than 1 per cent of the total stellar mass into potentially useable energy. A much deeper (by at least an order of magnitude) reservoir of useful energy is contained in the gravitational field of a stellar remnant (e.g. a white dwarf, a neutron star, or a black hole), even without already envisaged stellar engineering.[33] So the end of the Main Sequence lifetime of their star should be a boon to advanced sophonts, not a disaster! A highly optimized civilization will be able to prolong the utilization of its local astrophysical resources to truly cosmological timescales.[34]

The consequences for our conventional (that is, predominantly empire-state) view of advanced societies have been encapsulated in an interesting paper by Martin Beech:[35]

> A star can only 'burn' hydrogen for a finite time, and it is probably safe to suppose that a civilisation capable of engineering the condition of their parent star is also capable of initiating a programme of interstellar exploration. Should they embark on such a programme of exploration it is suggested that they will do so, however, *by choice rather than by necessitated practicality*.

In brief, the often-quoted cliché that life fills all available niches is clearly a non sequitur in the relevant context; interstellar colonial expansion should not be a default hypothesis, which, sadly, it is in most SETI-related and far-future-related discourses thus far. *Choice, rather than necessity*, in Beech's wise words, is what needs to be taken into account in any consideration of interstellar migrations.

A truly extreme version of the migration hypothesis was recently suggested by Robin Spivey in an intriguing paper dealing with the relationship of cosmological neutrinos to the considerations of habitability.[36] One of the conclusions was that, in the truly long term, the environments most conducive to the continued existence of intelligent beings are rich clusters of galaxies. The Virgo cluster is located about 16.5 Mpc away—and it is possible that we don't see any older sophonts in the Milky Way because they embarked on an *intergalactic migration* soon upon reaching technological sophistication and the correct understanding of cosmology. Apart from the issue of the general feasibility of intergalactic travel (if it's not in a miniaturized Armstrong–Sandberg probe version[37] but requires moving at least some of the supporting infrastructure), this cluster-bound relocation hypothesis offers a clear suggestion for empirical testing: we should look at the Virgo cluster galaxies for traces and manifestations of astroengineering, and in a general direction towards Virgo when searching for possible propulsion signatures.[38]

Speaking of which, why settle down at all? The following, literally logistic hypothesis about the generic future of advanced technological civilizations was proposed by de San in 1981 and, while it is not primarily about resolving Fermi's paradox, it does propose a possible way out of it:[39]

> **Eternal wanderers:** The generic fate of advanced technological civilizations lies in a nomadic existence in fleets of world-ships, emancipated from any permanent stellar or planetary habitats. They have no use for planetary systems, except possibly for refuelling and resupplying (even that could, arguably, be done elsewhere, without all the trouble of getting inside deep gravity wells; e.g. in the giant molecular clouds or star-forming regions). Therefore, such advanced sophonts are utterly uninterested in visiting Earth or colonizing the Solar System. Since their world-ships, even if powered by nuclear fusion or antimatter annihilation, produce weak detection signatures, there is no Fermi's paradox.

Among the motivations suggested by de San are safety from planetary catastrophes, including impacts or nearby supernovae, as well as the high quality of life which can be achieved in artificial, entirely designed habitats; the latter was the motivation suggested originally by Gerard O'Neill in his famous proposals for self-sufficient space colonies.[40] However, none of these seem persuasive enough to justify such a radical cultural convergence of lifestyles. This hypothesis violates non-exclusivity about as much as **Stop Worrying and Love the Bomb** does—with the additional twist that the absence of propulsion signatures observed so far further suppresses its likelihood.[41]

Finally, the most radical of related scenarios is one of *aestivation*, which also represents a kind of migration, but in time, rather than in space:[42]

> **Great Old Ones:** The first—and, correspondingly, the most advanced—Galactic civilizations hoard resources and put themselves into a state of aestivation (often misleadingly called *hibernation*) until the astrophysical environment becomes more conductive to high-efficiency information processing. In particular, the temperature limitations on computation as per Eq. (7.1) could be relaxed by simply waiting for a sufficiently long time—while gathering a hoard of material and free energy for subsequent prolonged efficient computations. While they are in a state of aestivation, their detection cross section is extremely low (although presumably some sensory and defence automation is active), so there is no Fermi's paradox.

At first glance, this hypothesis sounds paradoxical, since everything that happens in the universe produces entropy and degrades resources; thus, one can seemingly only lose by waiting. However, even rough calculations show that this is not necessarily the case: inside gravitationally bound systems, such as the Local Group of galaxies, major ongoing entropy-producing processes, like star formation and gravitational collapse, will not seriously tear into the reservoirs of free energy available to advanced sophonts until the distant future. In a sense, the **Great Old Ones** hypothesis can be regarded as being complementary to **Living on the rim**: while the latter implies relocation *in space* in search for the more efficient computing, the former implies relocation *in time* for the very same reason.

Of course, a 'minor' issue of the mechanism for such extremely long aestivation remains. The reliability of any system over cosmologically long periods of time should not be assumed lightly. In the fictional world of Schroeder's *Lockstep*, the periods of aestivation are much shorter and unfold within an essentially non-threatening environment.[43] Aestivation on a timescale of billions of years or longer presents a completely new order of problems, which we cannot ever hope to completely resolve. A critic might argue that, even if we accept the motivation beneath **Great Old Ones** as valid, the working of the hypothesis relies too much on almost-miraculous unknown technology. It looks too much like the naive *oh, they're invisible!* reply to Fermi's paradox. Besides, the environment could fluctuate to such an extent on such long timescales—including the emergence of upstart civilizations like humanity!—that the controlling mechanism should arguably be intelligent in its own right (posing further difficulties). Ensuring multiple redundancy of the controlling mechanism will require a substantial investment of resources in it—and whether such investment would be warranted in the long run, only future quantitative investigations and numerical modelling will show.

## 7.5 Sustainability

The sustainability solution of Jacob Haqq-Misra and Seth Baum is also related to the compact, highly efficient version of advanced extraterrestrial civilization I have dubbed the 'city-state' model.[44] In a sense—and in sharp contrast to the wild speculations of Section 7.4—the following is the most conservative among at the least remotely viable hypotheses in this book.

> **Sustainability:** Any fast-growing interstellar expansion is unsustainable at all times. Faster expansion ends in quick collapse, while slower modes are sustainable, but persist only on timescales much larger than the Fermi–Hart timescale and comparable to the Lineweaver timescale. So, we are currently living in the epoch of colonization of the Galaxy (i.e. in disequilibrium), but it proceeds much slower than it is naively assumed. In the long run, only cautious and frugal civilizations are likely to survive and leave detection signatures. Since those signatures would emerge only very slowly, the possibility that we have overlooked them so far is realistic.

This is a modern rendition of an old idea, first suggested by two of the SETI founding fathers, Michael Papagiannis and Sebastian von Hoerner.[45] Haqq-Misra and Baum envision a situation in which large-scale interstellar expansion is infeasible due to sustainability costs (and perhaps dysgenic factors, similar to the ones in Schroeder's adaptationist hypothesis), so that the prevailing model is a compact, 'city-state' sophisticated technological civilization, possibly slowly expanding, but at rates negligible in comparison to the expansion in either Newman–Sagan–Bjork regimes (i.e. those with no self-replicating probes) or Tipler regimes (i.e. those with self-replicating probes). This situation can be regarded as a placeholder for various more detailed evolutionary scenarios involving social, economic, or political collapse, or perhaps active resistance (so there is some crossover with **Interstellar containment**; see Section 6.6). It is worth noticing that there are *two* distinct components to the detection/contact cross section under **Sustainability**: one generated by sustainable, 'working' extraterrestrial civilizations along the mainstream evolutionary trajectory, and the other generated by 'failed' expansionist outliers. The authors of the hypothesis were quite aware of this circumstance; they actually suggested searching for traces and manifestations of 'failed' civilizations which have *already collapsed*, as a fruitful SETI strategy! There is, however, no first principle to decide what the ratio of the two contributing cross sections is. This belongs to the realm of cultural dynamics which is, as emphasized, completely uncharted territory. If the part of the cross section generated by failed expansionists is large enough, however, it obviously defeats the whole point of the hypothesis—we would have expected to detect such civilizations long ago, either via traces of their direct presence in the Solar System or through their recent astroengineering manifestations.

Some fine-tuning seems required again. If we focus on economic and social factors, we need to admit the need for 'passing the buck' here again. While it is a truism that we know little about economics and sociology above the planetary level (in spite of some heroic recent attempts[46]) and therefore there are likely whole levels of causal structure unknown to us which could constrain the history of advanced civilizations, there is something deeply unsatisfactory about leaning on our ignorance in this manner. The distinction between distal and proximal causation should be kept in mind at all times in our considerations—while it might be the case that fast-mode expansion could be the distal cause of a civilization's collapse, we should still prefer to have an account based on more proximal causes. In other words, among civilizations, there may be a form of very-high-level selection, where profligate spenders lose and frugal types win in the long run—although the success would not be obvious, at least not to primitive and short-term observers such as ourselves.

These hypotheses meet with the same criticisms based on (i) non-exclusivity and (ii) the lack of astroengineering detection signatures, as previously considered. One could even argue that slow expansion or the complete lack of expansion (as in the strict city-state model of advanced technological civilizations; see 'Logistics, Cultural Attractors, and City State versus Empire State' on p. 208) would actually make the second issue more acute. For example, a civilization characterized by striving for efficiency within a single planetary system seems reasonably better motivated for building a Dyson sphere around its star than a civilization devoted to rapid expansion over hundreds of systems. Ditto for stellar uplifting or other large-scale transformative processes.[47] A similar reasoning applies to other modes of astroengineering when they are regarded as profound changes in the physical environment enabling more efficient utilization of resources: if you are eager and ready to seek resources elsewhere, over interstellar distances, you will be more weakly motivated to long-term plan and rearrange your local resources for improved efficiency. If sowing and reaping over a large area (or even buying produce at the market!) is fine with you, your incentive to invest money and time into erecting a greenhouse will correspondingly diminish.

## 7.6 Metabolic Problems and Digital Indulgence

Fermi's paradox seems a somewhat unlikely topic for medical doctors. It is not in vain that we have devoted so much attention to the need for overcoming anthropocentrism—and, in a wholly different and admittedly beneficial sense, anthropocentrism is the physician's bread and butter. So, what better testimony of the depth and breadth of Fermi's lunchtime puzzle can we entertain than to encounter (after so many astrophysical, planetological, biological, etc., speculations) a health-based hypothesis for resolving it!

Three British biologists and physicians, led by Alistair Nunn, have recently proposed a novel solution to the apparent absence of extraterrestrial intelligence and its manifestations, using an extrapolation of hugely adverse trends present in contemporary human civilization.[48] Even if we take this hypothesis with a grain or two of salt more than we do some of the others, its potential impact upon our own future should be enough for us to consider it seriously.

> **Galactic stomach ache**: The evolutionary trajectories of all sophonts have an attractor—the state in which a civilization's development is arrested as the result of an explosion of degenerative medical conditions. We may be witnessing the very first steps of such a process in human civilization right now, with pandemics of obesity, diabetes, and other metabolic disorders, which ultimately originate in mitochondrial dysfunction, in most of the developed world. Having removed most of the stress due to our physical and biotic environment, we have with it removed low-level beneficial stress (known as *hormesis*). Already, the exponentially growing economic costs of maintaining health in face of these degenerative disorders are huge in comparison to investments in space research and exploration, not to mention utilization of extraterrestrial resources. If such trends continue and are typical, humanity could end up in a state in which almost all material resources and all creative energy are expended on the maintainance of a comfortable lifestyle free of external stressors, leading to a plateau in the development of cognition, and its subsequent diminishing. Modern human civilization may have already passed beyond the 'window of opportunity' to launch a dynamic and persistent attempt at colonizing the Solar System and fully utilizing its resources for further advancement. If the interplay between biology and culture is similar in most other Galactic biospheres, in particular in terms of the loss of hormesis before engaging in significant colonization of space, we may infer that most sophonts are locked in the trap of hedonistic imperative and cognitive decay. Hence, they stay on their home planet or very close to it, with very few surplus resources; hence, they are incapable of astroengineering or more ambitious cosmic communication; hence, as time passes, they are more likely to succumb to any other external or internal threat; hence, there is no Fermi's paradox.

This hypothesis could be also understood as a particular variant of the self-limiting scenarios given in Chapter 6, or even as a long-term (d)evolutionary loss of cognitive capacities, as in **Permanence**, which is discussed in Chapter 5. While those generic links remain—and will be further considered in Chapters 8 and 9—there is still a distinct flavour to **Galactic stomach ache**, as well as an additional economic aspect which eludes purely evolution-based or purely culture-based hypotheses. Nunn et al. point out that, already today, the annual cost of obesity in the USA ($2.1 \times 10^{11}$ dollars) is more than an order of magnitude larger than the cost of the national space research and exploration programmes ($1.77 \times 10^{10}$ dollars); *healthcare bills might have already caused our failure in space!* And this wide gap tends to increase as the time passes.

Importantly, besides this purely economic argument, Nunn et al. allege an entirely biological, functional relationship between environmental stresses and the emergence *and retention* of intelligence and other complex cognitive traits. They write:[49]

> Our suggestion is that the development of a comfortable environment, wherever it occurs, could lead to similar physiological dysfunctions as we observe across the current human population... life evolved, and now maintains its order, in response to a constantly changing and challenging environment. Remove the stress and life begin to lose its order, resulting in spiralling oxidative stress... advanced civilizations essentially become very comfortable and remove all stressors, which leads to spiralling health care costs and potentially, either a plateau in average intelligence, or even a fall. Clearly, neither of these would support the development of interstellar space travel.

Now, while we might argue that this reasoning is unduly anthropocentric and violates non-exclusivity, and even speculate that the authors are using an intentional 'cosmic' hyperbole in order to emphasize the lesson for us here and now, there is some food for thought in their study. First of all, the lesson is crucially important and should be much more publicized.[50] Further, while **Galactic stomach ache** indeed violates non-exclusivity, it is by no means certain that it violates non-exclusivity *more* than many older and more conventionally discussed hypotheses referenced in this book. Take, for example, **Stop Worrying and Love the Bomb** (Section 6.4); is there any compelling argument that nuclear weapons, and division of the world into conflicting ideological camps, are *less exclusive* than metabolic disorders? And hardly anybody denies the seriousness of the danger of nuclear annihilation/nuclear winter, do they?

While we are at it, where exactly has all the enthusiasm for 'the final frontier' of space gone in the last couple of decades? While the conventionally measured quality of life has monotonously improved over that period, we have actually witnessed a dramatic *shrinking* of our space-faring capacities, as argued by Robert Zubrin, among many others.[51] In the same time, we have also witnessed the explosion of pseudoscientific nonsense *about* space in a wider public (spearheaded by preposterous Moon-landing conspiracy theories, revived flat-Earthers, Raëlians, and other UFO cultists), with it being distributed more cheaply than ever by fatter-than-ever people.

Whoever thinks the ideas behind **Galactic stomach ache** are *too* lightweight would do well to also consider the following. To a large degree, the spread of humanity over the land surface of our planet and its cultural history has been governed by health issues. The Athenian plague, the late-Roman world pandemic ('the plague of Justinian'), the Black Death in the fourteenth century, the diseases which followed the conquistadores into the New World, the Spanish Flu of 1918—all of these have played roles in the outcomes we see today, probably more so than any military or political leader, or explorer. It was a contingency of evolution, not a pre-ordained deterministic fact, that smallpox was brought to the Americas by Spanish sailors and adventurers, and not the other way around. We may only speculate how a Eurasian pandemic of smallpox at some point after 1492 would have played out in such a counterfactually reversed world, but it is clear that the resulting Western civilization (if it existed at all) would not have been remotely similar to what we observe today. On a smaller scale, health issues have played some role in the demise of all previous civilizations on our planet. In many cases, it was exactly a particular lifestyle, diet, or custom which contributed to the downfall.

The data invoked by Nunn and collaborators show this is still the case—much better documented, of course—today, except that today's premier health issues are to an even larger degree the straightforward consequences of the predominant lifestyles and cultures. And the latter are the straightforward consequences of our mastery over most of the immediate physical environment. There is simply nothing random or accidental about unhealthy diets or habits, or pollution of air and water, as well as an ever-increasing desire for often pointless and frivolous comforts and pleasures.

Nunn and co-workers consider explicitly only obesity, diabetes, and metabolic disorders, but one can add coronary artery disease,[52] allergies of all sorts, autoimmune diseases, and Alzheimer's disease, as well as any other number of degenerative syndromes and illnesses. A poor lifestyle increases the risk of just about every disease, mostly due to an inflammatory component that appears to be related to mitochondrial dysfunction. While one might argue that *radical* medical breakthroughs will lead, in time, to a decrease in medical costs, and even—as the techno-optimists hope—the eradication of some of these syndromes, the issue is by no means certain. And one should keep in mind that *new* disorders and illnesses might appear, triggered by, among other things, global climate change, increasing pollution, or the misuse of biotechnology or nanotechnology.

The scenario of Nunn et al. is closely connected with the ideas of the evolutionary psychologist Geoffrey Miller and the media/art expert Wendy Ann Mansilla and her collaborators.[53] They have argued, first and foremost, that the development of technological civilizations such as the present-day human one and the near-future human one leads to the consumption of pleasure technologies which gradually discourage and displace (in terms of resources) the development of technologies required for the colonization of space, astroengineering projects, vigorous astronomical research, and other activities increasing the detectability cross section. But of no less importance are non-obvious feedbacks contained in the media themselves (the digital media of today, the Internet of things tomorrow, etc.), and their interaction with our evolved pleasure circuits; these further enhance this frivolous tendency and possibly make it, after a particular threshold is crossed, permanent. Miller uses the particularly evocative term *fitness faking* to describe the phenomenon in which personal hedonistic ideation or truly virtual reality completely trumps our physical and ecological environment. This destructive process is reinforced to the extent that there is a striving to increase adaptive value (or fitness, in evolutionary terms) solely with respect to the former:

> Technology is fairly good at controlling external reality to promote our real biological fitness, but it's even better at delivering fake fitness—subjective cues of survival and reproduction, without the real-world effects. Fresh organic fruit juice costs so much more than nutrition-free soda. Having real friends is so much more effort than watching *Friends* on TV. Actually colonizing the galaxy would be so much harder than pretending to have done it when filming *Star Wars* or *Serenity*.

An extreme extrapolation of the trend to be 'amused to death' (cf. Roger Waters and Pink Floyd) is clear:

> We are already disappearing up our own brainstems. Freud's pleasure principle triumphs over the reality principle. We narrow-cast human-interest stories to each other, rather than broad-casting messages of universal peace and progress to other star systems.
>
> Maybe the bright aliens did the same. I suspect that a certain period of fitness-faking narcissism is inevitable after any intelligent life evolves. This is the Great Temptation for any technological species—to shape their subjective reality to provide the cues of survival and reproductive success without the

> substance. Most bright alien species probably go extinct gradually, allocating more time and resources to their pleasures, and less to their children.

In a similar vein, Mansilla et al. point to empirically established attentional biases and intrusive mental concepts linked to the pleasure-giving patterns appearing in digital media.

Finally, one could view the topic from another angle. During the nineteenth century and the first half of the twentieth century, a form of evolutionary theory called *orthogenesis* was quite popular, especially among German and other continental naturalists. Espoused by big names such as Eimer, Haake, Cope, Berg, and Schindewolf (who was also the first to propose that cosmic explosions could cause mass extinctions in Earth's history), orthogenesis was historically the major rival to Darwinism on the opposite side of the functionalism/formalism divide.[54] Darwin himself called it rather dismissively (but very aptly from the present point of view!) 'descent theory' in the beginning, and Schindewolf called it 'typostrophism' in its last blossom, but it amounts to the same idea: a global morphological type contains a large but *ultimately finite and hardwired* potential for variation. The pool of possibilities will be manifested through one or more internal channels which direct the evolution of a species from its beginning to its inevitable end. Against Darwin and subsequent adaptationists, orthogeneticists believed that variation is not isotropic and, although they did allow for natural selection shaping some of the traits of species, they ascribed the development of the highest relative trait frequency to directional change along the preset channels of a given, discrete morphological form. Archetypical examples given for orthogenesis included cases where evolution clearly led organisms towards extinction by the mechanical continuation of once adaptive and subsequently maladaptive traits: for example, the ever-increasing length of teeth in sabretooth cats, since this, after the teeth exceeded some optimum size, clearly impeded the efficiency of feeding. In one of the last serious analyses of orthogenesis, before the tide of the Modern Synthesis swept it to the proverbial 'trash heap of history', the philosopher of biology Marjorie Grene wrote:[55]

> Furthermore, Schindewolf agrees with the older paleontologists that within each type, once it had appeared, there is a progressive, orthogenetic development. In fact, there is a rhythm analogous to that of birth, maturation, and senesence: the sudden appearance of a new type, its orthogenetic advance, and finally a stage of the breaking up of types which usually leads to extinction.

While orthogenesis is no longer a viable *biological theory* for explaining changes in the living world, this does not mean that some of the empirical evidence amassed by the orthogeneticists of old is easy to explain or that the effects which were ascribed to half-mystical orthogenetic mechanisms could not be mimicked by other, truly occurring processes. And it certainly does not mean that some version of it could not be applicable to cultural evolution. We do not understand the mechanisms of cultural evolution very well, but there is no doubt that—in contrast to biological evolution—it incorporates an important element of old Lamarckism: the inheritance of acquired characters. Cultural traits are acquired and, barring great physical or historical catastrophes, are transmitted to future generations via a variety of media. So, if Lamarckism gets a pass in the domain of cultural evolution—why not cultural orthogenesis as well?

It might well be the case—similar to the situation in **Permanence**—that *our* kind of cognition is ill-suited to grand feats of space colonization and astroengineering. Miller hints that some particular human cultures and lifestyles might be better adapted to survive the 'Great Temptation' and eventually establish contact with similar strands of other Galactic sophonts. But one could think of this as too easy a solution: it might be that the entire hardwired neural structure is, in the long run, ill-adapted to large tasks outside of the home planet. Maybe the Galaxy (will) belong to hive-mind intelligences[56] or something as yet unconceivable.

## 7.7 A Logistic Résumé

The internal unity of logistic hypotheses rests on the inherent challenges of maintaining an advanced technological level and avoiding the pitfalls of 'unlimited' expansion and consumption of resources. As previously argued, there seem to be at least two attractors in the space of evolutionary trajectories of advanced civilizations, namely the city-state model and the empire-state model. While a fully fledged empire-state model leads quickly to a Kardashev's Type 3 civilization (or the 'Galactic Club' as an aggregate of civilizations), thus leading us directly into paradox, it seems possible to 'bend the rules' a bit and remain in the attracting field of the empire-state model, while at least decreasing the amount of paradoxical consequences. This could be possible as a result of a porous, discontinuous presence (**Persistence**), the use of a specialized habitat (**Red Empire** and **Brown empire**), the complete rejection of fixed habitats (**Eternal wanderers**), or a long delay of the expansionism (**Great Old Ones**). Alternatively, a slightly different operationalization of the city-state model would further decrease the detection cross section in **Living on the rim** or **Galactic stomach ache.**

Bending the rules has its price, as always: most of solutions in this category can only answer weaker versions of the paradox. **Red Empire** and **Brown empire** are particularly vulnerable in this regard, since there is no visible rarity of such stars in the vicinity of the Solar System, no particular reason to suppress messaging and astroengineering—and, in general, no reason to expect a small detection cross section. Some degree of cultural uniformity, which is, as previously argued, in collision with non-exclusivity principle, is required for other hypotheses to work as well, although perhaps least for **Persistence**, since this assumption could be partially outsourced to the astrophysical properties of the distribution of resources. In some cases, deeper analysis of some of these hypotheses via the use of game-theoretical tools would be highly desirable; this remains a task for the future astrobiological research.

There are obvious borderline regions with other classes of solutions. In particular, the adaptationist solution (**Permanence**; see Section 5.3) can be interpreted as a version of the baseline idea of the specialized habitat hypotheses (**Red Empire** and **Brown empire**), as can the requirements stemming from **Sustainability**. Catastrophic **Resource exhaustion** (Section 6.5) is just the other side of the coin of **Sustainability**; and **Sustainability** itself might require some explanatory help from **Interstellar containment** (Section 6.6). While I shall discuss these cross-connections in the Chapters 8 and 9, it is important to keep in mind that it is general principles and assumptions that matter, and these have multiple intersections in the multidimensional space of parameters. In navigating the complex maps of that landscape, we need to keep in mind that particular authors put more or less emphasis on one side of the theoretical construction than on the other.

*A time to break down, and a time to build up.* This concludes the broad list of explanatory hypotheses. While it has been clearly impossible—especially in a wieldy book format—to consider *all* suggestions appearing in about three-quarters of a century of the history of Fermi's problem, I am satisfied with achieving a representative cross section of the explanatory hypotheses literature, which has grown too large for any single survey.[57] The important taxonomical principle—that each resolution of the paradox corresponds to violating (or at least relaxing) at least one additional philosophical assumption—has held throughout. Our next step, therefore, is to bring all of the players onto the scene at once and then try to establish who fares the best.

CHAPTER 8

# The Tournament

## *How To Rate Solutions and Avoid Exclusivity*

Fermi's paradox is a prototypical example of the problem involving *unknown unknowns*. The former US Secretary of Defence Donald Rumsfeld was certainly no astrobiologist, but he correctly described this sort of difficulty in the following, somewhat infamous, statement:[1]

> Reports that say that something hasn't happened are always interesting to me, because as we know, there are known knowns; there are things we know we know. We also know there are known unknowns; that is to say, we know there are some things we do not know. But there are also unknown unknowns—the ones we don't know we don't know. And, if one looks throughout history..., it is the [ones in the] latter category that tend to be the difficult ones.

A quintessential summary of Fermi's paradox is as follows: we have a report/evidence that something, otherwise expectable, hasn't happened so far.[2] This is Sherlock Holmes' dog that didn't bark, Magritte's missing men with the newspaper, Pascal's frightening silence instead of the harmony of spheres, Lem's Master's voice without the Master, Borges's invisible Congress of the World, and Lovecraftian colour out of space, which is not a colour at all. Something is amiss—and many have voiced their views, as surveyed in four preceding chapters: astronomers, physicists, biologists, philosophers, engineers, geoscientists, and authors—even the occasional medical doctor, computer scientist, or spy. Now we are up for an assessment of this battlefield of ideas.

So, which is the best solution? Clearly, the answer to this question depends on the adopted set of value criteria. In this chapter, I shall give a tabular overview of the main classes of hypotheses, revisit the criteria of detectability in light of the major assumptions of each class, and try to show that the number of truly different elements or 'building blocks' of these hypotheses is smaller than hitherto thought. To completely unveil these building blocks, we need to ask *meta-questions*. How easily could a solution be generalized to the universe beyond the Milky Way? What minimal additional information do we need from SETI or other fields of research to clinch a case for one hypothesis or another? Is there a price to be paid in unrelated disciplines in adopting a 'Hypothesis X'? Such questions would flesh out the true similarities between seemingly varied ideas and 'remove the unnecessary parts' from the slab of marble, in Michelangelo's quip. The remaining part of the astrobiological landscape will ground a number of more or less radical research programmes aiming at elucidating, undermining, and falsifying some or all of the remaining hypotheses, to be outlined in Chapter 9.

## 8.1 A Table Too Large?

There is something archetypal in the human tendency to create long lists, as memorably argued by Umberto Eco.[3] We have surveyed a large number of hypotheses—here, I put all of them in a single table, by broad category and with (necessarily subjective) grades and comments (see Table 8.1). The main purpose of Table 8.1 is to have a reminder at hand. The grading is done along the lines set by David Brin in his seminal review, with the refinements discussed in Chapter 3. Again, individual items on the list are just placeholders for multiple versions of the same set of ideas; for instance, it doesn't really matter whether advanced sophonts destroy themselves under **Self-Destruction, Advanced Version**, predominantly through (i) bioterrorism, (ii) the misuse of nanotechnology, (iii) geoengineering gone awry, (iv) turning their home world into strange-quark matter, or (v) the invention and abuse of *krakatit*.[4] The consequences for the astrobiological landscape and Fermi's paradox remain the same.

Grades are given in the US system (i.e. A, B, C, D, and F, with +/– as a further fine gradation). Grades are, obviously, a subjective category; I have tried to list the arguments for and against in the appropriate section, although some particulars will be further analysed below. F is assigned to hypotheses which clearly do not work *as solutions to Fermi's problem*; if they could work, but require an exclusive conspiracy of causes or unbelievable collusion or address only weaker versions of

**Table 8.1** ***Hypotheses for resolving Fermi's paradox: A recapitulation***

| No. | Solution | Page | Grade | Comment |
|---|---|---|---|---|
| SOLIPSIST HYPOTHESES | | | | **Violate direct realism** |
| 1. | **Fermi's flying saucers** | 109 | F | |
| 2. | **Ancient flying saucers** | 112 | F | |
| 3. | **Special Creation** | 113 | F | Violates naturalism |
| 4. | **Zoo hypothesis** | 116 | D | |
| 5. | **Interdict hypothesis** | 116 | D+ | |
| 6. | **Leaky interdict** | 117 | D | Same issues as UFOs |
| 7. | **Planetarium hypothesis** | 120 | C | **Zoo hypothesis** on steroids? |
| 8. | **Peer hypothesis** | 121 | D+ | |
| 9. | **Simulation hypothesis** | 122 | B– | Practical metaphysics? |
| 10. | **The Paranoid Style in Galactic Politics** | 124 | F | Does not account for manifestations and artefacts |
| 11. | **Directed panspermia** | 127 | F | |
| 12. | **Bit-string invaders** | 132 | F | |
| 13. | **New cosmogony** | 134 | B | |
| RARE-EARTH HYPOTHESES | | | | **Violate Copernicanism** |
| 14. | **Early Great Filter** | 153 | B– | Prototype rare-Earth scenario |
| 15. | **Horizon to the rescue** | 155 | F | Sweeps the problem under the rug |
| 16. | **Gaian window** | 156 | A– | |
| 17. | **Permanence** | 159 | B– | |
| 18. | **Thoughtfood Exhaustion** | 164 | C | Science-centric as well as anthropocentric |
| NEOCATASTROPHIC HYPOTHESES | | | | **Violate gradualism** |
| 19. | **The Gigayear of Living Dangerously** | 172 | C | |
| 20. | **Astrobiological phase transition** | 174 | B | |
| 21. | **Stop Worrying and Love the Bomb** | 179 | D | Default view during the Cold War |
| 22. | **Self-Destruction, Advanced Version** | 180 | D+ | |

*Continued*

**Table 8.1** *Continued*

| No. | Solution | Page | Grade | Comment |
|---|---|---|---|---|
| 23. | **Introvert Big Brother** | 182 | C– | |
| 24. | **Resource Exhaustion** | 185 | B | Benefits from our myopic biases |
| 25. | **Deadly Probes** | 188 | B+ | |
| 26. | **Interstellar Containment** | 195 | F | |
| 27. | **Transcendence (general)** | 196 | C–/F | |
| 28. | **Transcension hypothesis** | 198 | B– | |
| LOGISTIC HYPOTHESES | | | | Violate economic assumptions |
| 29. | **Red empire** | 210 | D | |
| 30. | **Brown empire** | 211 | D | |
| 31. | **Persistence** | 213 | C+ | SETI friendly |
| 32. | **Living on the rim** | 217 | C+ | Crossover with 28 |
| 33. | **Eternal wanderers** | 219 | F | *Very* exclusive |
| 34. | **Great Old Ones** | 219 | C | Undeveloped |
| 35. | **Sustainability** | 221 | B– | |
| 36. | **Galactic stomach ache** | 223 | C | |

the problem, they have been assigned a D. Note that the grades are *not* assigned for their general likelihood to be true: for example, **Directed panspermia** could well be true but, as the resolution of Fermi's paradox, it fails. Oh, and the **Hermit Hypothesis** (Section 1.6) obviously gets a D–.

Above that, everything is much hazier, less clear, and necessarily more subjective. While it might sound unjust to those accustomed to grading 'on the curve', I stick to the reasoning that no hypothesis can really be assigned an A, since none has been shown to be a clear and obvious favourite for resolving the problem which is highly—the take-home lesson, even if everything else is ignored—non-trivial. There are no explanatory paragons of virtue around, so the best we can do is scramble for the second best: those which follow as many of the philosophical precepts discussed in Chapter 3 as possible, avoid passing the buck, and leave as little as possible to auxiliary assumptions. These hypotheses are assigned an A–. The rest are graded mainly according to the severity of violation of the non-exclusivity principle, the degree to which the hypothesis is developed, and the number of auxiliary assumptions it makes. Again, much of it is still necessarily subjective (the Linnaean taxonomy was as well, especially in its early days, and many chemists of the day faulted Mendeleev for the perceived fancifulness of the periodic table of elements), but can still be useful for comparison.

What has been probably underestimated thus far is the amount of contact and partial overlap between different hypotheses, even in different categories (see Figure 8.1). **Transcendence** hypotheses are obviously related to some of the logistic hypotheses, notably **Living on the rim** and **Sustainability**; it seems that the latter could much better be calculated and directed by postbiological superintelligences, rather than the beings still chained by ancestral evolutionary luggage. I shall consider the prospects and implications of postbiological evolution in more detail in Chapter 9. For the moment, consider how much real and apparent overlap interferes with building taxonomy of any kind of complex objects. Two possible approaches are discernible here, roughly analogous to the distinction—and occasional tension—between classical morphological taxonomy and cladistics in biology. We could insist on purity of our categories observed in the morphological sense, or we could try to judge which elements reappear, even in disguised or subtle form, in various items on the list and how tightly they are causally connected and integrated in explanatory hypotheses.

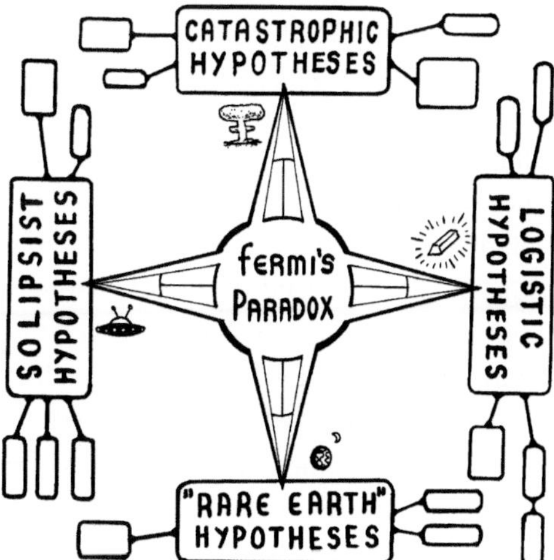


**Figure 8.1** Schematic redux of the main categories of explanatory hypotheses for resolving Fermi's paradox. Each category contains a number of at least remotely plausible solutions we need to prune.

*Source*: Courtesy of Slobodan Popović Bagi

At first glance, and on the purely morphological train of thoughts, the fact of contact/overlap might suggest that we need more inclusive hypotheses, fractionally combining some of the individual items shown in Table 8.1. For instance, we could formulate a **Hypothesis X** which conjoins **Gaian window** (50 per cent), **Stop Worrying and Love the Bomb** (25 per cent), and **Brown empire** (25 per cent) to explain all (100 per cent) of **StrongFP**.[5] Many such games could be played, obviously. While it is tough to argue against that in principle, in practice there are difficulties on both philosophical and practical level. On the philosophical level, it would amount to some wild backtracking: while we have expended some effort in trying to clearly separate the causes and factors in formulating explanatory hypotheses, we would be mixing them up again. And, while we might agree as a general matter that 50 per cent violation of the non-exclusivity principle (say) is better than 100 per cent violation of the same, it is highly arguable *exactly how much better* it is. One would be sorely tested to accept a sort of epistemic diminishing returns: once you allow for violating a general symmetry principle, the doors for increasing arbitrariness and 'anything goes' are wide open.[6] In addition, of course, there is a problem already mentioned in regards to the rare-Earth hypotheses in Chapter 5: is it likelier that a composite solution acts as completely efficient sieve and eliminates all possible causes for concern—or that our imagination simply fails and evolution finds a way around something we assume blocks its path?

In practice, most of these hypotheses have not yet been developed as precise quantitative models, nor have progressed beyond thought experiments. So, there is much work to be done 'on the ground', before we advance to the level of assigning weights to components within a composite hypothesis. Besides, the history of science teaches us another important point, which Philip Kitcher calls the division of cognitive labour.[7] Scientists tend to divide their labour to explore an unknown epistemic landscape. If the landscape is *very* unknown and mysterious, as is the case in astrobiology and SETI studies, then the enclaves in which new understanding is gained will be few and distant, and communication between them quite difficult, which is clearly seen in our present context, in the degree to which the hypotheses listed in Table 8.1 are *uneven* in terms of both their intrinsic depth and the amount of attention they have attracted from researchers. While obviously joining in the great modern conflux of astrobiology, they hail from different disciplines and carry imprints of particular scientific cultures or subcultures. Even without entering specific philosophical and ideological prejudices, it seems that some of the hypotheses were simply overlooked by researchers for quite innocent reasons (e.g. lack of imagination, lack of resources, the model was too complex to quantify in detail, the research would have caused negative publicity or interdisciplinary tension, etc.).

We need to work, however, with what we have. This has been usual and standard in science since the time of Galileo—and note how a notorious 'sceptical' double standard becomes acceptable only in astrobiology and SETI studies. As noted in Chapter 1, there are many wiseacres in and around science who would postpone investigation into extraterrestrial intelligence and Fermi's paradox 'until we know more'. Occasionally, it is murmured along the way that it is a 'wasteful' thing to do at present, with implication being that the resources are better redirected towards the proverbial 'finding a cure for cancer'. From the point of view of real history and philosophy of science, such an attitude is utter nonsense. Imagine somebody advising Darwin not to look into causes of variability of living beings 'until we know more' about the mechanism of heredity. Actually, some gradualists in the geosciences made fools of themselves by sending a similar message to the Alvarezes in the 1980s: it's wasteful to look into effects of things falling from the sky 'until we know more' about mass extinction episodes.[8] 'Sceptics' often hide their personal dislike for a particular *research programme* under the banner of 'we need to know more'—of course we do, but how exactly without doing actual research on the topic, approaching it from any conceivable angle, applying both reasoning and imagination to it? Without going into epistemological detail, it is safe to say that any student of Fermi's paradox should not pay much attention to this kind of complaint.

If some work is to be done at the 'meta-level', we might as well compare dividing lines. Hanson's 'Great Filter' is perhaps the simplest—and in some respects the most brilliant—example, since we need to bother with essentially only two versions of a Great Filter: the Early Great Filter and the Late Great Filter.[9] An Early Great Filter is essentially the eponymous solution of Section 5.2. A Late Great Filter is whatever can prevent the ascent of an extraterrestrial civilization along Kardashev's scale once the civilization emerges (this is essentially the entire neocatastrophic category, except for the **Transcendence**-style options, but also **Sustainability, Galactic stomach ache**, etc.).[10] It is more or less tacitly assumed that whatever could make it past both the Early and Late Great Filters would be readily detectable in a number of forms (e.g. colonization, manifestations, and/or artefacts)—so the absence of these is puzzling.

Kardashev's classification itself is slightly more complex tool, which is based on the distribution of energy resources and their control by a civilization. It has proven to be, in general, a rather reliable and useful vehicle—maybe not so much as intended by Kardashev in guiding observational SETI projects as in guiding the theoretical thinking about the role of intelligent life in manipulating and controlling its physical environment.[11] Kardashev's scale has motivated many, if not most, of the ideas about large-scale manifestations and astroengineering artefacts of Galactic sophonts, to be considered in the next section. Human civilization in 2013 was of Kardashev's Type 0.72. Arguably, it could already be detectable to advanced extraterrestrial astronomers from some of the nearby stars, either via our radio-/microwave emissions or through manifestations such as Earth's artificial nightglow (cf. Figure 1.2). As we consider larger and larger civilizations, ascending the scale of Kardashev types, they become more and more detectable out of proportion with resources controlled *unless* there is intentional suppression.

So, there might not be *that many* high-quality, non-exclusive solutions to **StrongFP** as it may seem at first glance. Long lists occasionally get too long for their own sake. The purpose of scientific theory is to compress the data: instead of memorizing the many particular instances of force causing acceleration proportional to the mass of a body, we need a single universal theoretical expression ($\vec{F} = m\vec{a}$) into which all specific instances are subsumed. In order to *hope* to ever obtain a meaningful astrobiological theory subsuming all individual instances of life and intelligence emerging over the Galaxy, we need to identify the relevant features firmly characterizing forms we are looking for. A critic might argue that this conclusion conflicts with what we have stated in Chapters 1 and 3 about the futile nature of the definitional project. The criticism would be decisive if not for the fact that we are interested only in *detectable* aspects of intelligent life, which allow us to neglect and sweep under the rug many complications.[12] In the entire overview of the battlefield, presented in Chapters 4–7, we dealt with the notion of the detectability of other Galactic sophonts and their creations. We need to further flesh out this notion now.

## 8.2 Manifestations, Artefacts, and Detectability

In one important sense, the manifestations and artefacts we seek within the expanded SETI mandate (or the Dysonian SETI) are just Copernican generalizations of what has been acknowledged recently as the human impact on our planetary environment on Earth. Currently ongoing debate on the precise definition of the *Anthropocene* directly relates to the circle of issues we face in the SETI context. The Anthropocene represents a proposed subdivision of geological time in which human activities started to significantly impact

Earth's system and all its subsystems (e.g. the atmosphere, the lithosphere, the hydrosphere, and various ecosystems).[13] Some authors directly talk about the *technofossil record* of human civilization.[14]

Ironically, the concept of the Anthropocene may contribute to an emancipation from anthropocentrism. Its recognition would mean admitting that the old prejudice about there being a strict separation of culture and nature is outdated nonsense, so there is strong scientific as well as ideological resistance to it in many circles. The fiercest opponents of the Anthropocene concept are those who believe in the 'free ride' of humanity on Earth's natural resources—the same anthropocentric cartel mentioned in Chapter 5. For our present purpose, the astrobiological ramifications of the Anthropocene are manifest: the very concept presents an 'official' acknowledgement that, *even within a purely naturalistic world view and within the 'hard' natural sciences, one has to take into account intentional manifestations, traces, and artefacts.* Before the onset of the Anthropocene debate, intentional influences—notably greenhouse gases emissions, fallout from nuclear tests and accidents, and various other forms of pollution—were regarded as aberrations, or errors to be corrected—an inferior form of meddling with nature. Not denying these aspects and, in particular and emphatically, not denying the threatening and very real truth of the anthropogenic global warming, it is important for astrobiology to regard the human influencing of our physical, chemical, geological, and biotic environment as a systematic, regular, law-like process—not 'meddling', not an aberration, but an unavoidable systematic process—a process through which more and more features of the physical universe obtain intentional origin and character as cosmological time passes. Indeed, this is a process which, unlike greenhouse emissions, should, in general, be welcomed, even if we consider its ethical aspects.

I cannot go further into the fascinating—and very much ongoing—story about the Anthropocene here. However, the pivotal question for any student of Fermi's paradox is:

> **If we can talk about the technofossil record of humans, why not about the technofossil record of some older and more advanced Galactic sophonts?**

What could such an alien technofossil record consist of? If humans detectably influence our physical surroundings on Earth, why not suppose that we will one day—if we avoid all global catastrophic risks—influence our physical surroundings beyond Earth as well, in the Solar System and the Galaxy at large? And what could *our* record by that time consist of? Clearly, things such as radio messages sent in various directions, probes such as the *Pioneers* and the *Voyagers*, artificial satellites in stable orbits, and the present-day nightglow of inhabited Earth are just very, very minuscule steps in this direction. Activities which have usually been thought about as first and necessary moves in human colonization of space—constructing permanent lunar bases, asteroid mining, moving parts of industry into Earth orbit or stable Lagrange points, putting many artificial satellites in orbit around all Solar System planets, sending first small interstellar probes—will each and *necessarily* increase our detection cross section. Any step *beyond* that might not be clear at present, but will (per considerations of Chapter 1, per Kardashev's scale, and per the logic of the Dysonian SETI) continue to dramatically increase the detectability of humankind, unless much intentional effort is invested in suppressing it. Each future generation will be able to review the technofossil record of its time and conclude that each additional layer of it makes humanity ascend further on Kardashev's scale and increase its detectability *without any desire, policy, intention, or conscious tendency to do so.* Critics may point out that the advancement of human technology led to the decrease of our contact cross section in some cases; most notably, the advent of cable TV decreased the emissivity of our planet in the bands previously used by major television stations and operators. This seems to be more an exception than the rule, as the economist William R. Hosek has emphasized:[15]

> Disembodied technology, such as education to improve the productivity of human capital, also contributes to output growth but it cannot substitute entirely for material objects. You cannot eat disembodied technology, wear it, or use it to travel from one place to another... Any society of intelligent beings must have a value system that permits it not only to select among competing uses of scarce resources in the short-term but also to select among resource uses that yield benefits in the short-term versus uses that yield benefits in the long term. Individuals with a finite life span will be inclined to prefer benefits that can be experienced more highly than benefits that can never be experienced and, because of unforeseen events, may not even be experienced by future generations. It is our finite and uncertain life span, and resulting time preference, that locks us into short-term perspectives and actions.

Overcoming these limitations of primitive civilizations is part of the definition of an advanced technological civilization or any higher Kardashev's type. Therefore, we are still entitled to ask: what part of the large design space accessible to future humans—or perhaps other Galactic sophonts right now—can be detectable over interstellar distances?

Firstly, a point we need to repeat and repeat, since the risk of sounding nauseating is way smaller than the damage already created by misunderstanding of this point in astrobiology and SETI circles: *all messages are manifestations, but not all manifestations are messages*. In Chapter 1, we have discussed reasons for belief—first elucidated by Kardashev—that the impact of sophonts on their physical environment is unavoidable. Obviously, sufficiently powerful technological civilization could, in principle, hide efficiently by expending significant resources and attention into encryption (cf. Section 4.5), stealth, and ultimately merging into the environment to the level of being indistinguishable from it by lesser minds (cf. Section 4.7). While the diversity of sophonts can be helpful—in the sense that very strange forms of cognition could easily pass unnoticed—the hiding could arguably never be easy for a technological community. Even the extremely exotic (by our anthropocentric standards) and certifiably paranoid inhabitants of the planet Quinta in Lem's *Fiasco* are, in the end, revealed. Morphological and cultural diversity requires, however, a matching diversity in method and approaches for detection. There are many stops and waystations on the road which the great Italian author Italo Calvino described as the 'transformation of ourselves into the messages of ourselves'.[16] This is not particularly new either: as historians of science teach us, since the nineteenth century, there have occasionally appeared proposals to engineer parts of Earth's surface in order to communicate with hypothetical extraterrestrial (usually Martian) sophonts.[17] Even the specific association of Fermi's paradox with astroengineering predates Brin's study. It was Carl Sagan in his *The Cosmic Connection*, which was originally published in 1973, who started his Chapter 33, entitled 'Astroengineering', with the story of Fermi's lunch.[18]

Secondly, we need to elaborate a point which serves as a sort of stumbling block for at least some students of SETI. *There is no sharp boundary between the traditionally understood SETI signal and the Dysonian manifestations, traces, and artefacts*. Usually we think about signals as something of short duration—fleeting and ephemeral—but this need not be so in the general case. Consider, for instance, the 'letter from the stars' (see 'Stanislaw Lem: *His Master's Voice*', pp. 18–19): Lem's fictional researchers estimated that the neutrino signal has been repeating itself in continuity for at least a billion years! Such persistent emission can certainly be taken as an artefact even on intuitive grounds, as much as we posit the invisible but clearly powerful beacon which emits the signal as an artefact. Such an extremely persistent emission might be rare or non-existent in the real universe, but almost any other conceivable signal—in whose intentional nature we might have confidence—needs some degree of persistence. The catch is in the qualification: the famous Ohio State University 'Wow!' signal did not *persist*—therefore, among other things, our degree of confidence in its intentional origin is sharply decreased.[19] All protocols suggested for *our* intentional signalling (sometimes known under the METI acronym) are formulated in such a manner that their repetition/persistence helps potential detecting sophonts to identify such emissions as intentional.[20] Intuitively, it is clear why this is so: natural processes generating (pseudo)random noise might, in the fullness of space and time, mimic any particular feature we associate with the intentional signal. However, the very definition of noise makes persistent or repeated simulations of this kind significantly less likely. In addition, transmitting beacons might be regarded as artefacts in their own right, especially if they utilize extreme astrophysical phenomena, like the system of transmitters orbiting a black hole and using its gravitational focus for efficient transmission, as suggested by astrophysicist Albert Jackson.[21]

Third, while I use the expression 'astroengineering feats' to denote both artefacts and manifestations, there is no sharp boundary between them. Tentatively, we can define artefacts as something substantial which persists in time, while manifestations are either one-shot or episodic phenomena which may not be substantial at all but exist in the abstract space of information or geometric forms. This is intuitively clear: a Dyson sphere lasts for a long time (in principle, a period of time comparable to the Main Sequence stellar lifetime, possibly longer[22]) and is detectable at any given instant of time, while the engine signature of an antimatter rocket is present only during the acceleration/deceleration/manoeuvring phase, and also depends on the unknowable (for a distant observer) spatial trajectory of the spaceship. It makes perfect sense to operationally distinguish between the two and label the Dyson sphere an *artefact*, and the emission of an antimatter drive a *manifestation* of an advanced technological civilization—and the more so as different observing techniques and procedures are required to detect these categories. Again, this will never be a clear-cut, waterproof distinction and should be regarded as provisional and operational only.

Therefore, among the *manifestations* of technological civilizations, we may count one or more of the following:

- **nightglow** created by artificial light sources on the night-side of the planets (what we generally consider light pollution; see also Chapter 1, Figure 1.2)[23]
- **stellar uplifting/'stellar wraps'** (as a manner for both obtaining usable material and stellar rejuvenation)[24]

- **antimatter burning** (for spaceship propulsion or other industrial purposes)[25]
- **stellar waste dumps** (as side effects of industrial activity (e.g. for nuclear waste disposal) or even as intentional beacons)[26]
- **non-stellar hydrogen fusion** (for spaceship propulsion or other industrial purposes)[27]
- **atmospheric industrial pollution**, including climate-changing gases (an unintentional consequence of industrial development)[28]
- **laser-beam leakage** from light-sail propulsion (an unintentional consequence of interplanetary, interstellar, or even intergalactic transport)[29]
- **mining activities** (for building resources, or $^3$He for energy)[30]
- anomalous activities at **stellar focal regions** (research)[31]
- **shepherding of celestial bodies** like comets and asteroids into particular intentional orbits (for the best utilization of their resources, using them for changing orbits of planets, or other industrial purposes)[32]
- **modifying physical properties of stellar objects**, including cepheid variables, neutron stars, relativistic binary sources, or the Galactic centre (for various purposes, including computation, communication, and energy)[33]
- **wild cards** (manifestations following from motivations, goals, and capacities entirely unknown and unconceived by us so far)[34]

Some of these are graphically sketched in Figure 8.2. All could be thought of as straightforward extensions of the very same drive which led early hominids to create and use fire to expel bears and tigers from their caves or prepare better food. Utilizing natural resources in order to achieve better control over the environment or tangible advantages over one's rivals or larger profits, including social and intellectual capital, are motivations common to many living species on Earth and could be argued to represent an integral part of the evolutionary luggage created by natural selection. While we may speculate as to what extent postbiological civilizations (including possible posthumanity) will be liberated from such burdensome motives, or whether evolutionary mechanisms other than natural selection could create different motives and drives on other planets, it seems reasonably safe to conclude that at least *some* sophonts will have similar set of motivations and drives. In addition, as much as we need to take into account possible 'wild-card' manifestations, there might also be 'wild-card' *motivations* in play as well. Ancient human civilizations would have been quite surprised, perhaps, if they were to see large land areas set aside for airports near each of modern cities.

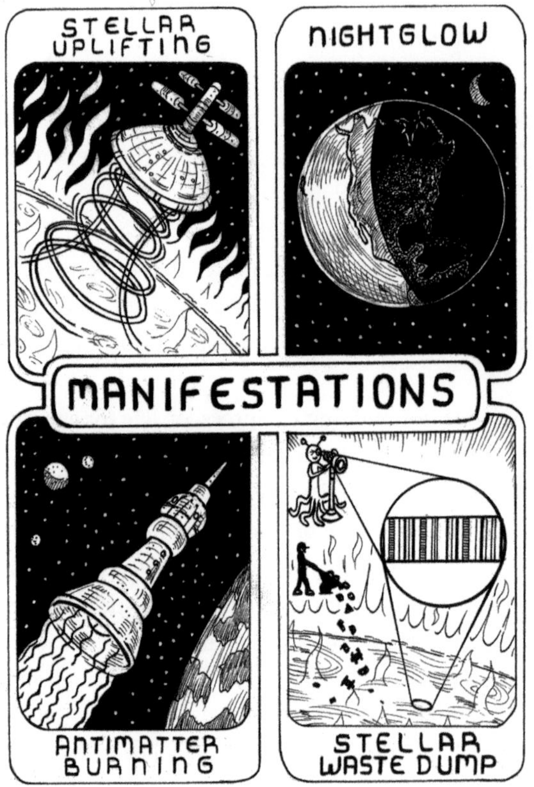


**Figure 8.2** Some examples of astroengineering *manifestations* of advanced technological civilizations. All of these are, at least in principle, observable across interstellar distances with essentially present-day astronomical technology.

*Source*: Courtesy of Slobodan Popović Bagi

Into this 'wild-card' category we may also put those manifestations which are of purely aesthetic, symbolic, or culture-specific value. One might think of them in analogy with great sporting events: a *very* careful extraterrestrial observer of Earth could detect particular activities in terms of communications, traffic, and so on, related to events such as Olympics or Super Bowls. (Other possible analogies include events such as V-Day parades and commemorations, May Day demonstrations and marches, UN General Assembly meetings, the Cannes film festival, large scientific congresses, etc.) These manifestations would be rather localized, transient, and perhaps periodic or quasi-periodic. Now, if we multiply this by a large enough factor (reflecting the location on Kardashev's scale) and realize that 'local' might mean a very large chunk of space when Type 2.x civilizations are concerned, we may get a vague picture of why such manifest-ations might be detectable even from interstellar distances. Note, however, that this category—lacking a known functionality—would be impossible to quantitatively model and predict from the viewpoint of an external observer.

As mentioned, we now follow a smooth transition to the realm of artefacts—or, as it has been often stipulated in older science-fiction literature, BDOs (from 'Big Dumb Objects').[35] Any future comprehensive list of astroengineering artefacts is likely to include the following:

- **Dyson spheres/shells/swarms** (usually for purposes of habitation, energy, or computation; see Figure 8.3)[36]
- **solar-wind-powered satellites** and other forms of transiting objects (for energy or habitation, or even fragments of Dyson swarms under construction; see Figure 8.3)[37]

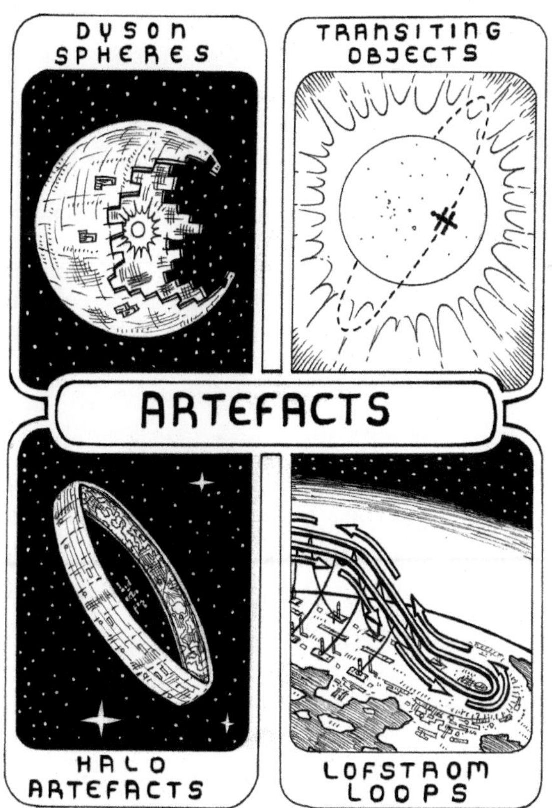


**Figure 8.3** Some examples of astroengineering artefacts suggested in the literature. While the exact detectability is dependent on too many concrete engineering choices, it is clear that at least some of these could be detected by twenty-first century observational techniques.

*Source*: Courtesy of Slobodan Popović Bagi

**Figure 8.4** A toroidal O'Neill colony, with an illuminating mirror, intended for the habitation of around 10,000 people. This artistic vision by the celebrated space artist Rick Guidice, was a product of several summer studies and workshops organized by NASA around the year 1975 on the topic of large-scale space colonization. It is a disturbing compliment to both space administrators and artists of the 1970s that very few renderings comparable in terms of both technology and aesthetics have been produced in the roughly 40 years since.

*Source*: Courtesy: NASA Ames Research Center

- **ring worlds, 'halo' artefacts, 'orbitals', or 'coronels'** (for habitation or for industrial or research development; see Figures 8.3 and 8.4)[38]
- **Lofstrom's loops, space elevators, Jacob's ladders**, and similar circumplanetary contraptions (for transport to and from planetary surfaces, for circumplanetary manoeuvres, and even for habitation; see Figure 8.3)[39]
- **artificial planetary rings** (for accelerating payloads, environment control, or habitation)[40]
- **shell worlds or supramundane planets** (for habitation, for industry, or for baryonic matter resources)[41]
- **'stellified' planets or brown dwarfs** (for terraforming large rocky satellites or nearby planets, or for energy)[42]

- **matryoshka brains or Jupiter brains** (extremely large computing devices; might look similar to Dyson spheres/swarms to external observers)[43]
- **Alderson discs** (for habitation?)[44]
- **solar shields, orbital mirrors, solettas, or shielding swarms** (for influencing climate, terraforming, protection from stellar flares or supernova/GRB radiation, illumination of planetary surfaces or orbital habitats, or as a part of solar power plants)[45]
- **'Dysonships'** and other non-relativistic vehicles (transport, migrations)[46]
- extragalactic **Planck-energy particle accelerators** (for research, defence?)[47]
- **black boxes** (not necessarily an artefact per se, rather than the way how alien artefacts and their region of space might look to an outside—and very distant—observer; included here because the study of Brian Lacki offers a clear and precise way of detecting them)[48]
- **wild cards** (artefacts following from motivations, goals, and capacities entirely unknown and unconceived by us so far)

There is no comprehensive study of detectability of these (and other) astroengineering feats so far, in spite of more than half-century since the original Dyson sphere proposal. Some recent studies like those of Paul Davies, Dick Carrigan, Morris Jones, Stuart Armstrong, and Anders Sandberg emphasize the necessity for new approaches to astroengineering artefacts, motivated by developments such as 3-D printing or molecular nanotechnology.[49] Some of the items on the list could be misidentified as natural astrophysical objects; for instance, artificial planetary rings may look *really* similar to naturally occurring planetary rings when observed from afar, even without any attempt of stealth. Shell worlds will, of course, look just like normal planets from a distance, although their sizes, densities, or surface features may be anomalous. On the opposite end of the detectability spectrum, Dyson shells/swarms, ring worlds, and large, rectangular transiting objects have no persuasive naturally occurring analogues. Quantifying this scale of possibilities is clearly a big project for future SETI studies and exploratory engineering.

(To this list we could add the self-reproducing von Neumann probes discussed in Section 6.6, if they are not especially stealthy or very small. The process of self-replication, presumably occurring at the outskirts of planetary systems, in cometary/asteroidal belts analogous to the Kuiper belt in our Solar System, should be counted as potentially detectable manifestation. However, if hostile von Neumann probes were to emerge through mutations and evolution, it makes clear strategic sense for them to be rather stealthy and hard to detect.)

The presence of a 'wild-cards' category on each list should be illuminating: there is no way of foretelling what engineering ideas will be around in communities of intelligent beings with different cognitive structures and more time at their disposal to develop powerful technologies. There is no way of foretelling, in fact, what novel engineering ideas will be around in 10, 50, 100, or 1000 years *on Earth* (if we do not destroy ourselves until then). All other specific processes and devices listed belong firmly to the category of technological problems that are currently unsolved but which we should expect to solve. In other words, they are more like flying cars than time machines. However, one might expect that there are *unknown unknowns* in this domain as well: engineering challenges we cannot and will not be able to conceive on Earth, but which could come naturally to sophonts evolving in some other place in the Galaxy. We can imagine such situations easily: sophonts evolving on a planet close to a neutron star or a black hole[50] might be facing astroengineering challenges related to extremely strong gravitational fields; such challenges would be both unclear and *unimaginable* to us, at least at present. A planet or a satellite with a magnetic field that was much stronger than that of the Earth—like the world shown (albeit unrealistically) in the movie *Avatar*—might be another such case, enabling feats of astroengineering which human thinkers are unlikely to come up with.[51] We do not know how to construct such artefacts—but we do not know why we should bother in the first place, since we are unlikely to understand their functions, utility, and purposes outside of their proper context. Thus, they are paradigmatic unknown unknowns.[52] We should not a priori reject such wild cards, since they might, for all we know at present, contain most of the detection cross section in the Galaxy.

The other side of the coin is *detectability*. We may not be able to give the precise and general definition of detectability yet—remember, formalization comes to a scientific field when it is completely understood or very nearly so!—but we are able to say a thing or two about its particular properties. Firstly, detectability should be a scalar, non-negative measure; a perfectly stealthy civilization (or part of a civilization, like a colonized planet or a set of resources or an installation) has zero detectability. One can hardly go less than that! Secondly, it has to be, in general case, a function of space and time, since variously distributed and managed resources have different detectability in different epochs. For example, the GTZ, as implied by **Living on the rim** (Section 7.4), should contain the most detectable targets, certainly more than the too hot and risky region inside the Galactic bulge. Third, detectability is a function of observational methodology, in particular of frequency/wavelength if we consider the electromagnetic spectrum as the main carrier of information. This easily generalizes to the energy densities/fluences/amplitudes if we consider non-electromagnetic ways of detection, like neutrinos or gravitational waves. If Dyson shells/swarms are easiest to detect in the infrared, this implies detectability is higher at wavelength of, say, 50 μm than on 10 cm (the radio-wave domain) or 900 Å (the ultraviolet domain) *for this particular type of civilization*. Fourth, as is clear from the same example, detectability needs to include a parameter or an index describing a type of civilization in some convenient taxonomy. This could be Kardashev's classification, or some more complex and elaborate scheme, but is necessary for detectability to be modelled and to convey really useful practical information.

Fifth, and most importantly from the practical point of view, we may reasonably argue that detectability must be *superadditive*. Namely, if $d_{\mathrm{m}}$ is our adopted measure of detectability of a civilization managing the set of resources $m$ (including the spatial volume controlled), we may argue that it holds that, on the average:

$$d_{m+n} \geq d_m + d_n \tag{8.1}$$

So, for example, a civilization managing five planetary systems is, on the average, more detectable than either five civilizations managing a single planetary system, or two civilizations, one managing three planetary systems, and the other managing two planetary systems. Obviously, communications and transport between the managed systems are 'extra' potentially detectable processes, from the point of view of an external observer. In the worst case, those could be undetectable, so that we would have equality in Eq. (8.1); but, in a realistic case, we expect inequality—even a very strong one.[53] Quantity does not make up for quality.

Further elaboration of detectability is certainly necessary if we are to be able to offer anything like a unified framework for searching for traces and manifestations.[54] Even with our present simplistic understanding, it is clear what properties of detectability mean for practical SETI: specific searches should make a *compromise* between the volume and duration surveyed on one hand, and the detectability measure on the other. Obsession with a large number of civilizations, including primitive ones similar to humanity, evolving in parallel (large values of *N*, in the Drake equation) should be toned down, and finding traces and manifestations of interstellar colonization and appropriate energy consumption should be given higher priority. In more than one sense, large *N* is a red herring. The spectrum of possible targets should be increased by novel, original, and creative theoretical work, conservative hand-waving notwithstanding.

Against the lists of potentially detectable astroengineering artefacts and manifestations, the list of performed/ seriously suggested research programmes in this area looks rather modest, if not outright poor. It may read something like this:

- a small number of SETA searches for artificial objects in the Solar System[55]
- searches for Dyson spheres and similar artefacts in the Solar neighbourhood, via infrared excess[56]
- the search for tritium lines in the spectra of nearby stars, as such lines indicate massive non-stellar fusion of hydrogen[57]
- the search for the gamma-ray signature of antimatter burning[58]
- searches for Kardashev's Type 3 civilizations in external spiral galaxies via the Tully–Fisher relation[59]
- the search for Kardashev's Type 3/Type 2.x civilizations in external galaxies, via infrared Ĝ surveys[60]
- the search for transiting artificial megastructures[61]

This is a short list. It is short in spite of the fact that some of these searches are among the cheapest observational activities ever conceived in astronomy. It is short for approximately 60 years of SETI history, even when normalized for low funding levels and the rather small number of researchers involved. (There may be a few items missing from the list, but not many of them; in this respect, it is drastically different from the lists of manifestations and artefacts.) The idea that manifestations and artefacts are crucial to the SETI prospects has been very slow in coming—in spite of the authority of Dyson and Sagan—due to a number of reasons, the most obvious being the fear of SETI practitioners of being taken for pseudoscientists, on a par with the dreaded ufologists. An alien artefact sounds too much like something out of *The X-Files* or von Däniken's scribbling, although it is not the pop-culture associations but the objective merits and demerits of an idea which should be judged. Of course, the SETI community has been living for decades in panicky fear of not being able to find funding and not being taken seriously, so such a reaction is understandable, to a degree. For the very same reason, SETA activities have been few, small, and generally shunned by the mainstream SETI community. While this fear factor is undoubtedly justified in a culture where conservatism is so often extolled as a virtue, it should not be overrated, especially in recent decades. In fact, the recent surge in interest for detectable signatures of astroengineering, mentioned in relation to the Ĝ survey or KIC 8462852, has been coming for quite some time. The sleeper has awakened, it seems.

To prove that, it is enough to notice that, by far, the bulk of references cited in this section are of very recent origin (i.e. post-2001). On the other hand, the frequency of papers on the old-fashioned SETI topics like radio bandwidths, interstellar scintillation, beacons, and so on, has remained rather steady over the same period of time. So, it seems that the interest in topics of astroengineering and its detectability is taking a dramatic upturn in both absolute and relative terms. We shall soon, if this trend continues, be in much better position to assess the detectability intervals for each item in the lists in this section under any particular observation method. That this goal now seems realistic, after decades of being disparaged as pseudoscience or simply ignored, should give us pause and make us reflect upon the changing epistemology of SETI studies.

## 8.3 No Small Prices To Pay?

There is an obvious selection effect in play here. If the answer to Fermi's question were trivially easy, it would not have been discussed in this book, or in many research and popular articles. Nor would it provoke such visceral negative reactions as it occasionally does among 'the learned' in the orthodox SETI community.[62] Similar to other 'Big Questions', like the mind–body problem in cognitive sciences and philosophy, the question of contingency versus convergence in evolutionary biology, or the origin of the fine-tuning of the observable universe in cosmology, there clearly are no simple answers. The complexity of the problem is unavoidable, and the silence is non-negotiable.

What we need is a short-listing procedure akin to the one used from ancient times in contests and elections. Apart from the considerations mentioned in Chapter 3, there are several elements of our analysis which emerge naturally from the set of hypotheses described. Consider the universe outside the Milky Way; according to the venerated cosmological principle, it should be homogeneous and isotropic, on the large scale, for each observer, wherever he or she is located. This homogeneity and isotropy should apply not only to physical properties like matter density, metallicity, and so on, but also to astrobiological properties.[63] Therefore, one question which might be posed to all who wish to resolve **StrongFP** is, to what extent can the 'Great Silence' within the Milky Way be generalized to the rest of the universe and, most notably, to extragalactic SETI observations?

Note that, while some solutions generalize naturally, others do not. Earth-like planets might be rare in the Galaxy, but it requires a completely new order of confidence/presumption to argue that they are rare within a huge number of external galaxies. While the existence and properties of the universe beyond the Milky Way are entirely irrelevant for adherents of the **Planetarium hypothesis**, the **Simulation hypothesis**, or **Special Creation**, it is certainly not the case with any hypothesis which assumes ergodicity in the broad sense discussed in Chapter 7. Consider, for instance, the cycles of rises and falls of interstellar civilizations, as implied by **Permanence** or even **Resource exhaustion**. There is no reason to believe that these cycles are any different in the Milky Way than in the billions of other (non-dwarf) spiral galaxies. So, even if there is no current Type 2.x civilization in the Milky Way, there might be many in external galaxies, and we should be able to observe signs of them, even accounting for the time delay between the generation of those signs and their reaching Earth. For instance, we should be able to observe manifestations of civilizations which were in their ascending phase 2.54 Ma ago in M31, or 68 Ma ago in NGC 3982. (Of course, as discussed in Section 8.2, at least in the case of some artefacts, the delay between generation and observation could span a much longer interval of time.) Preliminary—and *quite* insensitive—searches have failed to reveal such manifestations and artefacts so far; if that is confirmed by further more detailed and more sensitive work, this will be a strong argument against both **Permanence** and any other hypothesis which increases the *total cross section* for detection at least linearly with the volume surveyed. Since that volume is, effectively, other galaxies, the question we need to ask is, to what extent does the astrobiological history of other galaxies follow that of the Milky Way (irrespective of what the latter actually looks like)? Some tantalizing hints have recently appeared in the literature.[64]

Rare-Earth hypotheses are tough nuts to crack here, since it is, at least, *not obvious* how rare the conditions for complex biospheres, in fact, are. Is there one 'Earth' in a thousand habitable planets? One in a million? One in a billion? One per galaxy? Some extreme construals imply uniqueness 'in the visible universe'[65] but, in most cases, supporters are wisely silent on this issue. Studies conducted by the Potsdam team of astrobiologists have offered estimates of 1 in $\sim 10^5$ up to 1 in $\sim 10^7$ in the Milky Way.[66] This is a worthy set of potential targets for astrobiological research programmes—and is clearly *at variance* with the self-professed scepticism and even derision with which Ward and others wrote about SETI.[67] As argued above, Drake's *N* is a red herring—so, even a single other advanced intelligent community in the Galaxy would have justified all and every SETI searches, if that community were advanced enough. So, there is a key equivocation game at the heart of the **Early Great Filter**. It requires something like the **Gaian window** mechanism to patch—a local mechanism, uniformly decreasing probability of the 'Goldilocks' evolutionary outcome. (Global patches like **Horizon to the rescue** will not help much here.) Of course, additional help could be forthcoming from Galactic regulation mechanisms, like in **Astrobiological phase transition**, provided that fine-tuning issues could be resolved. At present, it is unclear whether this really works—more numerical modelling and simulations are necessary before a firm conclusion can be reached; however, the good news is that it is relatively clear what needs to be done on the theoretical level.

Notice that I used the clause 'at least linearly'. Rare-Earth hypotheses can actually make the detection cross section rise *faster* than the spatial volume under scrutiny. The original hypothesis of Ward and Brownlee does not ascribe some special property to the Milky Way in the overall set of spiral galaxies; ironically enough, one may speak about 'galactic Copernicanism' in this regard.[68] But, if we are—for the sake of the argument—alone in the Milky Way, and if the Milky Way is close to the mean or median in terms of the number of intelligent species it contains, this would still mean that there are spiral galaxies with dozens of civilizations, in a sufficiently large sample. Most of these civilizations are bound to be older than ours and so potentially observable. So, strangely enough, proponents of rare-Earth hypotheses might still seek some verification from extragalactic SETI observations.[69] Of course, many other explanatory hypotheses predict much stronger signals than that, so we must wait for more sophisticated models and more precise observations.

The bottom line here is that, from the perspective of extragalactic SETI, most of the discriminatory power could be applied against hypotheses which either reject the reality of the extragalactic universe (the **Planetarium hypothesis**, the **Simulation hypothesis**, **Special Creation**, and perhaps **The Paranoid Style in Galactic Politics** as well) or suppress the expansion of intelligent communities at fairly low level. In addition to the rare-Earth hypotheses, the latter include the neocatastrophic hypotheses, apart from the **Transcendence** group and some isolated cases from the logistic category: **Great Old Ones** and **Galactic stomach ache**. The last two have not been really well developed and we may need more work to establish for certain how weak an extragalactic signal they could produce. Altogether, this reiterates the impression that the overall most well-rounded solutions are likely to come from either the rare-Earth camp or the neocatastrophic camp.

There is another key theme spanning several distinct solutions: postbiological evolution/superintelligence/technological singularity. While a significant portion of Chapter 9 will be devoted to this topic, here we need to point out how this theme links several solutions: **Transcendence (general)** and the **Transcension hypothesis** are obvious cases, but consider for a moment the **Simulation hypothesis**. The technology necessary to create a simulation of even a single human being, not to mention the rest of the observable world, cannot exist in a vacuum. Such technology can only be the product of an advanced civilization capable of building advanced information-processing devices and, hence, in so far as we believe in physicalism about minds, artificial minds. As argued by Nick Bostrom and others,[70] there are several pathways to superintelligent artificial intelligence, but the most promising is a recursive self-improvement loop: the artificial intelligence improves its software a bit, thus becoming slightly smarter and getting the capacity for improving its software a further bit, and so on. So, from a human level of AI, we could gradually reach to a level of being somewhat smarter than even the brightest humans (say, AI+) and, from there, advance further very quickly towards a truly superintelligent artificial intelligence (AI++). This has important consequences for judging what comes first: large-scale simulations of the universe (including human-level minds) or AI++. It seems that, lacking extremely tight external controls, we should expect superintelligent AI++ to emerge before simulations such as those required by the **Simulation hypothesis**. Therefore, either simulations are being run by AI++ (for reasons likely to be as incomprehensible as those in **Special Creation** or **New cosmogony**!), or we should consider some version of **Transcendence (general)** to be inherently likelier.

Obviously, the temporal ordering of technologies impacts other proposed hypotheses. **Living on the rim** is many orders of magnitude easier to implement if the civilization consists of uploads of AI+s or even AI++s, since a digital structure could be moved from one side of Galaxy to another much more easily than a material industrial infrastructure together with embodied sophonts could be. And, of course, in contrast to the case with embodied sophonts, it is much easier to implement optimization at near-absolute zero for postbiological disembodied forms. This simple fact undercuts somewhat the famous 'biological scaling hypothesis' proposed by Freeman Dyson.[71] In addition, the same considerations are probably applicable to **Great Old Ones** (having to wait for low temperatures to occur in order to become more efficient is more appropriate for postbiological sophonts than for biological sophonts), although, at present, this hypothesis is too vague for certain conclusions to be drawn.

Solipsist solutions have a seemingly paradoxical property that our credence in them increases with further insight into the difficulty and complexity of Fermi's problem as hinted at by Conway Morris. In most cases of more mundane research questions, the exact opposite is the case. If scientists try to cure a disease or discover a hitherto unknown Precambrian fossil, and it turns out that the pathogen is more elusive or that the force of erosion is stronger than hitherto thought, this outcome is not considered to strengthen the case for a particular medical approach (say, cryotherapy) or a particular location to search for fossils (say, the Precambrian beds of Patagonia). This is because, whether in physiology or palaeontology, we retain the usual, detached, Archimedean stance: *our* properties are not pertinent to the particulars of the research problem in question.[72] However, as elaborated in Chapter 1, this is emphatically *not* the case in regard to Fermi's problem. Anthropic bias is very much operative in this case, as our choice of explanation will be directly affected by how many *other* observers are there, explicitly or implicitly.[73] While F-graded hypotheses in this category can be shrugged off, D-graded ones deserve at least a further word. In the era in which the **Zoo hypothesis** and its elaborations were suggested, there was no astrobiology in the modern sense, nor did we know much about extrasolar planets, not to mention statistical analyses of their age distribution and other properties. Even the stellar dynamics of the Milky Way—especially phenomena like radial diffusion—was quite mysterious. All developments in both astronomy and future studies (including exploratory engineering) indicate that stable 'zoo-like' solutions are not tenable. Not only would such sort of policing of astronomically large volumes of space consume much resources, but one could argue that the implied lack of detection signatures logically implies an inefficient utilization of resources, even within core regions of the policing (or 'interdicting') civilizations or their Galactic Club. The Prime Directive of *Star Trek* fame simply cannot and will not work for industrial and post-industrial societies. The very act of *seeing* the *USS Enterprise* or any related manifestations (e.g. teleportation) would represent meddling in cultural affairs. While one can contrive scenarios that would imply that we simply haven't looked in right places and haven't observed their manifestations and artefacts yet, this seems unnatural. Stealth is essentially in conflict with efficient, large-scale utilization of resources; it is feasible to be stealthy if you are just aestivating, as per **Great Old Ones**; but, if you are actively maintaining a zoo or an interdict, it is not.

**Peer hypothesis** and **Directed panspermia** just move the problem elsewhere, without really trying to solve it.

**New Cosmogony** leads us into a labyrinth. If it is to be something more than a sophisticated and flashy pseudo-theology, it needs to be subject to some form of empirical testing. But how could such a test work at all? The game-theoretic approach used by Lem as a dramatic device for propelling Prof. Testa to his Stockholm rostrum will not operate in reality, at least not until we gain vastly more knowledge on the emergence of the effective, low-energy laws of physics from M-theory (or whatever Theory of Everything turns to be correct). Such knowledge— one has to be sober at this point—might never be forthcoming. Is there any other way? Empirical evidence about *very early* abiogenesis somewhere else in the universe (testifying to the ancient origin of the Player civilizations) would favour **New Cosmogony** but is extremely unlikely to be found. Not only is such evidence unlikely to survive for billions of years, it would most probably be extragalactic, so that taking a 'closer look' at it would be impossible. If Lem is at least faintly correct that the Players would seek to modify our low-energy physics, the recognition of such evidence would be hardly possible, since it would have reflected—*in addition to all other crazyness*—different properties of matter itself. That said, one might speculate that, for instance, a discovery of a Dyson shell or a similar contraption at high red shift, corresponding perhaps to the epoch of the very first habitable planets, 9 or 10 Ga ago, would go some way towards aggravating **StrongFP**—and therefore, as outlined above, strengthening the solipsist category in general, and **New Cosmogony** in particular, as it is the best of the bunch. Such a discovery—made by whatever super-duper-advanced observational methods—would certainly make the existence of the Players much more likely. While that is certainly not an immediate observational prospect, it is always worth remembering how often observational astronomy has pleasantly surprised even the biggest optimists in the field by being able to detect something which was confidently predicted to be below the detection threshold of our best instruments.

The *real* problem with **New Cosmogony** is that, at least in one important sense, it is a swindle: it does not explain the huge temporal gap between the age of the Players and our own. Obviously, the distribution of ages of sophonts under **New Cosmogony** must be strongly bimodal, with one very early peak (for the Players) and another close to our own age (for us). Why are there no intermediate cases? We have seen from Lineweaver's work and its subsequent improvements that first habitable planets emerged in the Milky Way more than 9 Ga ago; this does seem to be enough time for a hypothetical civilization arising on those first planets a couple of billion years later to become a Player, but what has prevented habitable planets that are 6, 7, or 8 Ga old from evolving to a not-yet-Player but still very advanced and detectable civilization? There seems to be no astrophysical processes doing the explanatory work to account for the bimodality.

Or perhaps we are looking at things backwards: perhaps it was the same global regulatory mechanism that allowed for the **Astrobiological phase transition** which the Players had fortuitously survived in their early epoch. Any realistic regulation mechanism is stochastic in nature, so there is clearly a minuscule possibility of surviving it even in the early days of the Galaxy (or any galaxy). Obviously, sophonts arising so early in the Galactic history would be in a tremendously advantageous position to expand and create Kardashev's Type 3 civilization. That this has not happened—to the best of our knowledge—ought to tell us something. (The alternative is, of course, that Kardashev's types fail to include some of the important cases, which is what, in my view, Lem actually had in mind when he wrote *The New Cosmogony*.[74] His Players would correspond to Type 4+, manipulating resources of the entire universe. A sufficiently early arising civilization might actually jump quickly *over* the steps of Kardashev's ladder and advance to the Player status without having to go through the intermediate steps or leaving a recognizable trace in the Galaxy.)

Another important insight is that we have actually survived thus far, testifying that the idea that there exist sophonts that are both malevolent and superior is almost certainly unfounded. The Players of **New Cosmogony** might actually disregard us entirely—after all, do we really worry about *each* group of beetles or ants milling around our houses, schools, or factories? Players might cause our destruction inadvertently, but the very size of disproportion in any possible sense or metric makes our notion of malevolence rather inapplicable. They should—and Lem's narrator gives us exactly this key aspect—be thought of as the changing laws of nature, to which we cannot meaningfully ascribe either benevolence or malevolence. To do either would be hopelessly anthropocentric, not to mention superstitious. The same applies, obviously and even more forcefully, to the Director(s) of the **Simulation hypothesis**: the simulation might end, thus annihilating our existence, for any number of external and, from our standpoint, inconceivable reasons, such as running out of assigned computing resources or getting annoyed and switching the simulation off, but there is little sense in claiming that malevolent Directors are bent on destroying us. The same applies to the Creator within the **Special Creation** context. We can call the sophonts implied by these three hypotheses—the Players, the Directors, and the Creator deity/deities—*overwhelming*, since their physical and cognitive powers are beyond all standards of comparison. Without entering the dusty theological debates, we can at least state that overwhelming sophonts, if they exist, are not malevolent in the conventional sense of the word. If they were, we would not be here in the first place.[75]

Much more interesting for contemplation is the next lower tier of beings. The very powerful but not overwhelming sophonts implied by the **Zoo hypothesis**, the **Interdict hypothesis**, **Leaky interdict**, the **Transcension hypothesis**, **Living on the rim**, **Interstellar Containment**, and even perhaps the **Planetarium hypothesis** and **Great Old Ones** (i.e. control mechanisms) are likely to have been aware of our existence for quite some time, as well as have the physical resources to quell our advance. That this obviously has not happened so far is yet another indication of what we already know (at least since Chapter 1): that the naive view of astrobiological evolution, including maximal expansion drive and 'survival of the fittest', is incompatible with any Copernican solution to **StrongFP** (i.e. with any non-rare-Earth hypothesis). These sophonts, parenthetically, seem viable as SETI and perhaps even CETI targets, if and when they manifest any intention of communicating.[76] Although incredibly powerful by our current standards, the sophonts postulated by these hypotheses are, in principle, comprehensible through the shared physical world and (probably) the shared scientific insights.[77]

The only real danger from agents which could be adequately described as malevolent comes from **Deadly Probes**. On the one hand, this is just a single hypothesis out of many (on our adopted level of abstraction), and that very fact should encourage us to feel 'at home in the universe'. On the other hand, that particular option is truly disturbing, since the analysis in Chapter 6 supports, at least on a preliminary and qualitative level, the conclusions of Brin's review, with due attention given to dissenting views such as the one of Kowald.[78] In other words, **Deadly Probes** represents an option which sacrifices the least number and the least important of our philosophical principles and expectancies about feasibility, for the biggest gain in explanatory terms. It violates only gradualism (which is anyway the weakest of the philosophical assumptions considered in Chapter 3), and it assumes the practical feasibility of von Neumann self-reproducing probes; even the latter is assumed in a weak-enough sense, since it is self-reproduction *imperfection* which is a source of risk, as it enables mutations leading to hostility towards biological sophonts, or sophonts in general. Even worse (from an anthropocentric or a SETI perspective), since the work of Armstrong and Sandberg, the hypothesis generalizes sufficiently well beyond the Milky Way, since the same manner used for exploration and peaceful colonization could also be used for the creation of extragalactic deadly probes.[79]

Negative synergic effects associated with **Deadly Probes** extend further, notably in the direction of **Resource Exhaustion**. If we regard the Galaxy as a wider ecosystem encompassing many billions of planetary ecosystems, then it is natural to expect that organisms which reproduce exponentially would fill it rather quickly. The same has, historically, happened with bacteria in almost all terrestrial ecosystems. Most of the useful resources will be used up by such organisms—and, in the case of non-renewable or very slowly renewable resources—will be actually *locked up in such organisms*. Such role in the Galactic ecosystems could easily be played by deadly self-reproducing probes. Actually, the situation might be even worse, since terrestrial bacteria live for only a short time, and dead bacteria are efficiently recycled through metabolism by their peers; on the other hand, the interstellar space is so vast that such 'cannibalistic' recycling of the resources locked up in von Neumann probes seems unlikely. In addition, secondary feedback mechanisms could well play in favour of such hostile self-replicators. In particular, if Galactic sophonts have already been burnt by the cosmic version of the 'tragedy of the commons', or they are afraid of being so,[80] this could make filling all available ecological niches with von Neumann probes much easier. Alternatively, as per **Resource Exhaustion**, sophonts with even a slight advantage over their peers could be sorely tempted to use von Neumann probes to 'flag' as many resources as possible, thus increasing the chances for either mutational or intentional corruption. And, even if no corruption ever occurs, the bubble of exhausted resources will expand quickly, reducing the probability of new technological civilization ever arising over a huge and increasing volume of space, as described in Chapter 6. This 'deadly synergy' delineates another cluster of plausible hypotheses for resolving our main puzzle.

## 8.4 The 'Great Filter' Redux

At the end of his inspiring popular book on Fermi's paradox, Stephen Webb suggests a 'patchwork quilt' solution which unifies a number (although not all—nor would that be possible) of proposed solutions. It boils down to a specific version of the rare-Earth solutions considered in Chapter 5. By eliminating individual factors contributing to the total contact cross section, Webb ends up with zero cross section, interpreting it as the lack of sophonts in the Galaxy, if not within the entire past light cone.

Somewhat ironically, Webb points in the correct epistemological direction, although his endgame is deficient (and, indeed, inferior to some of its own ingredients!). There is something essentially different about the rare-Earth category in comparison to all the others. That is due to the unique role of Copernicanism. As we shall see in Chapter 9, it is this philosophical principle that is in the line of strongest fire in the battles surrounding Fermi's paradox. In essence, the entire controversy surrounding Fermi's paradox can be read as the final, the most sophisticated, and the most complex challenge to the relentless Copernican relegation of Earth, the morphological and functional characters of the terrestrial biosphere, humankind, *and cognition of our type* to the status of mediocre typicality. While nobody dares to challenge the key role of Copernicanism in bringing about modern science and technology, the sappers beneath the mighty fortress have another thing in mind: in both the logical sense and the conventional physical sense, the future is *already existent*. Philosophers often call this the B-theory of time or, occasionally, the block-universe view of time; relativists have long ago established the view that there is no absolute past or future, and there is no physical basis for what we perceive as time *flow*.[81] If the future is to belong to humans or posthumans—in any case, *the offspring of this Earth*—then the issue is, in one atemporal sense, already firmly settled. And that could reasonably be the case only if one or another version of the 'rare-Earth' scenario is correct.[82] And so Earth, as the Tsiolkovskian 'cradle of the mind', is returned via indirect route to the central position of, if not the entire universe, at least the Galaxy, or the universe of values.[83] This vision of magnificent loneliness, of having all the resources for oneself, is arguably one of the main unmentioned drivers of this latter-day geocentrism and anti-Copernicanism.

This sets the stage for the outcome of our tournament. I shall not proclaim any particular hypothesis as a winner; instead, I shall point to the clusters which fared the best under the analytic weapons discussed in this chapter. The strongest players in the game would thus be in the following three clusters of related hypotheses:

**Early Great Filter–Gaian window–The Gigayear of Living Dangerously**

**New cosmogony–Transcendence (general)–Transcension hypothesis–Living on the Rim–Great Old Ones Deadly Probes–Resource Exhaustion–Sustainability**

Unsurprisingly, each of these clusters has a distinct flavour and a distinct set of research—and even ethical—prescriptions. While contact with extraterrestrial sophonts is meaningless under the **Early Great Filter** cluster, we should consider two very different forms of contact under the remaining two clusters of options. The **New cosmogony/Transcendence** cluster suggests that, while we should try to make the best of our available resources, we need to make understanding cognition, and especially very advanced modes of cognition, our highest priority. There might be some of the Transcendences out there who are willing to talk. In addition, understanding cognition may be of use in dealing with the extremely efficient sophonts of **Living on the Rim** and, perhaps, the control mechanisms of the sophonts described in **Great Old Ones**. Moreover, there might be a whole 'bush' of possible evolutionary trajectories available to humanity, out of which we will need to choose one—and what could be better for this than the direct observation of example outcomes? Finally, **Deadly Probes** might present us with a huge challenge and risk. We should be looking both for traces of deadly probe activity and for the remnants of civilizations such probes would have previously destroyed. We should put a high priority on the detection of any astroengineering signatures, especially those related to mining resources, presumably on small and cold bodies on the outskirts of extrasolar planetary systems. In addition, the dynamics of resources consumption in the Galactic context will soon be an empirical issue in observational astronomy. It needs to be monitored and important strategic decisions for the future of humanity taken on the basis of such observations and comprehensive astrobiological models.

What roads should we take from here? Which research directions and priorities look most likely to yield advances in our understanding of such a complex puzzle as Fermi's paradox? What fruits can we expect from further work, and how quickly will the field mature to the point where it is beyond any reasonable methodological criticism? I submit that there is a tremendous amount of work to be done and that there is reason for substantial optimism that we could make huge strides in answering this puzzle on the timescale of years or decades, at most. However, a necessary condition for such a shift would be increased boldness in tackling the most complex and delicate issues involved. While boldness cannot be instilled or even planned, it may often be *awakened*. It is to this wake-up call that the chapter before you is devoted.

## Subjective Favourites

In situations like this, as emphasized in the 'Introductory Note', some further personal disclosure is highly advisable. Therefore, I list here five hypotheses from Table 8.1 which I find to be the most thought provoking, original, intriguing, or aesthetically pleasing—as eminently distinct from *most likely to be true*. Some readers steeped in conventions of positivism and Victorian/Kaiser Wilhelm-style academic stiff-neckedness might object that it is inappropriate for a scientific text to endorse subjectivity (even if it can never hope to entirely avoid it), but skipping this little box would be clearly harmless for even the most obsessive reader. Newton allegedly stated that problems are more useful for understanding than solutions are, and Einstein extolled the virtue of 'intellectual love' (*Einfühlung*) for the subject of research. These nuanced feelings could, from time to time, be nudged by subjectivity, emotion, and human interest, as all scientists who have engaged in public outreach, from Flammarion, to Eddington, to Hoyle, to Sagan, to deGrasse Tyson, know very well.

So here are my favourites, without further comment, in the order in which they appear in this book:

1. **New cosmogony**
2. **Astrobiological phase transition**
3. **Deadly Probes**
4. **Transcension hypothesis**
5. **Galactic stomach ache**

CHAPTER 9

# The Last Challenge For Copernicanism?

The paradigmatic 'one-hit wonder' duo of (Denny) Zager and (Rick) Evans came into the spotlight for the first and the last time with the song *In the Year 2525*, which reached Number 1 on the Billboard Hot 100 for 6 weeks commencing 12 July 1969. The song was written by Evans in 1964, but only released in 1968, becoming popular the following year. With more than one hundred covers since then by various artists ranging from Nat Stuckey, to Laibach, to R.E.M., to Fields of the Nephilim, it has become one of the most popular music pieces coming out of the 1960s counterculture. Symbolically, it was at the top of the charts on 20 July 1969, when astronauts Neil Armstrong and Buzz Aldrin became the first human beings to set foot on another celestial body.[1] And it takes the long view, although not without acknowledging existential risks:

In the year 2525
If man is still alive
If woman can survive
They may find…

The lyrics are usually interpreted as warning of the dangers of technology, portraying a future in which the human race is destroyed by its own technological and medical innovations. The last stanza of the song suggests that mankind undergoes a continuing cycle of birth, death, and rebirth:

Now it's been 10,000 years
Man has cried a billion tears
For what he never knew

Now man's reign is through
But through the eternal night
The twinkling of starlight
So very far away
Maybe it's only yesterday

Cycles of history present a recurrent (pun intended!) theme of the history of ideas, as we have seen when discussing neocatastrophic and logistic hypotheses. Another perspective on the lyrics is possible, however. The truly important sense of time is not what is measured by clocks and calendars, but what is measured through the histories and achievements of intelligent beings. The depressing conclusion is valid until a community of intelligent and creative beings emerges which is capable of breaking the cycle of the 'eternal return'. This may take a long time as measured by the clock on the wall, but in the logical sense it is irrelevant—'maybe it's only yesterday,' although it took billions of years. Eternal return is, in a sense, self-defeating on long timescales. In the same manner as orthogenesis turned out to be powerless in face of the onslaught of natural selection,[2] so cyclical motion is fading when compared with any conceivable progressive alternative in the fullness of space and time.

This can serve as a metaphor of the astrobiological evolution on several levels. Even if individual instances of abiogenesis and noogenesis are either snuffed out or remain limited in one way or another, perhaps by missing the 'windows of opportunity', in the long run we expect the landscape to change. In this concluding chapter, I would like to restate the major conclusions of this study and, together with the directions for research previously proposed, suggest that a slight change in attitude towards SETI in general, and Fermi's paradox in particular, is appropriate. This will not come as a surprise, since similar calls for overhauling our SETI-related thinking have been heard in recent years from many quarters.[3] Still, prospects of resolving Fermi's paradox must count among the strongest motivations for doing so; while there are many other outstanding puzzles in the field of SETI studies, none is arguably wider in scope and more firmly connected with the epistemological and methodological basis of all the research in the field—and in many other sciences and arts.

The main points of the present book could be sketched in a cartoonish manner as follows:

- **Fermi's paradox is a fabulously complex and rich intellectual problem.** As few similar 'big issues' in the history of ideas, it touches upon many different areas and disciplinary troves of knowledge, encourages a healthy disregard for institutional and traditional boundaries, and provokes innovative, original, maverick, out-of-the-box, 'crazy' thinking. In sharp contrast to those regarding Fermi's paradox as a frivolous and unimportant puzzle or, even worse, as an obstacle to the progress of SETI, one which needs to be swept under the rug and ignored, I have shown that it is a serious multidisciplinary problem, generating a large amount of research and lively, fruitful intellectual activity. If anything, the research interest in Fermi's paradox has increased since the turn of the century, and as of 2018 this trend shows no signs of abating.

- **There are four major categories of solutions, corresponding to rejecting at least one major philosophical assumption.** The increasing intellectual activity has generated a large number of explanatory hypotheses and their variants, versions, reissues, covers, and so on. In order to make sense of this jungle of ideas, we need a workable taxonomy. Fermi's paradox is based on, among other things, four key philosophical/economic assumptions: scientific realism ('our observations of the Great Silence are correct'), Copernicanism ('we are not special'), gradualism ('no large discontinuities in astrobiological evolution'), and common-sense economic assumptions ('viable consumption of resources leads to detectability'). They look so large and so general that they are usually tacitly assumed, and yet abandoning at least one of them generates at least superficially plausible explanatory hypotheses.
- ***For many are called, but few are chosen.***[4] While dozens of explanatory hypotheses (if minor variations are taken into account, the number would be in the low hundreds) have been suggested, very few of them are plausible responses to the strongest versions of the paradox. The conclusion demonstrates that Fermi's paradox is similar to other great scientific questions posed by humanity: rational methodological analysis will act as a filter, creating fewer and fewer plausible answers and honing the hypotheses remaining in the pool, in the same way that natural selection acts in the course of biological evolution, until something resembling optimality is achieved. It has been an unfortunate prejudice which favoured weaker versions of the problem (the absence of visitors as opposed to the absence of manifestations and traces in general), that is, **ProtoFP** and **WeakFP**, instead of **StrongFP**. In addition, the key role played by the non-exclusivity principle reflects the most important filter any plausible hypothesis needs to pass. Recent

advances in many fields, from infrared astronomy to computational neuroscience, from theories of abiogenesis to virtual history, provided both new views of the puzzle and refined some of the old ones.

- **The puzzle is tightly connected to the future of humanity.** With a few exceptions, the choice of the optimal solution to Fermi's problem severely constrains the options for the future evolutionary trajectory of our own, human civilization. Stronger forms of this conclusion hinge on Copernicanism, and it is for exactly this reason that Copernicanism takes central place in our analysis. It is not true, however—and in particular it is not *necessarily* true—that those solutions which reject Copernicanism (the 'rare-Earth' category) cannot tell us anything useful about future human prospects; even the lack of particular constraints is still a relevant piece of information. Only the solipsist category, which is the least acceptable for independent reasons, decisively breaks the relationship between the correct solution of Fermi's problem and the most likely direction of the future evolution of humanity.

I wish to make three additional points—or exhortations—in this concluding chapter:

1. **In more than one sense, stronger versions of Fermi's paradox truly present the last—and the greatest—challenge to Copernicanism.** Our location in space and time, our relation to the cosmological, geological, chemical, and biological evolutionary processes which brought about our existence, our consciousness, subconsciousness, and the rest of our psychological and cultural make-up, even our relationship with our own artefacts and machines—elucidating all these revolutionary achievements which led to our 'decentring', 'demeaning', or 'humiliation', as prescribed by Copernicanism, is incomplete unless we put ourselves in the context of *independently evolved minds.* But this cannot be the whole story since, without taking into account Fermi's paradox, we could—just as some of the 'founding fathers' of SETI did—sit and relax waiting indefinitely for the eventual success of one listening programme or another. With Fermi's paradox in its stronger versions, it is not enough to relegate the issue to uncertain future; we need to seriously face the unpleasant conclusion that we should expect to have already encountered traces and manifestations of older Galactic sophonts.

2. **Radical ideas deserve more scrutiny.** This applies not only to the whole complex of ideas coming under the banner of postbiological evolution, but to those even more drastic deviations from our oversold 'common-sense' ideas about advanced technological societies. The huge diversity of biological morphospace and the even 'huger' diversity of postbiological design spaces clearly suggest not only that we need to emancipate from merely following and investigating existing large-scale trends, but also that we need to consider alternative, 'phantom,' and 'crazy' approaches. Some examples of these, like the extremely distributed computing hypothesis, are given in this chapter, while the connection with some of the explanatory hypotheses considered in Chapters 4–7 will be reiterated.
3. **There is a significant opening for reforming SETI projects directly inspired by Fermi's paradox.** Reasons for requesting reform and modernization of the SETI enterprise (a request that is rather ironic for a field which was originally regarded as the 'future incarnate') are many and important, and several of them have to do with Fermi's paradox. Such reform would mean putting SETI studies on more solid foundations in epistemological and practical terms, as well as reaching for a long-term view. Pleading for a renewed SETI—and reinvigorated enthusiasm for it—is tantamount to taking the cosmic vision of the future of humanity seriously. SETI needs an unashamedly pragmatic approach and it needs to stop being on the defensive and stop being afraid of politicians, journalists, and/or fools. Progress in this area is strongly contingent on further interdisciplinary and intercultural collaboration. The place of mind in the universe is such a tremendously complex and multifaceted issue that there is no escape from it into allegedly safe disciplinary pens and no hiding behind the fortified walls of departments, institutes, and other arbitrary administrative units. Fermi's paradox and its many provocations can serve as long-brewing catalyst for a kind of SETI renaissance in the same manner other famous antinomies of science, such as Maxwell's demon or Russell's paradox, brought about flourishing in their fields of study.

As this journey approaches its end, I shall refrain from favouring particular solutions and instead point out to broad research programmes which are likely to bring new insights into the problem. In addition, I shall offer some outlines for serious scientific discussion of the future of SETI enterprise and of astrobiological research in general.

## 9.1 Copernicanism Once Again

'The search for alien intelligence is an exercise in the Copernican principle', writes Paul Davies, one of the deepest scientific thinkers of today.[5] As discussed in Chapter 1, Fermi's paradox goes some further steps along the same alley, asking questions about the whole Galactic (at least) *set* of civilizations, not just individual potential contactees. A single, localized, incidental SETI success would not resolve the more general issue of **StrongFP**. The non-exclusivity principle acts as a sharp tool, excising vast chunks of the parameter space while asking questions about the uniformity of the structure or behaviour of any conceivable set of advanced technological civilizations. So, what conclusions can we finally draw about the lessons of Copernicanism for Fermi's paradox?

Evolution brings about cognition, which can recognize a subset of the parameter space containing cognition—and search for instantiations of that subset in the physical universe. One can look at this huge evolutionary loop in another way, prompted by immortal metaphors and symbols from art. M. C. Escher's famous 1956 lithograph *The Print Gallery* contains a weird self-reference which earns it a particularly detailed and inspirational discussion in Hofstadter's seminal *Gödel, Escher, Bach*.[6] In the lithograph, a young man in a gallery is looking at a print of a Mediterranean seaport, and among the buildings in the seaport is the very gallery in which he is standing. Is this what the universe really looks like? Hofstadter writes:[7]

> Now are we, the observers of Print Gallery, also sucked into ourselves by virtue of looking at it? Not really. We manage to escape that particular vortex by being outside of the system. And when we look at the picture, we see things which the young man can certainly not see, such as Escher's signature, 'MCE', in the central 'blemish'... Thus we, on the outside, can know that Print Gallery is essentially incomplete—a fact which the young man, on the inside, can never know.

While it requires a tremendous effort of abstraction and imaginative reflection to reach such an image, I suggest that much of our troubles with anthropocentrism originate in the lack of yet another layer of observing/modelling we need to superimpose on *The Print Gallery* to really account for the place of our cognition—and *any* cognition sufficiently similar to ours—in the universe. Even if our cognition is rather generic and typical, vagaries and opportunities of evolution, both biological and cultural, might have created a narrow, quite atypical filter. Since it is seldom explicated—and certainly not advertised, outside of esoteric philosophy of science circles—I shall call it the *implicit filter*. When interposed between us and the evolutionary loop, this additional layer will create an illusion of exceptionality, a false 'bird's-eye' (or, more controversially, 'God's-eye') view. The 'blemish' is still here, but it acquires a new meaning. If the implicit filter is sophisticated enough, the meaning could be sophisticated as well, obtain structural significance and—crucial from the point of view of this book—we could not a posteriori recognize it as a blemish, or something out of place or extraordinary.[8]

The implicit filter is symbolically shown in Figure 9.1. We have encountered it many times in this book, notably in discussions of **Gaian window**, **Permanence**, and **Sustainability**. Humans use very specific tools (including very specific *cognitive* tools) to obtain representations of the world, including the world of art and creativity. To claim that the paradoxical nature of some aspects of our representation is entirely up to the world itself or entirely due to our inadequate cognitive toolset is simplistic hubris, demonstrating that we fail to understand the question. There is no real beginning, nor the end to the chain of cognition; in logical terms, all possible evolution has already happened. We might only discuss whether cosmology allows for a large-enough world for all those evolutionary trajectories to be realized. With the recent advent of various forms of the multiverse and the ideas of creating a universe in the lab, it seems that the cosmological help is indeed forthcoming.

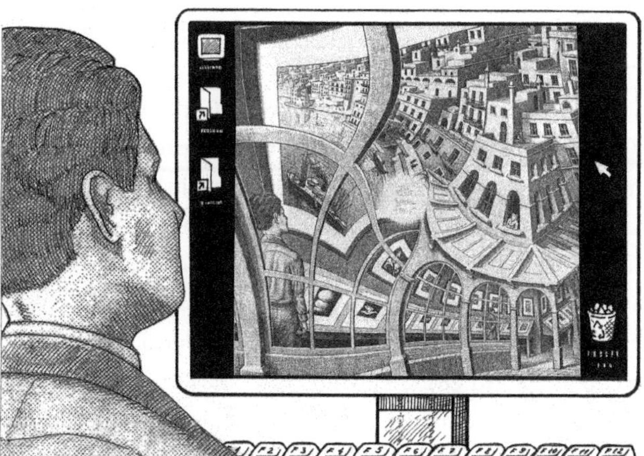

**Figure 9.1** A symbolic representation of *yet another layer* of understanding and perspective we need in order to face fundamental cosmological and astrobiological issues.

*Source*: Courtesy of Slobodan Popović Bagi

In Section 9.3, I shall discuss briefly the new and promising science and technology of distributed cognition. For the moment, we need only one of its basic tenets: *our nervous systems do not form representations of the world; they can only form representations of interactions with the world.* Even observation itself as the baseline interaction conforms to this essential truth. In modern, quantitative science, we look at the world through discrete numerical models which facilitate explanation and prediction. Our choice of modelling techniques and approaches, our selection of phenomena to include or neglect, our manner of dealing with the issues arising from information processing, our placement of results in the overall explanatory and predictive schemes—all these play a role within the implicit filter. The entire methodological complex creates multiple observation-selection effects which are, as we have seen in discussing many suggested hypotheses to explain **StrongFP**, major obstacles to our insight.

The implicit filter acts not only at the level of methodology, however; it acts on the evolutionary level as well. The greatest odyssey in humanity's history, its *cosmic destiny*, both literally and metaphorically, all that 'prophets in the wilderness' like Nikolai Fedorov, Konstantin Tsiolkovsky, Olaf Stapledon, Arthur C. Clarke, and Richard Buckminster Fuller preached about, cannot be ever really comprehended without providing the answer to the key puzzle, articulated by Fermi 70 years ago: *Where is everybody?* There is no escape from hard questions. No matter how much do we tend to avoid, procrastinate, and wiggle out of such problems, using all sorts of rhetorical arsenals, perceived loopholes, quid pro quos, guilt by association (with pseudoscientists or political ideologues)—at the end of the day, we must face the simple fact that either we are missing something crucial or we need to abandon Copernicanism and accept one or another anthropocentric view.

This dilemma is brutal. The relentless march of Copernicanism has repeatedly threatened and destroyed our cherished myths and prejudices in the course of the last (almost) five centuries. The famous passage attributed to du Bois-Reymond and repeated with adequate addition by Sigmund Freud tells us about Copernicus's, Darwin's, and, eventually, Freud's blows to human illusions of grandiosity within the cosmological domain, the biological domain, and, eventually, the mental domain, respectively. The last domain had been a bit shaky anyway, and it can be propped up and given truly universal meaning only within the set of different possible, non-human minds—hence the importance of the artificial intelligence and SETI fields for this particular philosophical narrative.

This dismantling of cosy anthropocentric illusions has, clearly, produced a counter-reaction and countless attempts to turn the wheel of history back, or at least arrest it in place. As discussed in Chapter 5, the anti-Copernican cartel is, paradoxically, livelier than ever now in the twenty-first century, when we are on the real threshold of our possible cosmic destiny. There is a view prevailing in much of the Web, the media, and academic literature that we have already become a navel-gazing, self-absorbed civilization, without much chances of developing a sustained cosmic presence and industrial bases all over the Solar System.[9] We have encountered some of the arguments and memes to that effect when discussing **Self-Destruction, Advanced Version, Sustainability,** and **Galactic stomach ache.** This is very dangerous path, since it impedes the best—and ultimately only—prospect for humanity to achieve its cosmic destiny. And the issue has very practical significance. This has been formulated nicely by Hoyle:[10]

> Many are the places in the Universe where life exists in its simplest microbial forms, but few support complex multicellular organisms; and of those that do, still fewer have forms that approach the intellectual stature of man; and of those that do, still fewer again avoid the capacity for self-destruction which their intellectual abilities confer on them. Just as the Earth was at a transition point 570 million years ago, so it is today. The spectre of our self-destruction is not remote or visionary. It is ever-present with hands already upon the trigger, every moment of the day. The issue will not go away, and it will not lie around forever, one way or another it will be resolved, almost certainly within a single human lifetime.

The same was warned about by George Gaylord Simpson in his (in)famous criticism of the then nascent SETI projects.[11] It is quite possible that our cosy anthropocentrism prevents us from seeing the large negative adaptive value of the intelligence we possess. As the current generation is likely to live to the 500-year jubilee of *De revolutionibus orbium coelestium* in **AD** 2043, it is important to take stock of what Copernicanism has achieved and what it means to our scientific outlook. Here, it makes sense to see—and urgently—what it means in an astrobiological context.

Firstly, we need to abandon our *Galactic provincialism*. As put succinctly by Milan Kundera: 'How to define "provincialism"? As the inability (or the refusal) to see one's own culture in the *large context*.'[12] In the case of our study of Fermi's paradox, the relevant larger context is—at least—the Galactic one. We need to put our own evolution, both biological and cultural, into a larger context and thus reject Galactic provincialism. We need to begin *thinking in Galactic terms*, something very few researchers, philosophers, or even artists are equipped or willing to do. This is a grand task ahead of us as a species, and astrobiology and

SETI studies are well-positioned to give a significant contribution—but not exclusively; many disciplines, including 'soft' sciences dealing with societies and cultures, can and have to do valuable work. This was clear even at the time of the SETI 'founding fathers',[13] but is conveniently forgotten from time to time. (This is not really a one-sided affair: rather, consistent strands of actors in social science and humanities have consistently opposed space research, space programmes, and, indeed, the very long-term vision for humanity for decades.[14]) That exactly is the substance of an *extended mandate*[15] of astrobiological research: the three canonical questions usually cited in connection with the astrobiological revolution from 1995 until today[16] should be understood in an inclusive, evolutionary, and Copernican manner. We are still far from such an understanding, although a vague outline has been emerging in recent years. The fact that a large number of the studies which I cited in this book were published in the last decade or two clearly testifies that the winds have changed.

There is a huge task ahead, however: the task of finishing the Copernican revolution. While the original Copernican revolution led to reappraisal of our position in the universe in terms of spatial coordinates, the Darwinian evolution led to reappraisal of our position in regards to the terrestrial biosphere, and the cosmological Big Bang revolution led to reappraisal of our position in cosmic time[17]—the ongoing astrobiological revolution has potential to lead to reappraisal of our position in terms of universal complexity. In more than one respect, through modern astrobiology we are facing some of the deepest and oldest problems dealing with the role of life and mind in the universe at large. In contrast to all previous epochs, when answers to these questions were necessarily speculative, we are now living at the cusp, at the tipping point—in the very moment when firm empirical resolution of these ancestral big puzzles is, for the first time ever, in sight. We should not regard ourselves as pinnacles of complexity in the universe; instead, we should reason as if we were near typical for our given epoch.

As mentioned in Chapter 5, there is a powerful and widespread anti-Copernican cartel holding sway over most of the human culture and society at present. Among other things, it could be argued that anti-Copernicanism is, to a large degree, responsible for historical problems encountered by the real-world SETI projects. This cuts both ways: both extreme and unwarranted scepticism towards the existence of potential SETI targets *and* a loose, unserious, anything-goes view linked to excessive optimism and even UFO enthusiasm. Indeed, how one could account for the irrational persistence of so much of the UFO mythology without at least partially crediting an implicit belief that

Earth and humanity are still somehow special and particularly interesting for alien spaceships and abductors? And it becomes more aggravating when we contrast this mythology—and countless spin-off conspiracy theories and other *X-Files*-like paraphernalia—with the clear, cold, scientific truth about billions of Earth-like planets all over the Galaxy. It is only anthropocentrism, that most powerful drug of the human mind, which is capable of sustaining such a 'bowl of lies' (cf. Leonard Cohen) for a prolonged time. So, ironically, the UFO-conspiracy buffs have more in common with religious fanatics and misguided postmodern humanities academics than either would care to admit. But, at both extremes, the road for serious SETI studies is closed: both believing (i) that Earth is unique in the empty universe, and (ii) that the universe is full of alien species which all come to Earth as a place that is especially interesting to crash on and abduct people and cows are ultimate stop signs for scientific search. The damage done to the honest, open-minded research from *both* sides is rather obvious.

If we zoom in even further, the tension between proponents of minimal cosmic engagement and those envisioning the true cosmic future of intelligent beings has been present since the beginning of serious SETI thinking, although it has been partly submerged and hidden for various reasons. For example, Kuiper and Morris in an influential paper in *Science* in 1977 mildly criticized the Project Cyclops report for assuming that all interstellar travel needs to include fuel for the return travel.[18] It may sound like a technicality, or even trivia, but do reconsider such a tender criticism today, in an age of general recess of cosmic thinking. In this day and age, thinking about *interplanetary*—not to mention interstellar—logistics sounds to most people as bored luxury at best, or delusional irresponsibility at worst.[19]

In this important and timely sense, the job of the Copernican revolution cannot be finished without at least a modicum of success for SETI programmes. In other words, we need to work hard on finding the middle way between the Scylla of extreme scepticism, and the Charybdis of fanciful, alien quasi-religiosity. The success needs not consist of a dramatic bombshell about receiving 'The Message' from our 'brothers in the mind' or 'kindred spirits', as represented in much of the bad pop culture about this topic, and unfortunately accepted by part of the orthodox SETI community. Even less should we hope and strive for a cosmic revelation (*The* Revelation?) to give us insight into the deepest question about existence, universe, and everything else. A success in the true sense would constitute understanding of the general shape of the astrobiological landscape of the Milky Way—the capacities and potentialities for transformation on a large scale. Instead of a lacklustre and passive 'waiting for the miracle', the success might instead be a gradual awakening to the fact that the universe is, in many places, at the transition point, a cusp in its history, in an epoch in which 'natural' is becoming supplemented by 'intentional'. The success might be similar to what Alastair Reynolds describes in his ultrashort science-fiction story-essay 'Feeling Rejected':[20] astronomers coming to terms with the multiplicity of indirectly detected advanced civilizations, even to such a comic extent that peer review tends to reject submitted papers dealing with *just a single technological civilization* of the Kardashev 1.x type. It's not newsworthy anymore!

Let us not forget contingent facts of the history of ideas: after each revolutionary overturn, we have found ourselves fruitfully accepting our deeper embeddedness in nature, coming to terms with naturalism and opening innumerable new vistas for further understanding. So, all the sound and fury of the regressive forces of anthropocentrism are ultimately deprived of any coherence—there is no 'end of the world', no existential or morale catastrophe waiting for us if we acquiesce in what is, anyway, enlightening, liberating, and true.

There is another reason why Fermi's paradox is the last and the greatest challenge for Copernicanism; I have hinted at this reason at many points in this book and it is tightly interwoven with the logistic hypotheses outlined in Chapter 7 and the astroengineering projects described in Chapter 8. If practical SETI searches so far have mainly been anthropocentric, searching mostly for beings like ourselves—the exceptions being rare Dysonian studies like the Ĝ survey or others cited in Section 8.2.—then their outcomes are, unfortunately, less significant than is usually and naively thought.[21] If a fisherman uses the net woven in such way that it catches only fish larger than 50 cm, he should not be terribly surprised if he returns hungry most of the time. The subsample of civilizations similar to our present civilization may not, for many reasons previously argued, be representative of the whole set (or the high-complexity region of the astrobiological landscape[22]). We should also ask, what does it really mean to be similar to ourselves—not as we are now, but as we are to be in the most typical moments of our *total* history? This is by no means obvious, and the proponents of 'humanist anthropocentrism' may have an escape clause here: it is possible, for instance, that mankind (or something sufficiently similar) persists for a billion years from now, in which case it might be the case that many or all possible civilization configurations will be sampled by our descendants, including those radically different from what we perceive today. In other words, some or most of our descendants will be *alien* enough to be effectively comparable to hypothetical present-day aliens. In 1930, inspired by the futurist thinking of J. B. S. Haldane, Olaf Stapledon sketched his magnificent vision of interplanetary humanity which is genetically engineered for living on different planets and moons, in his *Last and First Men*. The process was dubbed *pantropy* by another classic author of science fiction, James Blish.[23] In this Stapledonian vision, it is still justifiable to subsume many plausible and apparently heterogeneous forms of Galactic sophonts into something which is human-like *in general*, including many generations of future humans or posthumans. And this leads directly to the key highlight of human evolution and arguably the most important topic in all of future studies.

## 9.2 The Importance of Being Postbiological

In several places in this book, postbiological evolution has been invoked either as giving the context of some of the proposed solutions (e.g. **Transcension** or **Great Old Ones**) or as a process undermining some of the underlying assumptions (e.g. **Galactic stomach ache**). In addition, it is one of the strongest points of contact between astrobiology on one side and future studies, as well as bioethics. From one point of view, it is already here. The ongoing debates on human enhancement via drugs, prosthetics, implants, brain–machine interfaces, and so on, are just the tip of the iceberg, the early beginnings of what is very likely to become the central topic in humanity's future. If we zoom out just a little bit and take a slightly wider view in both a temporal and a structural-parameter sense, we shall see a strong trend towards the substitution of biological mechanisms for more durable and reliable intentionally created mechanisms. In the long run—but *not* as long as evolutionary or astrophysical timescales!—this can only lead to postbiological civilization.

Many authors have pointed this out. For instance, Paul Davies, in his short, but quite instructive and provocative *Eerie Silence*, writes:[24]

> In a million years, if humanity isn't wiped out before that, biological intelligence will be viewed as merely the midwife of 'real' intelligence—the powerful, scalable, adaptable, immortal sort that is characteristic of the machine realm. Thereafter, machine intelligence will accelerate in power and capability until it hits fundamental bounds imposed by the physical environment, whatever they might be... By the same token, we can expect any advanced biological extraterrestrial intelligence to long ago have transitioned to machine form. Should we ever make contact with ET, we would not be communicating with... humanoids, but with a vastly superior purpose-designed information-processing system.

Clearly, the consequences for practical SETI projects and, indeed, the very methodology of SETI are tremendous—and it is quite indicative that they have been noted only recently.[25] Coupled with the ideas of interstellar colonization and astroengineering, postbiological evolution changes the entire game: we need not—and indeed should not—target habitable planets and circumstellar habitable zones in our SETI searches. Instead, we ought to focus on regions with the greatest amounts of resources, including metals and energy, as well as low working temperatures, as the best locales for optimized computation. Surveying warm, wet places would not make much sense. SETI circles should also be very, very interested in developments in fields such as hypercomputation, quantum computation, and artificial intelligence (the interest being conspicuously lacking so far, with a few honourable exceptions), since these are likely to offer fresh insights into the parameter space describing advanced machine intelligences of our own postbiological future, as well as postbiological evolution elsewhere.

Strategic assumptions behind both most SETI searches and thinking about Fermi's problem are also undermined by the possibility of postbiological evolution. As discussed in relation to hypotheses such as **Transcension hypothesis** or **Living on the rim**, the concept of expansion to 'fill all ecological niches' might be a fiction imposed on us by our own biological nature and evolutionary past (the **Deadly Probes** might be an exception to this, to be more studied). If many human behaviours—and, by extension, the behaviours of other biological sophonts—are strongly influenced by the evolutionary baggage of the species, it is only reasonable to conclude that a transition to the postbiological realm will obviate such behaviours. The same applies to many cultural characteristics. If disciplines such as sociobiology or evolutionary psychology claim to explain many aspects of the human psyche and behaviour, then those aspects would become marginalized or extinct if humanity passes into the postbiological realm. (While, of course, this does not mean that we shall lack explanations for psychological or cultural characters, as these shall be provided by forthcoming *post-sociobiology* or *postbiological-evolutionary psychology*.)

However, if we zoom out still further, we can see a catch. In the longer run, the dichotomy between biological and postbiological might become immaterial, and even false. This has been clearly expressed by a contemporary science-fiction writer:[26]

> The difference between biological, biochemical, electronic, or neuro-electronic information systems, at that level of civilization—no difference, is it? Once you

> can rebuild yourself from the molecular level up, and out of any substance you fancy, soft or hard, stored as a pattern in a mainframe or spun out into any form of matter need calls for—no such thing as machines you can properly call by that name. It's all alive. Or all dead.

The emphasis is correctly put on the 'level-of-civilization' parameter. While there might be other prerequisites for such *merging of biological and technological* organization, it does seem to primarily depend on the overall civilizational level—as indicated by proxy, for instance, by Kardashev's type or perhaps position on the Barrow scale.[27] The last sentences of Wright's passage suggest a viewpoint similar to the one embodied by **New cosmogony**: there is no point in trying to discern biological from postbiological, as much as there is no point in trying to discriminate between natural and artificial after some—very high—threshold civilizational level.[28]

Thus, we conclude that the process of influencing physical environment might, at least for a subset of all sophonts appearing within a cosmological domain, be represented by a function with a maximum, such as an inverted parabola or a cycloid. In the beginning, nascent sophonts exert negligible intentional influence upon their natural environment; if stages of civilization, such as the hunter-gatherer, agricultural, and pre-industrial stages, have meaningful analogues in sophont societies, the situation would correspond to those long epochs. As time passes, the population, on the average, will increase, and new resources will be discovered, with increased ecological impact. If that does not in itself cause social collapse and other possible types of extinction events, we may expect both the spatial scope and the functional depth of sophonts' influence over nature to increase dramatically with the advent of space travel and subsequent colonization of space. The opposite trend will, however, start to manifest itself as well: many advanced technological processes will either utilize natural evolutionary innovations developed and perfected over billions of years of evolution, or will be intentionally designed to *minimize impact* on the natural environment, for both practical and ethical (and possibly aesthetical or other culture-specific) reasons. We can observe a seed of this trend in current human attempts to recycle as much waste as possible, use renewable energy sources as widely as possible, and even use evolutionary algorithms and other forms of biocomputing. Negotiating between these two large-scale attractors is the likely destiny of any intelligent community ever evolving in the Galaxy: larger and larger perturbations of nature on one side of the equation versus stronger and stronger attempts to emulate nature or achieve more and more efficient symbiosis with it. Terraforming, the seeding of distant planets with spores of life, and astroengineering projects which increase habitable space and biological productivity, such as O'Neill habitats, are some of the manifestations of this opposing tendency; attempts to show the rather plain truth that *exploration and colonization of the universe are essentially ecological endeavours* are precursors of those more ambitious enterprises.[29] I shall consider another manifestation, widely distributed computing, in Section 9.3.

This can also be put in the context of *Big History*: recent attempts at giving an overview of history from the Big Bang to the present. Big History examines long time frames using a multidisciplinary approach based on combining numerous disciplines and methods from science and the humanities.[30] While Big History has demonstrated an increase of local complexity in the course of cosmological history, it has also demonstrated a shift in the material substrate within which this complexity is manifested. There is no reason to expect the biological substrate, as exemplified by the terrestrial biochemistry based on proteins and nucleic acids, to be the last and final word. On the contrary, it seems natural to expect other substrates to begin to play a role in the unfolding of the universal drama. And the digital substrate we perceive nowadays, which can—as testified by extensive results in the field of artificial life[31]—support at least as great a complexity as the natural biological substrate, offers huge new prospects in this regard. As put by Cadell Last in a comprehensive review of Big History as related to future studies:[32]

> Throughout the development and evolution of our local universe there has been an interconnected growth of complexity from physical, chemical and biological systems, as well as cultural and technological systems.... In this way the universe gives the appearance of internalizing its future potentiality within a network of billions of biocultural nodes that in aggregate represent a phenomenon capable of producing yet another level of complex organization. This perspective does not succumb to the trap of anthropocentrism as I am not arguing that humans are 'reclaiming centrality'. Instead I am making the philosophical argument that humans could represent an important process in the context of the growth of local complexity that is part of a much larger 'multi-local' cosmic phenomenon...it is entirely plausible that cosmic evolutionary theory has application on a universal scale, with other analogous levels of local complexity developing via a type of 'universal culture'.

So, Big History effectively offers a conciliatory way of joining the (still) non-Copernican fields of learning, such as the social sciences and the humanities, to the mighty river of Copernican science, a river whose basin maps out the possible cosmic future of humanity. The latter is the very future outlined by Fedorov, Tsiolkovsky, Stapledon, Wells, Clarke, Fuller, O'Neill, and the modern-day transhumanists. The opportunity ought not to be passed up lightly—there aren't that many of them in this day and age.

## 9.3 Radical Departures: Distributed Computing, ~~Star~~Universe Maker, and the *New Cosmogony* Revisited

Postbiological evolution, as previously discussed, has gradually become an 'almost mainstream' issue. Even the most rabid bioconservative opposition acknowledges the importance of the subject matter.[33] In contrast, truly radical issues rarely get serious attention and scrutiny, due to the dual threat of being perceived as sensationalist, pseudoscientific, and fantasy-prone, on the one side, and inhuman, dangerous, and outright evil on the other. Similar to what has been discussed in connection to astroengineering manifestations and artefacts (see Section 8.2), one needs a safe passage between Scylla and Charybdis in order to at least build foundations for a reasonable and serious discourse on these issues. In this section, I consider *examples* of these new, radical themes, with an eye to the future: these have not been seriously studied so far, but are likely to become so in the near future. Here, they serve as placeholders for innovative, bold, radical thinking and a call for detailed, quantifiable work of tomorrow.

While I write this sentence, with my peripheral vision I observe some lavender plants growing on my windowsill and a sparrow flying above, and I hear the persistent humming of my desktop fans. Why is the latter, rather than the former, thought to be more involved in the cognition necessary for these words to emerge? In discussing various solutions, we have seen that the structural elements of many approaches to Fermi's puzzle revolve around information technologies, in particular the large-scale networking which, in the course of the last couple of decades, 'brought us all closer together' as technophiles and Internet enthusiasts would claim. Hypotheses presented under the titles of **Simulation hypothesis**, **Transcension hypothesis**, **The Paranoid Style in Galactic Politics**, **Bit-string invaders**, and perhaps even **Thoughtfood Exhaustion** could not have been formulated, or perhaps even made intelligible, in the pre-Internet era. If we perceive this connection as an expression of a zeitgeist, we should not stop there. Consider what many experts regard as the proverbial 'next big,' namely the *Internet of things*, alluded to above; we may use the definition of the US National Intelligence Council as a working one:[34]

> The 'Internet of Things' is the general idea of things, especially everyday objects, that are readable, recognizable, locatable, addressable, and controllable via the Internet—whether via RFID [radio-frequency identification], wireless LAN [local area network], wide-area network, or other means.

While this has been described in cyberpunk fiction many times over, it seems that we are approaching technological level necessary for its realization.[35] And it has a huge transformative potential for human life and culture, and even, as Melanie Swan has persuasively argued, for the definition of personhood. The concept of *quantified self* represents not only a necessary step for any postbiological evolution, but also a prerequisite for implementing cognitive processes in any substratum and any context.[36]

For our present purposes, the key issue is that the Internet of things will enable the further distribution of computing. Coupled with the ideas of postbiological evolution previously discussed, this leads us directly into questioning the conventional concept of extraterrestrial intelligence, whose absence seems so puzzling. The cognitive scientist Edwin Hutchins offered a detailed description of distributed computing and cognition, with many vivid examples, in his classic book *Cognition in the Wild*.[37] The properties of the system are at least as much determined by the nature of the media in which the representations are realized and by the pattern of connection between individual components as they are by the properties of those components. The very title evokes what we may encounter in SETI studies: a signal of cognition hidden—intentionally or not—in the vast cosmic wilderness of noise. If the cognition is distributed widely enough, we may not be able to detect it—which is the point made by Lem and Hoyle, as previously discussed in connection with **New cosmogony**. But if we wish to pursue the idea further, there is more to it, applicable not to a single hypothesis, but to many explanations for Fermi's problem. As put by Lem in his discursive *Summa Technologiae*:[38]

> The Intelligence we shall discover one day will possibly be so different from our ideas of it that we shall not even want to call it Intelligence...It does not look like intelligent activity to us because man favors a heroic attack on the surrounding matter. But this is just a sign of our anthropocentrism.

Not 'attacking the surrounding matter'—but *working with it* or *working through it* in a form of widely, cosmologically distributed cognition; that is a signpost of not only different, but actually and justifiably *more advanced* intelligence. The desiderata of ethics and efficiency seem to coincide here.

Why work through *some* of the matter, however—and not *all* of it? A closely related issue is the creation of entirely artificial universes in a lab—not simulated universes, but those as physically real as our observable universe.[39] This could be regarded as the ultimate form of astroengineering, as was discussed in Chapter 8. The artefact would be an entire universe, with its own full-package physics (not just bits and pieces of local physics, as **New cosmogony**). It might not be observable—or at least especially conspicuous—from afar, since most ideas of this sort imply that it will be surrounded by an event horizon; in other words, it will look like a black hole from the outside. And black holes are notoriously classically boring objects: the famous 'no hair' theorem of general relativity implies that black holes are completely described by only three numbers: their mass, their angular momentum, and their electric charge. It follows that it is very difficult, to say the least, to distinguish between 'natural' and 'artificial' black holes. For example, in Section 8.2, I mentioned the stellification of substellar objects as a possible astroengineering project.[40] If such a process has been used to artificially create black holes, we may expect there would be a sizeable population of artificial black holes, some of which could contain engineered universes—but we would not be able to establish that from afar.

Such 'basement universes' could be conceived as scientific experiments, but they also could be by-products of other activities and concerns we cannot hope to envision at this point.[41] Recall Darwin's parable about a dog and Newton's mind. A distinguished British cosmologist, Edward R. Harrison, suggested in a 1995 paper[42] (which was regarded as a bit tongue-in-cheek[43]) that advanced technological civilizations would tend to create many basement universes—and that it is the surest way to ensure something like Lee Smolin's natural selection of universes and explain the observed fine-tuning of fundamental constants and cosmological parameters. Through this process, the habitability of the whole *multiverse* (or the ensemble of universes) would be fractionally increased. So, there might be a strong ethical incentive for creating new basement universes, if it is at all possible.

Another way of formulating the same idea is a fascinating hypothesis posed by James Gardner, which he dubs the *selfish biocosm*.[44] Gardner has developed a grand picture of a cycle of cosmic creation in which highly evolved intelligences with a superior command of physics spawn baby universes engineered to be able to give birth to new intelligent life. Within the context of the selfish biocosm, the ability of our present universe to support intelligent life as well as it does (via fine-tunings of fundamental constants and cosmological parameters) is not an accident but the result of a succession of increasingly 'cognition-friendly' universes emerging within the multiverse, in a process of both (evolutionary) intelligence-guided experiment and engineered (developmental) design. In one sense, the selfish biocosm of Gardner gives further incentive to the study—if not the corroboration—of **New cosmogony**-type hypotheses.[45] And not only that: in contrast to traditional metaphysical or theological versions of the design hypothesis, we would be fully entitled to seek some empirical evidence for it. Strangely enough, as mentioned in Chapter 4, some proposals have been shyly offered recently, in particular those involving CMB, as well as the lattice QCD calculations and the related highest-energy cosmic ray distribution.[46] This is in itself an incredibly strange and intriguing development, which demonstrates how even classical metaphysical questions could become legitimate subjects of scientific inquiry. While this development may yet fizzle out, the deep epistemological questions it raises are going to stay with us, even if—due to another outburst of bureaucratic scholasticism or any other reason—they are ignored for an arbitrary amount of time.

## 9.4 Fermi's Paradox as a Successful Provocation

The journey is often more important than the arrival. The history of Fermi's paradox, as an extremely complex and rich intellectual problem, could be understood as a continuous strand of *provocation* of some of our cherished notions and beliefs. As Popper famously said about lectures, the best aspect of a paradox is its challenging and provocative nature. As we have seen, the challenge was largely successful so far. But the provocation is a means, not an end in itself. The end is building a synthetic view, the emergence of an overarching explanatory structure, drawing an outline of the high-complexity, self-aware sector of the total astrobiological landscape. The end is understanding the place of mind—any kind of mind, liberated from the Galactic provincialism and parochialism—in the cosmological scheme of things; not just historically and chronologically, but on the most general *logical level*: what can minds do in the fullness of space and time, free from any constraints except logical coherence and the basic physical law? How *really* big is the design space in which advanced intelligence might tinker once it is liberated from the shackles of its biological evolutionary origin? What *we* could hope to achieve in order to prevent some other young Galactic sophonts from naively asking 'Where is everybody'?

Is there any chance of building a unified framework for resolving **StrongFP** and related problems? As I have tried to argue so far, this problem is not truly separate from the problem of building a sufficiently general theory of astrobiological evolution, a form of 'Galactic (neo)Darwinism'. It is entirely reasonable to consider intelligence/consciousness/tool-making/etc. as a natural outcome of the same biological evolution elsewhere as they have been on Earth. This is still valid even if there are entirely different modes of life, incommensurable to what we know from our terrestrial experience and historical record. There are things which are simply universal enough; recall Darwin's letter to C. J. F. Bunbury, dated 9 February 1860:[47]

> With respect to Nat[ural] Selection not being a 'vera causa'; it seems to me fair in Philosophy to invent any hypothesis & if it explains many phenomena it comes in time to be admitted as real. In your sense the undulatory theory of the hypothetical ether (the undulations themselves being not recognised) is not a vera causa in accounting for all the phenomena of Light. Natural selection seems to me in so far in itself not be quite hypothetical, in as much if there be variability & a struggle for life, I cannot see how it can fail to come into play to some extent.

Keeping in mind that, in Darwin's time, it was still usual to use capitalized 'Philosophy' as synonymous with 'science', we infer an important methodological lesson from the first sentence; it is the same one underscored by Ludwig Boltzmann in his famous—and often misunderstood—saying that 'there is nothing more practical than a good theory'. It is widely applicable to astrobiology and SETI studies, where researchers are often afraid of bold and daring hypotheses, although such hypotheses could explain a variety of phenomena. From the present point of view, however, still more interesting is the second part of Darwin's argument, namely that, if we wish to be intellectually honest, we have no choice than to accept that some bottom-line processes *must* play an explanatory role, though it might not be *the* explanatory role or contain all of the explanation. It is not a matter of a researcher's 'free will', 'attitude', or 'disposition' to include a process in the explanatory scheme—it is mandatory.

The lesson for those who attempt to resolve Fermi's paradox is clear. As we have repeatedly seen throughout our journey through the space of explanatory hypotheses, it is unlikely that there is just a single, well-defined, all-powerful explanatory mechanism. It is—just as in the case of evolutionary mechanisms operational in terrestrial biology—the question of relative frequencies. Just as in biological evolution most phenomena could be explained by an adaptationist model (as discussed in Section 5.4), while in no way demeaning the existence and relevance of genetic drift, horizontal gene transfer, and other evolutionary mechanisms, so here. It is most plausible that a single, non-exclusive hypothesis accounts for *most* of the explanandum in **StrongFP**, while some of the other acceptable options play auxiliary or local roles. Such a conclusion could be reached, however, only after careful examination of the high-complexity sector of the overall astrobiological landscape. In a sense, we need a real, comprehensive *astrobiological theory* in order to classify the explanatory hypotheses by their likelihood. Thus, we return to the words of Lévi-Strauss cited in the beginning; after touring many exhibits in Chapters 4–7, we better appreciate their wisdom. While erecting such a comprehensive theory might sound like a tall order at first, it is important to note that, embodied in the rare-Earth hypothesis, we have an example—however superficial, oversimplified, extreme, and often misleading—of such an emerging paradigm. Better and more quantitative theoretical frameworks are likely to come.

Winds of change blow stronger still, however. As mentioned briefly in Chapter 1, I would argue that, at present, we still have too *few* hypotheses for resolving **StrongFP**! This seemingly ludicrous assertion should be measured against our very scarce knowledge of the whole half of the problem—the xenosociological half, that is. To the examples from astronomy cited in Chapter 1, we could now freely add many examples of 'Big Questions' from what are essentially social sciences and culturology. How ergodic (Section 7.1) are pathways taken by Galactic sophonts? How frequent are totalitarian stages/phases/movements in the evolutionary trajectory of an average intelligent community? Is postbiological evolution a necessary phase in the history of all (or most) intelligent species? These questions will strike most observers as far worse than merely difficult; they seem formidable or, indeed, radically unsolvable. Our desperation in face of such questions reflects the relative lack of theoretical insight in the domain of cognitive, social, and historical sciences. Even extreme pessimists can hardly doubt that the situation in this respect—barring global civilizational collapse—will improve. And even a small improvement in absolute terms can be a relative boon in comparison to the dearth of interactions with social sciences/humanities so far. Even a minor fractional step forward can certainly bring about new explanatory hypotheses and create whole new cross-disciplinary fields of learning.

The relevant metric is not comparing the number of hypotheses, but comparing the number of hypotheses per relevant scientific field, per discipline, or even per theme. Since we have at least touched upon so many themes in the course of this book thus far, it seems clear that, on this metric, the number of hypotheses listed in Table 8.1. is by no means outrageous. At least this applies to the present. As we learn more, the shore of the 'ocean of unknown' lengthens, to paraphrase Newton's famous sentence and, while we may imagine (on some highly abstract level) that it will eventually contract, this era is not yet in sight.

Thus, we have another prediction: there will be many new explanatory hypotheses for Fermi's paradox in the near future, as the astrobiological revolution progresses and exploratory engineering goes farther and farther. However, they will be, at least in principle, amenable to the same taxonomy as applied here: we shall have to suppress scientific realism, Copernicanism, gradualism, or additional economic assumptions (or some combination of the four) in order to obtain even a remotely plausible explanation. The status of some of these philosophical assumptions might change in the future—although these things happen more slowly and less frequently that usually thought—but they will still strongly delineate the space of ideas. If they turn out to be consequences of deeper principles, such principles will, in turn, offer finer taxonomical distinctions than those presented here. The approach itself will still be validated.

## 9.5 The Failure of Conventional SETI as a Philosophical Failure

In one of the most moving scenes in the history of film, the culmination of Ridley Scott's 1982 legendary *Blade Runner*, the leader of escaped androids, Roy Batty (superbly played by Rutger Hauer), first saves the life of the cop (Rick Deckard, played by Harrison Ford) who tries to kill him, and then, just seconds before he dies from a programmed death (the 'fail-safe'), delivers a famous monologue on how his memories of the outer space are about to be lost 'like tears in rain'. The metaphor sounds particularly poignant in the rain-soaked atmosphere of the dystopian Los Angeles. Roy's speech alludes to the huge and (for those humans who remain Earth-stuck) unimaginable variety of cosmic experiences: 'I've seen things you people wouldn't believe. Attack ships on fire off the shoulder of Orion. I watched C-beams glitter in the dark near the Tannhäuser Gate.'

Deckard—the representative of extremely anthropocentric state power—who watches silently, seemingly experiences a *catharsis*, or revelation of the intrinsic value of life and mind—not just human life and human mind, but all life and all mind. His anthropocentric cocoon is smashed by the experience of mercifulness and revelation left—almost as a testament—by his intended victim. What Roy described *has already changed the human condition*; it is a done thing. Only the Earthbound masses, kept in order by conservative and repressive laws and police mechanisms, are still in the dark.

The ending of *Blade Runner* represents an artistic triumph in the fight against anthropocentrism. Robotics and artificial intelligence are fields that are, in any case, very close to SETI studies, since they all try to discern requirements for and properties of *non-human cognition*. Many scientific, philosophical, and artistic truths pertaining to artificial intelligence could be easily transposed into SETI, and vice versa. And, as we have previously seen, artificial intelligence and postbiological evolution are bound to play key roles in many of the explanatory hypotheses for Fermi's puzzle.

There is a deeper undercurrent in the symbolical representation of the wild, indescribable variety of the real universe, which human language is poorly adapted to. As well as we can ask ourselves, *Where is everybody?*, we can imagine that other intelligent observers ask the same question, completely oblivious of our ephemeral existence. Whoever finds such a picture depressing should remember that this is how many reacted—and continue to react to this day—when faced with the hard truths of Darwinian evolution. But the possibility and plausibility of such a state of affairs should not depress us, but motivate us to search—and search harder. And the search needs to be expanded beyond the confines of orthodox SETI searches.

Why is orthodox SETI so thoroughly *un*equipped to deal with Fermi's paradox? Clearly, the main reason is the cognitive dissonance induced by the simultaneous desire for success in establishing contact, on the one hand, and the cavorting that takes place with the anthropocentric cartel, and the acceptance of some of its poisonous memes, on the other. Recall Davies's words at the beginning of Section 9.1. We have seen that the rejection of Copernicanism leads into the acceptance of one of the rare-Earth hypotheses described in Chapter 5, in which case SETI, at least as far as the orthodoxy has conceived, makes no sense, since there are no viable targets. But the rejection of Copernicanism is also tantamount to the overt or covert subscription to one or another form of anthropocentrism—and vice versa: accepting anthropocentric thinking as valid entails the rejection of at least some aspects of Copernicanism. As far as resolving Fermi's conundrum is concerned, this leads to rare-Earth hypotheses—and a consequent lack of viable SETI targets. So, we are facing inconsistency.

A defender of the hallowed orthodoxy might ask where one can find this collusion of orthodox SETI with anthropocentrism. This one is actually rather easy, and critics—both destructive *and* constructive ones—have pointed to the answer in some cases for literally decades. The (in)famous criticisms of George Gaylord Simpson and some of the other evolutionary biologists rested in large part on the perceived anthropocentrism of orthodox SETI, as conceived by the 'founding fathers'.[48] The very methodology of orthodox SETI is unabashedly anthropocentric: only beings very similar to us—*present-day us*, that is—in the astrobiological landscape will tend to use radio waves to communicate among themselves and over interplanetary and interstellar distances. We listen to targets chosen explicitly from Sun-like stars and, in recent years, planetary systems, with the implicit idea that we should listen to possible SETI signals from Earth-like planets. This is firmly and squarely anthropocentric.

It is also ironic, however, in light of the stated views of the 'founding fathers' (esp. Sagan, Shklovsky, and von Hoerner) that *failure* of SETI would support anthropocentrism. This has been used, in turn, by some of the vocal *opponents* of SETI to equate it with allegedly unreasonable and uncritical application of Copernicanism. Such usage, for example by Barrow and Tipler in their encyclopaedic *Anthropic Cosmological Principle*, has become decisively outdated with the advent of the astrobiological revolution in the mid-1990s.[49] An in-depth analysis will easily show, however, that it is possible to retain both seemingly contradictory statements. This is so since some of the founding fathers—especially Sagan—conceived of SETI as being much wider than its orthodox formulation and practice. For them, it was quite natural to search for Dysonian astroengineering or for artefacts or ecologies drastically different from the terrestrial one.[50] It was only later that mainstream SETI views hardened and became the orthodoxy in which groupthink and dogma are so pronounced.[51]

Although Fermi's original question (not to mention Tsiolkovsky's or Stapledon's versions) is a full decade older than Project Ozma and the beginning of scientific SETI, it was not widely known until Michael Hart's paper in the mid-1970s.[52] And, by that time, the orthodoxy had already been hardened and heavily fortified. Hart's work—together with his research into narrow circumstellar habitable zones[53]—was conventionally received as part of the anti-SETI agenda and usually discussed only *in a negative context* by the SETI community. For the present purpose, it is immaterial that Hart's views were extreme in his insistence on the uniqueness of Earth and humankind; it is the reaction of the orthodox SETI community which indicates a deeper, structural flaw. Instead of constructively engaging with the published critical arguments, many orthodox SETI circles prefer to keep giving monologues about the merits of patient radio listening, and litanies on the lack of funding for the continuation of patient radio listening.

In a supreme irony, even orthodox radio SETI has been, in fact, searching for *really advanced* technological societies, although this fact has been rather craftily concealed. The Northern Irish physicist Sir David Bates pointed out in 1978—in a paper thoroughly ignored by both proponents and opponents of SETI!—that relying on omnidirectional radio beacons by orthodox 'listening' SETI projects does not make much sense under their own set of rules.[54] Namely, such beacons would have to be designed with call-response times on the order of $10^4$ years or more in order to have a reasonable choice of other parameters; and Bates is unnecessarily reticent when claiming that such a timescale would be unacceptable to governments anywhere in the Galaxy! Even on a purely technical level, the construction and maintenance of a system over such a long, in human terms, interval of time are no mean feats. It is quite debatable to what extent the construction of a machine that would last for $10^4$ years in a reliable manner is comparable with the construction of a Dyson sphere or an orbital ring. When we add cultural and political angles into play, the claim of orthodox SETI about 'looking for beings similar to ourselves' sounds even hollower. It almost goes without saying that building such durable institutions and structures qualifies as a macro-project in its own right!

Of course, there are various possible ways around Bates's argument. If aliens know we are here, they may opt to use a tight beam directed at the Solar System; this could increase efficiency, but keep in mind that this is also quite a distance from 'beings like us', since we are still decades and probably centuries away from being able to detect the presence of primitive sophonts on the surfaces of distant exoplanets. And we might hope to intercept their own internal communications—which requires quite a notch bolder leap of imagination and unwarranted optimism.

Even worse, from time to time there occur 'accidents' in which the credibility of SETI is seriously damaged[55]—while in 'normal' times the benefits derived from SETI *even without ultimate success* are sparsely and poorly advertised. Before the recent reignition of the search for manifestations and artefacts of higher-Kardashev-type civilizations, as documented in Chapter 8, the only highly visible example of SETI producing something of wider interdisciplinary interest was the SETI@home initiative, which sparked important work in widely distributed scientific computing.[56] This is too little for half a century of activities by some of the best minds of this planet—and too little in comparison to the grand vision of the 'founding fathers' themselves. When discussing the failures and woes of practical SETI, we should *never forget* that this was an endeavour originating with thinkers and visionaries who were bold enough to discuss the ecologies of Jovian planets' atmospheres, put Project Cyclops on the drawing board, contemplate a 'Galactic Club' of all advanced sentient beings, and consider the possibility of civilizations using the energy outputs of whole galaxies or QSOs.[57] Those were titanic visionary steps which encountered much resistance in their time; that much is understandable. What is eminently *not* understandable is the continued resistance—fortunately somewhat crumbling in recent years—in the twenty-first century.

There can be no doubt that the failure of SETI so far has mostly occurred on philosophical and methodological levels. Some of the pioneers warned about this long ago: the best example is Kardashev's sharp rebuke, quoted in Chapter 3, first given in 1985, and which he repeated multiple times in various versions throughout his writings: the key statement 'Extraterrestrial civilizations have not yet been found, because in effect they have not yet been searched for.'[58] Only very, very recently, with the breakthrough of the Dysonian thinking as discussed in Chapter 8, has this climate finally begun to change, and the ice to melt.

What has to be done? There is just one more exhortation at the very end of this book. Clearly, the prejudices of the past should be overcome. Fallacies supported only by emotional appeal to our human exclusivism, preposterous prejudices spread from behind barbwire academic and disciplinary borders, short-sighted nonsense propagated by the media—all this noise clogging the channels should be scrubbed and cleared. There is a huge amount of work to be done, but there is even more: there is a sense of excitement, a sense of drama, an unmistakeable smell of adventure, which is not so common these days in any field of human endeavour. Even in science, there is a marked lack of true enthusiasm and of the pure, selfless joy of discovery for its own sake and for the sake of universal knowledge, not for peer-reviewed-publication points, copyright-protected patents, or vague patting-on-the-back by politicians and industry or media moguls. The culture of 'infotainment' effectively suppresses diligence, hard work, and critical thinking—and it did not completely bypass astrobiology and SETI science. While the Saganian 'pale blue dot' message is more pertinent and more important than ever since *Homo sapiens* evolved a short 200,000 years ago, since we are facing the biggest risks and the highest uncertainties—it seems that we are more and more stuck with his other message: the necessity of being a 'candle in the dark'.[59] We should stop lamenting the alleged end of the space age—and become actively involved in reviving and reinvigorating space programmes and cosmic thinking on the global level. Even when that is achieved, it will not be the beginning of the end—but it might be the end of the beginning of the greatest of all journeys, the wildest adventure of all adventures.

*Abandoning past prejudices is not enough.* There has to be a positive momentum, an awakening, a new spirit of open-mindedness towards original, speculative, even crazy leaps on both the theoretical and the empirical stage. After all, this is what science has stood for in its most brilliant moments: courage, conviction, and the spirit of great adventure. These qualities might, in the final analysis, withstand the test of cosmic time, repulse the last challenge to Copernicanism, and ultimately break the Great Silence.

# ENDNOTES

## INTRODUCTORY NOTE

1. Subsequent, more precise studies give even larger temporal differences, as will be discussed in Chapter 2.
2. For an idiosyncratic sample, see Gindilis et al. (1969); Beck (1972); Ambartsumian, Kardashev, and Troitskii (1986); Davies (1995, 2010); Aczel (1998); McConnell (2001); Wilson (2001); Cohen and Stewart (2002); Duric and Field (2003); Sheridan (2009); Schulze-Makuch and Darling (2010); Bennett and Shostak (2011); Cranford (2015); and Ashkenazi (2017).
3. See also Chapter 7 of Ćirković (2012).
4. Ćirković (2012), esp. pp. 12–26, 203–15.
5. For contrasting views, see e.g. Brown (1993) and Norton (2004), on the opposite sides of this debate. On the relationship of paradoxes and thought experiments, see Hawthorn (1991), Sainsbury (1995), and Havel (1999).
6. Gould (1996), p. 39.
7. Gardner (1979).
8. Freudenthal (1960); the name is an abbreviation of the Latin phrase *lingua cosmica*: the cosmic language.
9. e.g. http://www.hasbro.com/common/instruct/Battleship.pdf, last accessed 5 November 2017.
10. And more of that style of philosophical argument can be found in Ćirković (2012).
11. This, for instance, is my reading of Lem's novels *Solaris* and *Fiasco*, although the latter one arguably implies an even worse situation: one in which CETI can bring about substantial *negative* value. Some of his other fictional works, notably *Eden* and *His Master's Voice*, masterfully present scenarios in which CETI can create some positive values, but they are meagre, uncertain, and quite expensive—giving the whole idea a somewhat anticlimactic and deflationary character, in sharp distinction to the overwhelming bulk of pop-cultural references.

12. Parasitic SETI searches are those which analyse the data obtained through other, more conventional astronomical observation programmes and therefore do not utilize separate telescope time, needing only human and computing resources (the latter could even be distributed all over the world via SETI@home and similar devices). See, for example, Bowyer et al. (1983), Tarter (2001), and Harris (2002). Similarly, archival SETI projects are searches for signals of possible artificial origin in the existing astronomical databases (e.g. Borra 2012). The whole approach could be regarded as being analogous to the famous cynical reply of Michelangelo to a silly question of an outsider about 'how exactly' he managed to create magnificent sculptures such as *David*, *Madonna and the Child*, *Brutus*, and the rest. *It's simple*, allegedly stated the master, *you just take a slab of marble and remove all the unnecessary parts.*
13. Interview on the amazon.com website at http://www.amazon.com/The-Eerie-Silence-Renewing-Intelligence/dp/0547133243 (last accessed 1 December 2017).

## CHAPTER 1

1. E. Jones (1985b).
2. In the further course of this book, I shall use proper astronomical units for distances, namely parsecs (pc) and their decimal multiples (kiloparsec (Kpc), megaparsec (Mpc)); a detailed discussion of the relevant spatial and temporal scales will be given in Chapter 2. Here, popular units are used not only because they give a more straightforward insight into the substance of the argument but also because, in Fermi's time, the use of parsecs was not as entrenched in astronomy and related sciences. For conversion purposes, 1 parsec = 3.26 light years; thus, the average distance between stars at the location of the Solar System is ~1 pc.
3. We shall discuss some of these arguments later; for the moment, see, for example, Forward (1986), Vulpetti (1999), and Long (2012).
4. Lytkin, Finney, and Alepko (1995); Young (2012).
5. For reviews at various levels of complexity and technical detail, see Brin (1983), Webb (2002, 2015), Verma (2007), and Ćirković (2009).
6. Hart (1975); Viewing (1975).
7. Gray (2015) is a fresh recent example. Various writings of Frank Tipler, notably Tipler (1981a, b, c). Barrow and Tipler (1986) exhibit the same tendency to oversimplify and misinterpret the historical record.
8. Shklovsky (1962).
9. Kardashev (1964). For an example of one of Shklovsky's papers from this time, see Shklovsky (1973).
10. Lem [1964] (2013), esp. Chapters 3 and 7.
11. Mitton (2005), pp. 171–2.
12. Hoyle [1964] (2005), p. 47. Hoyle rarely used written notes in his lectures and relied on his magnificent memory most of the time, so such mistakes were unavoidable

occasionally; it might also be interpreted as the confirmation of the half-mythical status the original lunch had obtained quite early.

13. Here belong those reactions to the paradox which proclaim it as a *Mu* question (in the Buddhist tradition, the one that should be *unasked*, for being unanswerable within its own conceptual framework; cf. Hofstadter 2000). Some of the hypotheses considered in Chapter 4 are certainly close to this category, while perhaps narrowly avoiding being full-fledged calls for unasking of the problem.
14. e.g. Pigliucci (2010), pp. 39–40, and Pigliucci, private communication to the author on 11 April 2011. Gray (2015) promotes the same literal reading, with an additional twist in form of the assertion that Fermi had in mind unfeasibility of interstellar travel only.
15. Ćirković (2012), pp. 178–81; p. 233.
16. For the overview of classic pluralism, see Crowe (1986) and Dick (1996, 1997).
17. Osterbrock (2001).
18. Mason (2008). For the Van de Kamp affair, see Croswell (1988).
19. e.g. Sandels (1986); Haines (1999).
20. e.g. Taylor (1976); Herwig (1999).
21. Without putting too big emphasis on particulars, one should not entirely discount Fermi's own personal experiences and his passive acquiescence with Italian fascism for 16 years, all the way up to promulgation of the *Manifesto of Race* and corresponding discriminatory laws by Mussolini in 1938, which directly threatened his family. See the excellent study of Di Scala (2005). As many other scientists on the Manhattan Project and in the subsequent era of hydrogen bomb research, he did occasionally show appalling insensitivity to the problem of the abuse of science and technology for military and ideological purposes. However, near the end of his unfortunately shortened life, the great man expressed serious concerns about humankind's capability of living with nuclear weapons in the long term (Cronin 2004). Since he shared such a pessimistic outlook with some of the key pioneers of SETI, like Shklovsky or von Hoerner, I shall return to this important topic when discussing the self-destruction family of hypotheses for resolving Fermi's paradox, in Chapter 6.
22. e.g. Partington (2003).
23. This is, one ought to emphasize, immensely more disturbing and ethically suspicious than anything one could have expected in Fermi's time when racism, colonialism, and various 'destinies' still hang heavily over people's hearts and minds. For further reasons to reject modern revamps of those evils in the cosmic context, see Cockell (2008) and Ćirković (2008).
24. While a detailed discussion is postponed to Chapter 6, it is worth noting that, while we resist the temptation of a 'Whiggish' history of ideas, clear continuities and causal relations are worth establishing in order to understand the context in which contemporary concepts emerged. Therefore, while 'transcendence' for early cosmists was certainly a different concept from the one in works of modern-day future

thinkers like, for example, Kurzweil (1999, 2005), Chalmers (2010), or Smart (2012), their worldviews are not entirely dissimilar either.

25. In spite of this strict qualification, I do not doubt that, sooner or later, some weird call for the 'postcolonial' SETI studies will be issued. Some indications to that effect already exist, for instance, in Massimo Pigliucci's rhetorical misconstrual of Fermi's problem as 'expecting Spanish conquistadores' from space (Pigliucci 2010, p. 39). Plato (or the pseudonymous author of the *Seventh Letter*) and, in a rudimentary form, Empedocles, Xenophanes, and Parmenides before him already knew that people were disturbed by the arbitrary conventions of language assigning words to things, and not by things themselves, but who nowadays has the time and peace of mind to read all those 'pompous Greeks'? And, lo and behold, for such bizarre criticism see, for example, Dittmer (2007), Williams (2010), and Slobodian (2015). For the general debunking of many misguided postmodernist notions about science, see Koertge (1998).

26. Hart (1975); Tipler (1980, 1981b, c). See, however, J. T. Wright (2018) for an up-to-date critical discussion of this assumption.

27. Boyajian et al. (2016). Some additional observations are reported by Lisse, Sitko, and Marengo (2015) and Marengo, Hulsebus, and Willis (2015); compare with J. T. Wright and Sigurdsson (2016); J. T. Wright et al. (2016).

28. e.g. by Lem [1964] (2013). Apart from the all-important review of Brin (1983), see also discussions in Kardashev and Strelnitskij (1988), Lipunov (1997), Dick (2003), Ćirković (2009), and Armstrong and Sandberg (2013).

29. Kent (2011). An attempt to delineate the meaning of 'local' is given by Cartin (2014).

30. Webb (2015). For example, the idea that extraterrestrial sophonts have developed different mathematics from us (Webb's #33; see also Ulvestad 2002; Kukla 2008, 2010) might be relevant from the point of view of *communication* (or its absence) with them. But their artefacts, like space probes, Dyson spheres, or other manifestations, are part of the same *physical* world we observe through our telescopes—at least as long as we stick to the scientific realism about those observations. Therefore, the absence of such artefacts or traces still count as an unresolved problem—which would be only honest and fair to call Fermi's paradox (in a stronger form) as well! For a similar recent example, see Lampton (2013).

31. Kardashev (1964). In brief, we are dealing with the following basic types: Type 1 – a civilization manipulating energy resources of its home planet; Type 2 – a civilization manipulating energy resources of its home star/planetary system; Type 3 – a civilization manipulating energy resources of its home galaxy. The equivalent formula suggested by Sagan would be $n = 1{+}0.1 \log(E/10^{16}\,\mathrm{W})$, where $n$ is the Kardashev type, and $E$ is the total power used by civilization. It agrees within an order of magnitude with Kardashev's initial estimates, and Type 2.0 corresponds to a civilization managing total energy emitted by the Sun (1 $L_\odot$) and it allows for the fractional Kardashev's types. For a review of various approaches to extension and modification of the scale, see Ćirković (2015).

32. By local here, I suppose the standard physical meaning: if any amount of energy, when transported from A to B has to pass through all intermediate spatial points, such transport is local. This property is the opposite of the one manifested in teleportation.

33. For a review of this classic understanding of habitability, see Lineweaver and Chopra (2012). Nonetheless, one should not be dismissive of 'wild' speculations such as the intelligent Black Clouds of Sir Fred Hoyle (1957) or strong-force-based life inhabiting neutron stars (Forward 1980), which, after all, very justifiably test the epistemological limits of our astrobiological understanding (cf. Benner 2009). We simply know too little of those possibilities to be able to reach any serious conclusions; see also Cleland and Chyba (2002); Tirard, Morange, and Lazcano (2010); Hengeveld (2011); and Cleland (2012). So here and in the rest of this book, unless stated otherwise, I shall be dealing only with life based on chemistry, not necessarily identical to the terrestrial one, but sufficiently similar in the sense that it requires planetary systems as its basic habitats. Some further considerations in this regard will be given in Chapter 5, in connection with 'rare Earth' hypotheses.

34. Hart (1975).

35. I use this as an expression for direct contact, without presuming that the concept of personhood, whether in the biological, psychological, or philosophical sense, has any universal validity. However, a precise specification of our statements, taking Copernicanism and other philosophical premises into account, leads easily to new insights into the substance of the problem (this is, after all, what makes Fermi's paradox interesting from a philosophical standpoint). In the particular case, abstracting away the notion of personhood may lead us either into reflecting upon the universality of the Darwinian evolution (pp. 159–64), or into considering various 'transcendence' scenarios (see Chapter 6, esp. Section 6.7).

36. Allègre, Manhès, and Göpel (1995); Bouvier and Wadhwa (2010).

37. In view of this circumstance, it is occasionally suggested that we also need a Search for Extra-Terrestrial Artefacts (SETA) programme as well (Freitas and Valdes 1980; Arkhipov 1996, 1997; Kecskes 1998). Although we shall discuss this possibility below, it is worth noticing that this is a special case of a more generally understood unorthodox SETI programmes advocated under the title of the Dysonian SETI (Bradbury, Ćirković, and Dvorsky 2011; Zackrisson et al. 2015).

38. Although these will be discussed in more detail in the following chapters, I shall give some prototypical references here, just for the flavour. For radio messaging, see Drake (2011), as well as the vast majority of the SETI literature; for local artefacts, see Freitas and Valdes (1980) and Forgan and Elvis (2011); for astroengineering feats, see Dyson (1960a, b); Arkhipov (1996, 1997); Weiler (1998); Annis (1999b); Arnold (2005); Beech (1990, 2007); J. T. Wright, Griffith, et al. (2014); J. T. Wright, Mullan, et al. (2014); J. T. Wright et al. (2016); and Griffith et al. (2015).

39. Lem (1984, [1964] 2013).

40. Brin (1983).

41. In atomic, nuclear, or particle physics, the mutual cross section for two particles is the area which is transverse to their relative motion and within which they must meet in order to interact with each other. It clearly depends on the properties of the interaction involved, as well as parameters such as particle energies. The probability for any given reaction to occur is in proportion to its cross section.

42. Here I assume several things taken for granted in any serious astronomical project, which still require a bit of elaboration for non-astronomical audiences (cf. Carlotto 2007). Our hypothetical alien observer should be observing in various wavelength bands, notably those corresponding to optical and near-infrared parts of the electromagnetic spectrum. This can be justified by appealing to the types of stars likely to harbour life as we know it (in biochemical sense): since those stars emit mostly optical and infrared radiation (with an important, but relatively small addition of part of the ultraviolet domain), creatures evolving in their vicinity would have gained fitness increase from being able to perceive exactly those parts of the spectrum. Although human astronomers rather quickly—in historical terms of hundreds of years since Galileo started telescopic astronomy—achieved some modicum of coverage of other parts of the spectrum, notably radio, microwave, X-ray, and so on, the vast majority of all our modern astronomical data still belongs to visible and infrared parts of the spectrum; and this situation is likely to persist, at least in the decades, if not the centuries, to come. This is not to necessarily say that an alien astronomer observing in another band would not come to the conclusion that Earth is inhabited by intelligent beings; after all, the essential point of most of *our* SETI projects is the idea that the chances of remote detection are, indeed, larger in the radio part of the spectrum than anywhere else! However, an even weaker conclusion is enough for our present purposes. In addition to this condition based on *spectral coverage*, another has to do with *angular resolution* of observations; it is this latter requirement which mandates using large space-based telescopes, preferably in the interferometry mode. Finally, the *temporal resolution* of observations must be such that the night glow of Earth's cities, traffic routes, oil platforms, and so on may be detectable within the natural time limit set by Earth's rotation (or the duration of the terrestrial night). The challenge these requirements pose for any astronomer is formidable, but the advances in both instrumentation and, even more, the quality of reduction and processing of astronomical data—which can compensate, in our own limited experience, for many deficiencies of the actual observing runs—have been formidable as well, especially for the last ~100 years. The fact that this timescale of astronomical improvement is quite minuscule in comparison to the Fermi–Hart timescale (which is, of course, minuscule when compared to the age of the oldest habitable planets in the Galaxy) is another way of stating the superiority of **StrongFP** over its rivals.
43. Coulter et al. (1994); Dick (1996); Garber (1999); Waldrop (2011).
44. M. Pigliucci, private communication to the author on 11 April 2011; Gray (2015).
45. I shall deal with problems specific to interstellar travel—and hypotheses resolving Fermi's paradox on that basis—in Chapter 7. That much greater weight given to detectability in this narrative should not be construed to imply that interstellar travel is easy or uncontroversial. But the important thing which often escapes students of SETI is that, *in the specific context of the paradox*, any relevant outcome of interstellar travel can be simply subsumed under detectability without any loss of generality.
46. Contingency of history does not mean that there were not good *local* reasons for doing A instead of B; what is important to understand is that those reasons are

themselves contingent on a large number of factors, such as the availability of resources, human capital, personal preferences, the vagaries of funding, and so on, which are usually just opportunistic happenstance.

47. Lem [1968] (1999). For excellent secondary sources on the novel, see Swirski (2000, 2006), Csicsery-Ronay (2013), and Glaz (2014).
48. e.g. Horvat, Nakić, and Otočan (2012).
49. At least for the moment and for the foreseeable future at timescales smaller than $t_{FH}$.
50. Webb (2015).
51. An incomplete list is given by Rauch (1998).
52. By astrophysicist Malvin Ruderman: 'For theorists who may wish to enter this broad and growing field, I should point out that there are a considerable number of combinations, for example, comets of antimatter falling onto white holes, not yet claimed' (Nemiroff 1994). The resolution of the GRB origin problem in the mid-1990s has been one of the great triumphs of extragalactic astronomy—and it has important astrobiological ramifications as well, which I shall discuss in Chapter 6.
53. For instance, it seems that there is a clear delineation between causative processes causing longer, >30 s, and shorter, <10 s, GRBs; the same seems to be valid for high-red-shift, as opposed to low-red-shift, Lyman-alpha lines in QSO spectra, the dividing epoch being something about $z \cong 1.3$. For GRBs, see Woosley and Bloom (2006) and Gehrels, Ramirez-Ruiz, and Fox (2009). For the origin of Lyman-alpha forest lines, see, for example, Rauch (1998), Ćirković and Lanzetta (2000), Yoon et al. (2012), and Rahmati et al. (2013).
54. Although philosophers have established fine lines of distinction between these (and other) conceivable realist positions (Miller 2016), I will use them as denoting the same thing in the rest of the text.
55. Besides the vast realm of the *X-Files*-inspired pop culture, see, for instance, Cooper (1991) for a 'believer' rant, and Barkun (2003) for a serious social-science analysis of the alien conspiracy movements.
56. For some points of entry, see Popper (1972), Nozick (1981), Lyons (2005), Maudlin (2007), and Turner (2007).
57. Kragh (1996, 2004, 2007).
58. I have tried to make the case for this analogy in Chapter 2 of Ćirković (2012).
59. See Eldredge and Gould (1972), Raup (1991), Huggett (1997), Palmer (2003), and Ćirković et al. (2008).
60. e.g. Zuckerman and Hart (1995).
61. #75 in the expanded edition, Webb (2015).
62. At least for the information emitted thus far; we could, though, more for reasons of technological efficiency and ecological conscience than interstellar paranoia, evolve into a fainter source.
63. Webb (2002), pp. 85–8.

64. See pp.119–20.

65. Bracewell (1975).

66. Note that this is quite different from the mundane view expressed by Stephen Jay Gould in the following terms: 'I have enough trouble predicting the plans and reactions of people closest to me. I am usually baffled by the thoughts and accomplishments of humans in different cultures. I'll be damned if I can state with certainty what some extraterrestrial source of intelligence might do' (Gould 1987, p. 405). The difference is exactly related to the desideratum of non-exclusivity sketched above, which will be discussed further in Chapter 3. We are *certainly* (pun intended!) not dealing with certainty: we deal with chances or likelihoods, in an attempt to make the explanation as non-exclusive as possible.

67. Egan (2008a, b). The citation is in Egan (2008b), p. 53.

68. Since absolute hermits make a poor story, *some* meagre interaction is described in the novel and the story. Due to—doubtless intentional—Copernican irony on Egan's behalf, it turns out that the hermits share DNA biochemistry with humans, in contrast to about a dozen other biochemical foundations for cognition evolving independently in the Galaxy.

69. For singletons, see Bostrom (2006).

70. Lineweaver (2001); see also Lineweaver, Fenner, and Gibson (2004). The improved calculations have been published by Behroozi and Peeples (2015), as well as Zackrisson et al. (2016); these only aggravate the problem, as will be elaborated in Chapter 2.

71. Cf. Williams, Kasting, and Wade (1997).

72. Noble, Musielak, and Cuntz (2002); Asghari et al. (2004); Kopparapu, Raymond, and Barnes (2009); Jaime, Aguilar, and Pichardo (2014).

73. Gillon et al. (2017).

74. Gonzalez, Brownlee, and Ward (2001); more details in Chapter 2.

75. e.g. Mojzsis et al. (1996); Bell et al. (2015); Dodd et al. (2017).

76. Lineweaver and Davis (2002).

77. Cavicchioli (2002).

78. Lahav, Nir, and Elitzur (2001); Ehrenfreund et al. (2004); Bada (2004).

79. e.g. Chernavskii (2000).

80. See, for instance, Schaller (1997); Bostrom (2003a, 2009, 2014).

81. Crawford (2009); Long (2012).

82. Rose and Wright (2004).

83. Badescu (1995); Badescu and Cathcart (2000); Badescu, Cathcart, and Schuiling (2006); Korycansky, Laughlin, and Adams (2001); McInnes (2002).

84. Sandberg (2000); Armstrong and Sandberg (2013).

85. Annis (1999b); J. T. Wright et al. (2014a, b); Zackrisson et al. (2015).

86. Leslie (1996); Bostrom and Ćirković (2008).

87. Taleb (2005), p. 39.

88. Ćirković and Bradbury (2006); Davies (2010); Bradbury et al. (2011).

89. Tipler (1980, 1981b); c.f. Valdes and Freitas (1980); Barlow (2013); Starling and Forgan (2014); Kowald (2015).

90. Ćirković and Bradbury (2006). I owe the concept, as well as so much else, to the ingenuity and thoughtfulness of the late Robert J. Bradbury (1957–2011).

91. Kardashev (1964). Kardashev's classification has proved to be a surprisingly useful tool for thinking about advanced Galactic societies and will be often used in the further course of this book (see also my detailed review in Ćirković 2015).

92. Roberts (2015), p. 2.

93. Lamb (2001), p. 4.

94. Kragh (1996).

95. Ćirković (2012).

96. Large distance—comparable with the distances between spiral arms of the Milky Way, for example—is desirable in this simplified scenario in order to avoid contamination with additional astrophysical information which could be more easier to obtain for our stellar neighbours, to minimize the possibility of the discovered civilization having the same ultimate origin as humans via some form of panspermia, as well as not to contradict negative outcome of the SETI projects actually undertaken in human history thus far. We still suppose that the signal originates within the Galactic Habitable Zone of the Milky Way (see Chapter 2). It has been elaborated in the voluminous SETI literature how one could hope to recognize the artificial origin of a signal without presuming our capacity to decode it into a specific *message* (e.g. Shklovsky and Sagan 1966; Michaud 2007). Moreover, there could in principle be no message at all—the signal might be a navigational or timing beacon of some sort, or just an unintentional waste product of extraterrestrial technology. Note that without actually (i) confirming that the signal is a message and (ii) deciphering it, all our conclusions must have the form of *inference to the best explanation*.

97. Failure to grasp this not-so-complex point leads to many misrepresentations of Fermi's paradox. In a recent example, Pigliucci (2010) attempts to dilute the acuteness of the paradox by way of a thought experiment with a single extraterrestrial civilization which exists elsewhere in the Milky Way but which is still unaware of humanity's existence on Earth. Such a civilization will have its own cultural counterpart to Fermi's paradox, a situation which the author claims undermines the strength of the paradox. However, this interesting attempt of undermining fails, because what we naively expect on the basis of general assumptions underlying Fermi's paradox (listed in Section 1.4. and discussed at length in Chapter 3) is not a single or a few extraterrestrial enclaves of intelligence coexisting with us now, but rather a *statistically significant ensemble* of such sites, mostly giga-anni older than

us, filling all or most of the Galaxy with its presence and artefacts. We do not need to presume, of course, a multiplicity of intelligent species—a single Kardashev Type 3 civilization would suffice. For further discussion of the statistical nature of the paradox, see Maccone (2010).

98. Unfortunately, outbursts of this kind of pseudo-theological thinking in the context of SETI have appeared in the literature (e.g. Drake 1976)—and have actually damaged the reputation and image of the whole enterprise (SETI sceptics actually *revelled* at such incidents, e.g. Tipler 1980, 1981b; Basalla 2006). See also the discussion of various types of SETI scepticism in Ćirković (2013).

99. This does not mean that there are no exceptions to this general rule. Notably, some of the solipsist hypotheses of Chapter 4 may not be applicable; for instance, if the universe is a simulation or a 'planetarium' created *for us*, then the overset of our peer civilizations does not exist in any epoch and our future is obviously decoupled from the Great Silence. Consolation remains small, considering other aspects of such a hypothesis. Another way of looking at it is that such simulated universes are not at all Copernican, so the key step of invoking our typicality must fail. On the other hand, UFO believers should not worry about Fermi-based gloom over the future of humanity; if *their* hypothesis turns out to be correct, there is no probabilistic reason why humans could not one nice future day use not-so-stealthy aircraft to conspire with governments of some younger and bewildered species on a distant planet! This would be just another example of the mirroring effect.

100. For points of entry into truly huge literature on this subject, see Matheny (2007), Bostrom (2003b, 2009), Epstein and Zhao (2009), and Baum (2010).

101. Fuller [1969] (2008), p. 60.

102. Obviously, the relationship between Fermi's paradox and future studies is mutual: if we have a particular reason to believe some **Y** about the future of humanity to be true, then, under Copernicanism, the same **Y** is likely to be applicable—probabilistically—to other intelligent species in the universe as well. This is the reason why some technological optimists vis-à-vis our future are sceptical towards SETI (e.g. Tipler 2003; Kurzweil 2005; Bostrom 2009); but see also Halley (2012), Ćirković (2014), Voros (2014), Frank and Sullivan (2014), and Miletić (2015).

## CHAPTER 2

1. There is another sense in which Shakespeare's meaning is very appropriate: *our* particular future, as humans, is almost inextricably linked to whatever is the likeliest explanation of the paradox, as explained at the end of Chapter 1. Some further discussion of this point will be given in Chapter 9.

2. In particular, Chapters 2 through 4.

3. For our present purpose, the relevant field is *classical* cosmology, that is, discipline dealing with the origin, properties, and evolution of our observable cosmological

domain (the 'universe'). In other words, I shall avoid discussing issues such as *quantum* cosmology, cosmological inflation, the existence of multiple cosmological domains (the 'multiverse'), and so on. While the latter are relevant to astrobiology in the wider sense they are largely irrelevant to our attempts at explaining any version of Fermi's paradox. Minor exceptions to this are dealt with in notes to Chapter 4.

4. The three numbers are the Hubble constant ($H_0$ or $h$), the total cosmological mass density $\Omega_m$, and the cosmological constant (or dark energy, in the modern parlance) $\lambda$. For overviews, see, for example, both textbooks by Steven Weinberg (1972, 2008). The difference between Weinberg's books is an excellent proxy for the distinction between classical and modern cosmological discourse. Further modern presentations are Peacock (1999) and Liddle (2015).
5. North (1965, 1994); Kragh (1996, 2007); Peebles, Page, and Partridge (2009).
6. York et al. (2000); Strateva et al. (2001); Strauss et al. (2002).
7. e.g. Penrose (1989); Treumann (1993).
8. Ade et al. (2014). For earlier best-fit WMAP results, see Bennett et al. (2013).
9. Loeb (2014).
10. Stapledon (1937); Olson (2015); Lacki (2016).
11. e.g. Maddison, Kawata, and Gibson (2002); Bovy et al. (2014).
12. Gerhard (2002); Binney and Tremaine (2008).
13. The Holmberg radius is technically defined as the length of the semimajor axis of the isophote corresponding to a surface brightness of 26.5 photographic magnitudes per square arcsecond (e.g. Binney and Merrifield 1998).
14. See Binney and Tremaine (2008) and Weinberg (2008). There is perhaps a minuscule amount of massive neutrinos (comprising currently unpopular hot dark matter, or HDM) as well in the halo, but CDM comprises at least 90% and probably more than 99% of the total halo mass.
15. Exceptions might occur in the distant future; calculations by Adams and Laughlin (1997) show that most of the remnant stellar population will 'evaporate' from the Galaxy on timescales of about $10^{19}$ years. Charlie Stross's supremely intelligent novella 'Palimpsest' depicts our Solar System—conveniently astroengineered for long-term survival—being intentionally hurled out of the Milky Way for a trillion-year-long intergalactic voyage through ever-emptier universe (Stross 2009). Interesting events can still occur on the outskirts of spiral discs; in particular, see Section 7.4.
16. There is some amount of controversy surrounding the question of the astrobiological properties of spectral type M stars (red dwarfs), which are the most numerous sort of stars altogether (comprising about 80% of the stellar population by number if the 50-parsec sphere centred on our Sun is anywhere near typical). I shall return to this issue in Section 7.2.

17. Lineweaver and Grether (2003); Cassan et al. (2012).
18. Stevenson (1999); Badescu (2011); Strigari et al. (2012); Wagner et al. (2016).
19. See, for example, Liu and Chaboyer (2000) and del Peloso, da Silva, and Arany-Prado (2005).
20. The literature on this subject has tremendously grown in recent years. See, for instance, Leitch and Vasisht (1998); Shaviv (2002a, b); Gies and Helsel (2005); Pavlov et al. (2005); Overholt, Melott, and Pohl (2009); and Beech (2011b).
21. Bouvier and Wadhwa (2010).
22. Binney and Tremaine (2008), p. 167.
23. Some of the newer contributions to this long-standing debate are Bailer-Jones (2009), Feng and Bailer-Jones (2013), and Filipović et al. (2013).
24. Bond et al. (2013).
25. Adams and Laughlin (1997).
26. Gonzalez, Brownlee, and Ward (2001). For elaborations, see Gonzalez (2005).
27. Marochnik (1983). For a discussion of the co-rotation radius, see Binney and Tremaine (2008) and Goncharov and Orlov (2003).
28. Marochnik (1983), p. 74.
29. Balázs (1986, 2000). His ideas have circulated for some time, since Marochnik (1983) contains a reference to an unpublished manuscript (dated 1982) by Balázs.
30. Lineweaver (2001).
31. Tadross (2003); Bovy et al. (2014).
32. Obviously, here we neglect exotic habitats, like Europan subglacial oceans or Titanian liquid-methane lakes. See also Johnson and Li (2012).
33. Lineweaver et al. (2004). For excess metallicity, see Peña-Cabrera and Durand-Manterola (2004).
34. Adams et al. (2004). See also Buccino, Lemarchand, and Mauas (2006).
35. Behroozi and Peeples (2015); Zackrisson et al. (2016); Vukotić et al. (2016); Forgan et al. (2017).
36. There is a difference between definitions of a 'terrestrial planet' in Zackrisson et al. (2016) and an 'Earth-like planet' in Behroozi and Peeples (2015), but the corrected median timescales in these two studies differ by less than 0.5 Ga, which is quite satisfactory at this stage.
37. In the rest of this book, I shall use the original Lineweaver timescale as the most conservative, while keeping in mind that the discrepancy with the naive view of Fermi's problem only increases with modern studies.
38. I shall return to this issue—which is quite important for both SETI and future studies—in Chapter 7.
39. Ćirković and Bradbury (2006).

40. A quote from Stanislaw Lem is highly illuminating here: 'Instrumental technologies are required only by a civilization still in embryonic stage, like Earth's. A billion-year old civilization employs none. Its tools are what we call the Laws of Nature' (Lem [1971] 1999, p. 208). I shall return to this key idea in more detail in Section 4.7.

41. Horneck and Rettberg (2007); Barrow et al. (2008); Shostak (2009).

42. e.g. Prantzos (2008).

43. About the discovery of some of those hypervelocity stars, which can leave not only their region of formation but even the whole of the Milky Way, see Brown et al. (2005, 2007).

44. Laughlin and Adams (2000).

45. e.g. Wallis and Wickramasinghe (2004); Napier (2004, 2007).

46. Sagan and Salpeter (1976); Benner (2009, 2010). For a somewhat wider concept of *conceivability* (of evolved biospheres), see Janković and Ćirković (2016).

47. Gonzalez (2005).

48. Suthar and McKay (2012); Spitoni, Matteucci, and Sozzetti (2014).

49. Lineweaver (1998); Gudmundsson and Björnsson (2002); Olum (2004); Olson (2015).

50. See, for example, Mash (1993), who gives a critical account of the application of Lucretian argumentation to the SETI case.

51. Sakharov (1975). As he was in the internal exile at the time and was not allowed to receive the prize in person, the lecture was read by his wife, Elena Bonner Sakharova.

52. For a classical treatment of cosmological horizons, see Weinberg (1972) and Hawking and Ellis (1973).

53. Almár (1992).

54. The best classical discussion is Sir Hermann Bondi's famous and idiosyncratic *Cosmology* (Bondi 1961).

55. Wesson, Valle, and Stabell (1987); Wesson (1991).

56. Of course, this even more forcefully applies to hypothetical higher Kardashev's types (e.g. Galantai 2004).

57. This would be in accordance with the essential parallelism of these two fields, a theme I have elaborated upon in Ćirković (2012), especially pp. 49–55.

58. A rather short list of the relevant references consists of Shostak, Ekers, and Vaile (1996); J. T. Wright et al. (2014a, b); Griffith et al. (2015); Zackrisson et al. (2015); and Gray and Mooley (2017).

59. This has been suggested as one of the major advantages of the Dysonian SETI by Bradbury et al. (2011). A fictional example of such ultra-astroengineering was mentioned in Carl Sagan's *Contact*, referring to the puzzling extragalactic radio source Cygnus A (Sagan 1985). A good introduction and justification has been offered by J. T. Wright, Mullan, et al. (2014).

60. The famous analytic approximation to the galaxy luminosity function is given by Schechter (1976). Numerical parameters of the Schechter function are given, for example, by the survey of Loveday et al. (1992); these do not differ much from the values obtained in other galaxy surveys at low red shift (Blanton et al. 2003).
61. Wesson (1990). For his work on Olbers's paradox, see Note 55 in this chapter.
62. cf. Gray and Mooley (2017).

## CHAPTER 3

1. Molière [1670] (2008), Act II, Scene 4.
2. In Molière's time, the heyday of the Scientific Revolution, philosophy still encompassed what would become 'special sciences'.
3. While it may seem paradoxical to claim such status for SETI, which has relatively recently celebrated a half-centennial, even people in the SETI community admit that this is a very young and immature field which is at the very beginning of its mission. In Ćirković (2012), pp. 148–83, I have discussed some of the reasons why the emergence of SETI in the late 1950s and early 1960s was in many ways premature and that only after the astrobiological revolution of mid-1990s were we in a position to put SETI correctly into its proper niche of dealing with the highest-complexity parts of the total astrobiological landscape.
4. The subject in the historical instance was Steven Weinberg, and it was beautifully elaborated in Wilczek's (1993) review of Weinberg's book *The Dreams of Final Theory*.
5. Some pioneering pointers in this directions are Puccetti (1969), Weston (1988), Mash (1993), and Lamb (1997, 2001).
6. For some of the troubles surrounding the definitional project in life sciences and cognitive sciences, see Cleland and Chyba (2002), Benner (2009, 2010), Tsokolov (2009), Tirard et al. (2010), Jagers op Akkerhuis (2010), Hengeveld (2011), Cleland (2012), and Trifonov (2011, 2012). Arguably, an even worse case of 'foundational trouble' is facing people working in more abstract areas such as string theory or quantum cosmology; while one could perform a psychological manoeuvre of pushing the issues of definition of life and intelligence outside of one's 'everyday activities', this is impossible to do with notions such as the universal wave function or such-and-such manifolds, which figure in everyday routines of any string theorist or quantum cosmologist. However, in spite of occasional criticism, those fields are largely successful on any reasonable scientific metric.
7. Hatcher (1982); Penrose (1989).
8. Butterfield [1949] (1997); Kragh (1996).
9. At least for us astronomers!
10. Kardashev (1985, 1986). See the quotation in Section 3.4.
11. The status of intelligence of marine mammals is still unclear (Browne 2004; Huggenberger 2008), while we still do not know with certainty whether the intelligent

Neanderthals truly comprised a separate species distinct from *Homo sapiens* (e.g. Hawks and Wolpoff 2001).

12. Jaynes (1976); Raup (1992); Nørretranders (1999); Koch (2004).
13. Lamb (2001), p. 96.
14. For interested readers, an excellent article in the *Stanford Encyclopedia of Philosophy*, written by David Papineau and available at http://plato.stanford.edu/entries/naturalism/ (last accessed 20 July 2017) provides a good entry point. See also Rudder Baker (2013).
15. 'I had no need of that hypothesis.' For historical details, see Gillispie (2000). In particular, it is important to understand that Laplace did not object to the existence of God per se, but instead rejected Newton's hypothesis that only occasional Divine intervention keeps orbits in the Solar System continuously stable; see also Stephen Hawking's comments at http://web.archive.org/web/20000708041816/http://www.hawking.org.uk/lectures/dice.html (last accessed 20 July 2017).
16. Kuhn (1957); Butterfield [1949] (1997).
17. See Crowe (1986), Gould (1987), Shermer (2002), and Kutschera (2012). Beside speculating on the life on Mars in a separate treatise (Wallace 1907), in his fascinating book *Man's Place in the Universe* (Wallace 1903), preceding even Tsiolkovsky's formulation of the proto-Fermi paradox for about three decades, he argued that naturalism cannot account for the fine-tuned structure of the universe. While his views would perhaps be closest to the 'rare Earth' ideas discussed in Chapter 5, there are important differences, discussed in Ćirković (2012), pp. 28–32.
18. The doctrine of *provisory methodological naturalism* (as exposed e.g. by Boudry, Blancke, and Braeckman 2010) represents the best framework for discussing both pretensions of conventional supernaturalists, like the proponents of intelligent design, and some borderline cases, like the simulation hypotheses to be discussed in Section 4.5.
19. Tegmark (1998, 2008).
20. For classic expositions of old religious pluralism, see Crowe (1986).
21. After all, its dubious virtue is that one can freely return to it if all else fails. At that stage, it might still have to compete, though, with the explanatory nihilism: the view that some deep questions simply have no answer and that phenomena to be explained are brute facts, neither requiring causal explanation nor capable of being explained in any non-trivial way. Fortunately, in the **StrongFP** case, we are very far away from such a state of desperation. See also the recent more general discussion by Fry (2012).
22. See, for example, Lyons (2003, 2005) and Rosen (2010).
23. Friedman (1974); Kitcher (1981).
24. See, for instance Des Marais and Walter (1999), Dick (2003), and Chyba and Hand (2005).

25. One should mention that, even if there is a viable alternative to Quinean realism, following prescriptions of something else than science (such an alternative has been hinted at, for instance, by Price 2007), such a metaphysical alternative will not invalidate the quest for an explanation of phenomena inferred assuming scientific realism. Therefore, such an alternative will not impact the review of hypotheses for explaining Fermi's puzzle in this book, except to the extent that some additional explanatory hypotheses could be formulated—as if we don't already have too many of them!
26. Merriam-Webster (http://www.merriam-webster.com/dictionary/mediocrity, last accessed 2 July 2017).
27. e.g. Carroll (2006); Hartle and Srednicki (2007); Page (2008); Gleiser (2010).
28. Here, under 'we', I consider *Homo sapiens*, as arguably the first terrestrial species possessing intelligence in our operational sense. There is no intention of denigrating either other peaks in animal intelligence (like giant squids or cetaceans) or our hominid ancestors and relatives.
29. Carter (1983, 2008, 2012); Wilson (1994); Livio (1999); Ćirković, Vukotić, and Dragićević (2008).
30. Scharf (2014), esp. pp. 36–8. For still more deflationary view, see Beisbart and Jung (2006).
31. With provisos regarding the status of early type and dwarf galaxies, as well as other physical constraints on habitability, as discussed in Chapter 2.
32. Shklovsky and Sagan (1966); Bracewell (1975).
33. Gustafsson (1998); Beer et al. (2004); Ćirković (2004a); Caldwell and Stebbins (2008); Robles et al. (2008); Krissansen-Totton et al. (2016).
34. Kardashev (1985), p. 497.
35. As well as stricter bounds obtained by Lineweaver (2001), Behroozi and Peeples (2015), and Zackrisson et al. (2016).
36. Taleb (2005, 2007); Almár and Tarter (2011).
37. e.g. Markley (2005); Raulin-Cerceau (2010); Raulin-Cerceau and Bilodeau (2012).
38. Obviously, this particular qualification cannot apply to a cosmologically distributed set of observers.
39. Bondi and Gold (1948); Hoyle (1948). For the best history of that crucial period of development of physical cosmology, see Kragh (1996); for a specific investigation into the role of gradualism—there called 'uniformitarianism'—see Balashov (1994).
40. Is it possible that a similar development might be taking place—as with most historical processes not clearly visible to the contemporaries—in astrobiology right now? Analogy is an elusive guide, but there are structural similarities between the historical pathways of the two disciplines, and the proponents of hypotheses we shall consider in Chapter 6 would probably answer in the affirmative!
41. Gould (2002); Shanahan (2004); Baker (1998).

42. e.g. Huggett (1997); Palmer (2003).
43. For the origin of the Moon, see Canup and Asphaug (2001) and Belbruno and Gott (2005).
44. For an interesting study of how even the vocabulary of geosciences has changed in the last quarter century or so, see Marriner, Morhange, and Skrimshire (2010). See also Baker (1998).
45. Raup (1999), pp. 35–6.
46. Glen (1994).
47. Gould (1989), pp. 129–31.
48. With exception made for relay races, where Bolt is also the world champion as a part of the 4 × 100 metres relay.
49. There is another analogy of relevance here: the one with the concept of *evolvability* in biological evolution (e.g. Pigliucci 2008). Increased evolvability improves the chances for survival of a lineage, although it might actually *worsen* the prospects of individuals and populations within the lineage. What will prevail on the balance can be very hard to see until we 'zoom out' and frame the issue a posteriori and in statistical terms (Janković and Ćirković 2016).
50. Brin (1983), p. 287. Please note that I am using the terms 'non-exclusivity' and 'non-exclusiveness' as synonyms (as suggested by e.g. *The Oxford Business Dictionary*), while being aware that in a *legal* context there is a slight difference in usage. Note that Brin himself uses the two phrases interchangeably as well (e.g. Brin 1983, p. 296).
51. Ibid. p. 296. 'ETIS' stands for 'extraterrestrial intelligent species'; $f_c$ is a factor in the Drake equation, representing the fraction of the intelligent species which develop detectable technologies.
52. See Ball (1973) or Deardorff (1986, 1987) for versions of Hypothesis B, to which I shall return in Chapter 4. I shall discuss self-destruction hypotheses in Chapter 6. In a sense, the principle of non-exclusiveness regulates what has been called 'cosmic irony', for example, in Lamb (2001, p. 56): 'It would be a matter of cosmic irony if civilizations turn out to be capable of destroying themselves at the very moment they achieve the ability to communicate with each other.'
53. For some points of entry into the discussion of contingency versus convergence, see Gould (1989), Ferguson (1999), Conway Morris (2003a), and Beatty (2006).
54. Barnes (2012), p. 531; emphasis in the original.
55. There are rough analogies to this situation in the history of science. For instance, functionalism with respect to mechanisms of biological evolution was rather an abstract philosophical assumption in the times of Lamarck and Darwin—who *both* were ardent functionalists. It was an excessively fruitful philosophical assumption, which by far most biologists in subsequent both Darwinist and Lamarckian traditions believed to be empirically true, but it could not be proved as such (in its proper domain, of course) until the advent of molecular biology. See, for example, Gould (2002).

56. e.g. Kasting (1988). Of course, there are corrections due to the greenhouse effect on other planets with atmospheres as well, notably on Earth, which owes at least part of its habitability to it.

57. See Chapters 6 and 7. The scenario is intentionally extreme.

58. Fry (1995), p. 389. See also Fry (2000, 2011, 2012).

59. This is the *extended continuity thesis* proposed in Ćirković (2012). Whether such an extension is legitimate remains an open question, too difficult to be tackled here. We mention in passing that at least one of the proposed solutions discussed in this book—the adaptationist hypothesis of Raup (1992) and Schroeder (2002)—explicitly denies this generalization.

60. Especially Haldane [1927] (2017).

61. Feynman, Leighton, and Sands (1964), vol. 2, pp. 41–2.

62. And pretty low-complexity biological thinking at that—hence those simplistic scenarios representing the expansion of intelligence through the universe as being analogous to the spread of a contagious pathogen through a population, for example, E. Jones (1981). Such an 'epidemiological' approach to Fermi's paradox represents a dangerous oversimplification; see Chapter 7.

63. Dick (2003), p. 69.

64. For wildly varying and often conflicting views, see, for example, Kurzweil (1999, 2005), Fukuyama (2002), Hughes (2004), Bostrom (2009), and Ferrando (2013).

65. Pigliucci (2010), p. 40, provides an example of this tendency.

66. e.g. Tremblay, Hartup, and Archer (2005). On the same trail is Robin Hanson's insistent claim that 'life will colonize' (Hanson 1998b).

67. This is a very tall order. It is a sobering thought that the seminal 1960 paper by Freeman Dyson on astroengineering (Dyson 1960a; see Chapter 8) was motivated in part by Malthusian concerns over the energy consumption and overpopulation problems, which he generalized to all technologically advanced civilizations anywhere. While the study itself is tremendously important—and very likely to become more relevant in the future, as we are getting a better hold on the possibilities and prospects of astroengineering—such concerns sound quaint or even naive today.

68. See also Parkinson (2005). Note that *this* apparent violation of Copernicanism is, indeed, (trivially) justified by our empirical knowledge on the prevailing conditions in the Solar System with respect to, say, temperature and pressure extremes, lack of atmospheric oxygen, and so on.

69. Tipler (1980).

70. See, for example, Gould (2003).

71. Drake (1962). I use here the form initially used by Brin (1983). The difference between various forms of the Drake equation would make an interesting case study in the history and sociology of science.

72. Casti (1989), pp. 343–68; Lamb (2001), p. 49.

73. Tipler (1980, 1981a, b, c, 2003); Mash (1993); Rescher (1985); Kukla (2001); Basalla (2006); Pigliucci (2010). What could justifiably be considered the real SETI theory has shyly began to emerge only very recently, with works such as Arnold (2005, 2013); Bounama, von Bloh, and Franck (2007); von Bloh, Bounama, and Franck (2007); Forgan and Rice (2010); Maccone (2011b); Vukotić and Ćirković (2007, 2008, 2012); Barlow (2013); Wandel (2015); Forgan et al. (2017); and Verendel and Häggström (2017). This is still very much fumbling in the dark, but it does show a genuine effort to connect with the relevant developments in astrophysics, astrobiology, evolutionary biology, complexity theory, artificial intelligence, and other relevant fields, as well as to overcome the simplistic approach of the previous decades.
74. An interesting analysis with a similar conclusion is given by Burchell (2006).
75. Walters, Hoover, and Kotra (1980). Other examples of specific criticisms of the Drake equation are given in Wilson (1984), Ćirković (2004b), and Ashworth (2014).
76. This does not mean, of course, that there are no problems with the Walters et al. (1980) study. It could be improved in several ways, notably by taking into account the improved detectability of colonizing efforts over interstellar distances. It is, however, a big step towards improved SETI theory, which *has remained virtually unnoticed*. For example, the NASA Astrophysics Data System (ADS) database records only fourteen citations of the Walters et al. paper in the whole 1980–2018 period, out of which four are by a single author (Claudio Maccone). The good thing is that the situation seems to be somewhat improving in recent years, since twelve out of these fourteen citations belong to the 2008–18 period.
77. Of course, Boltzmann thought at first that it would be possible to have locally temporally asymmetric theory via *Stoßzahlansatz* or the 'hypothesis of molecular chaos'; however, in the 1890s, he realized himself—in part as a consequence of objections raised by Culverwell, Zermelo, and others—that it was not possible, and so sought the solution in cosmological assumptions. See Boltzmann (1895), Steckline (1983), and Price (1996).
78. Ćirković (2003).
79. Of course, all this does not in any way demean the role of Frank Drake as one of the key 'founding fathers' of the whole SETI science and its *spiritus movens* for many decades. But his real claim to fame lies in conducting the Ozma project and other practical searches, and working tirelessly to promote the field, not in 'inventing' the mundane and simplistic equation.
80. Examples in this sense include Gleiser (2010); Glade, Ballet, and Bastien (2012); and Scharf and Cronin (2016).

## CHAPTER 4

1. Russell (1999); emphasis M. M. Ć.
2. Bioy Casares [1940] (1996); although unacknowledged, the link of Resnais's film to the novel has been repeatedly discussed by critics, for example, Beltzer (2000).

For a model of consciousness inspired by *The Invention of Morel*, see Perogamvros (2013); that a model of consciousness might be relevant for SETI, see Raup (1992). We shall return to this key point in Chapter 9.

3. e.g. Phillips (2012).
4. Consider the famous story of Leo Szilard stopping in the middle of busy London street upon realizing that '[Herbert George] Wells was right' and that a chain reaction of nuclear fission could be sustained in uranium atoms—with all the epoch-making consequences this insight eventually produced (Rhodes 1986, pp. 292–3).
5. Sagan and Newman (1983); Tipler (1980, 1981a, b).
6. e.g. ATLAS Collaboration (2013).
7. Recent research shows that the last pharaoh of the eighteenth dynasty died of malarial infection, probably following an accident; see Hawass et al. (2010).
8. Freeman and Lampton (1975); Campbell (2006); Carrigan (2012); Davies (2012); Vakoch (2014).
9. Conway Morris (2016).
10. Cf. Barrow (1999).
11. See, for example, Calude (2002), Gurzadyan (2005), and Chaitin (2006).
12. Although, of course, the threshold may change with time, and future cognitively enhanced *posthuman* observers could adequately explain artefacts of supercomplexity.
13. Some examples are the Ringworld (Niven 1970, and many sequels and prequels), the false Janus (Reynolds 2005a), or the black monolith (Clarke 1968; Kubrick [1968] 2006).
14. Lem [1968] (1999), pp. 22–7.
15. Hynek (1972). For a succinct historical review, see Chapter 6 of Dick (1996).
16. It is highly indicative that the number of serious publications arguing for such an explanation is quite small; see, for instance, Herbison-Evans (1977), Deardorff (1986), and Deardorff et al. (2005).
17. Friedman (2002).
18. Remember that the Fermi–Hart timescale given in Eq. (1.1) is based on the limitation of travel times across the size of the Galaxy. If we allow for faster-than-light travel, this timescale might be much shorter, making the paradox *more* pronounced. This is one of very rare points on our journey where immersive knowledge of science-fiction tropes and traditions can be a hindrance rather than a help.
19. Perhaps the best succinct *explanation of the underlying explanation* (no pun intended!) of the whole UFO phenomenon has been given by Carl Gustav Jung: 'The danger of catastrophe grows in proportion as the expanding populations impinge on one another. Congestion creates fear, which looks for help from extraterrestrial sources since it cannot be found on Earth. Hence there appear "signs in the heavens", superior beings in the kind of space ships devised by our technological fantasy.

From a fear whose cause is far from been fully understood and is therefore not conscious, there arise explanatory projections which purport to find a cause in all manner of secondary phenomena, however unsuitable. Some of these projections are so obvious that it seems almost superfluous to dig any deeper' (Jung 1959, pp. 12–13).

20. Sheaffer (1995), pp. 26–7.
21. Condon and Gillmor (1969); Klass (1974); Sagan [1973] (2000); Jacobson (2014).
22. Freitas and Valdes (1980, 1985); Steel (1995); Arkhipov (1996); Stride (2001); Haqq-Misra and Kopparapu (2012); Davies and Wagner (2013). The original justification for expecting artefacts in the form of exploratory space probes of advanced Galactic civilizations has been given by Bracewell (1970).
23. Much is often made of the impossibility of proving the negative, although this is mostly a red herring. After all, if the scientific world view rejected the existence of ghosts, witches, the 'philosopher's stone', and similar superstitions and widely held beliefs of centuries past, it did this in spite of the alleged impossibility of proving the negative. The relevant—loosely understood—parameter space for ghosts, for example, decreased with the advance of science until the concept became meaningless as an explanatory device; anything which had previously required the existence of ghosts for its explanation could be explained in the new paradigm without any reference to ghosts. Thus, while the current paradigm assumes no extraterrestrial intentional presence in the Solar System, it should be repeatedly questioned with better and better empirical tests. If there are really no artefacts of extraterrestrial intelligent origin in the Solar System, this empirical testing will clearly rule out the 'interesting' part of the parameter space of such artefacts sooner or later.
24. Freitas (1985b).
25. Note that this is a much wider concern than the one present in the science-fictional discourse about misuse of alien *weapons*: even completely peaceful artefacts, if originating in a sufficiently advanced technological environment, could be incredibly dangerous to undeveloped cultures. Nowhere is that, to the best of my knowledge, better explicated and discussed than in Lem [1968] (1999) and Strugatsky and Strugatsky [1977] (2007).
26. For example, the civilization of Rapa Nui (Easter Island) was recognized as such by European explorers on account of the quite conspicuous *moai* statues (Diamond 2005), and various Greek kingdoms in modern-day Afghanistan, Pakistan, and India were first reconstructed solely on the basis of their coinage (Tarn 1951). For epistemological problems surrounding archaeological research in unknown and complex contexts, see Kosso (2006); Lucas (2012).
27. Strugatsky and Strugatsky [1977] (2007).
28. For readers lucky enough not to remember those paragons of pseudoscience of 1960s and 1970s, Kolosimo was an Italian journalist and writer, an extreme communist, who argued that ancient texts like the Bible and even Homer's *Odyssey* retain witnesses' accounts of extraterrestrial spaceships and visitations. Von Däniken, Swiss hotelier and multiple convict, wrote a few dozen books arguing for extraterrestrial

visitations and the role of extraterrestrial technology in building many ancient monuments, from the pyramids of Egypt to the Nazca lines in Peru. A detailed criticism of these pseudoscientific claims can be found in Story (1976) and Colavito (2005).

29. Wallace (1903); Heffernan (1978); Kutschera (2012).
30. Crowe (1997); see also Dick and Lupisella (2010).
31. Wilkinson (2013) and many referenced therein; William Lane Craig, personal communication.
32. While this is an extremely important topic for any discussion of solipsist hypotheses, I cannot enter into it in any detail here; for a recent key result, see Fry (2012).
33. Available on the wonderful Darwin correspondence project (http://www.darwinproject.ac.uk/) as http://www.darwinproject.ac.uk/entry-2814 (last accessed 20 September 2017). Parenthetically, earlier in the same letter, Darwin amusingly mentions some of the criticisms *The Origin* had met recently, including one from 'Prof. Clarke of Cambridge', accusing the epochal book of 'consummate impudence'. Since many books in SETI studies are routinely labelled in similar terms (see particularly nasty examples in Basalla 2006), it is sobering to note that the manner of conservative criticism has changed little in a century and a half; if only there were more such impudent volumes!
34. Ball (1973); Fogg (1987).
35. Deardorff (1986, 1987).
36. On the other hand, one might ask whether such speculation is an example of *disconfirmation bias*, the tendency for people to extend critical scrutiny to information which contradicts their prior beliefs.
37. One example of such an extremely disruptive contact between morally very different cultures within the human/posthuman domain is given in Greg Egan's novel *Schild's Ladder* (Egan 2002).
38. Deardorff (1987).
39. A version of this idea has been the generator (pun/spoiler intended!) of the plot in Stephen King's novel *Under the Dome*, where a local, but strong, perturbation in our conventional physical reality is caused in the final analysis by private and low-level ('children') actors within an obviously extremely advanced technological society (King 2009). An appealing feature here is that actual nature of the intruding agents remains unknown—extraterrestrials? posthumans from the distant future? inhabitants of a higher-dimensional space?—as it should be effectively unknowable by observers belonging to the technologically inferior culture.
40. Bostrom (2006). For some relevant discussions of political aspects of future human/advanced extraterrestrial societies, see Crawford (1993), Ćirković (2008), and Cockell (2008).
41. Gould (1987), p. 405.
42. Vico [1725] (1968), par. 331.

43. It is hugely ironic that the lack of theoretical underpinnings of SETI is often decried by the very same authors who deplore the lack of more input from social/cognitive sciences in SETI debates—that is, the input from the fields which themselves lack broad theoretical foundations (Rescher 1985; Kukla 2001, 2010). See, however, prototypes of new and bold broaching of the topic in references such as Lemarchand and Lomberg (2009), de Sousa António and Schulze-Makuch (2011).

44. e.g. Webb (2002). However, even that is open to question, since in both hypotheses it is assumed that eventually, in the course of time, newcomer civilizations such as humanity could 'qualify' to become a full-fledged member of the 'Galactic Club' (Forgan 2011; Magalhães 2016). This could be construed as a form of falsification.

45. Baxter (2000). For a fictional description of this scenario, see Reynolds (2004).

46. Birch (1990). See also the discussion in Lamb (2001). For the astroengineering ideas, see, for example, Birch (1983, 1991).

47. As suggested by physicists since the early days of the cosmological inflationary paradigm, such as Sato et al. (1982); Farhi and Guth (1987); Linde (1988); Farhi, Guth, and Guven (1990); Linde (1990); and Crane (1994).

48. Bostrom (2003a).

49. *SimCity* is a placeholder for the long-standing series of computer games in which the player builds a complex structure (e.g. a city) from simple parts. For an interesting recent review on Cartesian doubts from a pen of a famous philosopher, see Smart (2004).

50. Crawford (2013).

51. Note that this is just a new twist on the old 'problem of evil'; this problem has been present in one form or another in all theistic accounts postulating a Creator (or Creators) of the world. It applies with equal force to the **Peer hypothesis**.

52. Of course, even higher moral standards are conceivable, where the obligation to actively help underdeveloped cosmic neighbours is regarded as an ethical norm—but whether such a state of affairs is realistic is highly uncertain. Human experience thus far suggests that entirely beneficial 'foreign' help is seldom, if ever, achieved in relationship between communities. Intense discussions on the topic of humanitarian interventions, led in circles of political science and international law studies, testify to the validity of this concern (e.g. Kinclová 2015).

53. See the interview with Neil de Grasse Tyson at http://www.theguardian.com/us-news/2015/sep/19/edward-snowden-aliens-encryption-neil-degrasse-tyson-podcast, last accessed 15 March 2017.

54. Hofstadter (2008). By an eerie coincidence, the lead essay is based on the talk Hofstadter gave in Oxford on 21 November 1963—*a day before* the assassination of President John F. Kennedy in Dallas, a historical event that gave rise to many of the most paranoid conspiracy theories ever devised.

55. Barkun (2003); Aaronovitch (2009).

56. e.g. Diamond (1999); Haqq-Misra et al. (2013); Korhonen (2013).

57. Lachmann, Newman, and Moore (2004).
58. As of April 2017, the NASA ADS lists only four citations of the Lachmann et al. (2004) paper, two of which pertain to SETI. A search via the more inclusive Google Scholar offers about twenty-eight citations, including multiple book editions, book reviews, and conference abstracts; again, slightly less than 50% of that count (twelve to fourteen) deals explicitly with SETI.
59. For a sampling of the voluminous literature, see Tough (1990); Finney (1990); Diamond (1999); Carrigan (2006); Michael (2011); Baum, Haqq-Misra, and Domagal-Goldman (2011); Musso (2012); Shostak (2013); Haqq-Misra et al. (2013); and Neal (2014).
60. Lem (1987), p. 135.
61. Crick and Orgel (1973).
62. e.g. Lovecraft [1936] (2005); Scott (2012).
63. This scene ('Strange Bedfellows') was deleted in the final version; it can be found on the 2012 DVD version.
64. Mojzsis et al. (1996); Lineweaver and Davis (2002); Bell et al. (2015).
65. Cleland and Copley (2005); Davies et al. (2009).
66. Note that the usual bioethical reasons for refraining from colonization did not apply (trivially) to an early, Hadean Earth—there was no native biosphere to be displaced!
67. Mautner (2004, 2005); Mautner and Matloff (1979).
68. Rummel and Billings (2004); Dunér et al. (2013).
69. We are witnessing the emergence of a new crossover field of astrobioethics from multiple viewpoints: see McKay and Marinova (2001), Cooper (2014), and Chon-Torres (2017).
70. Napier (2004, 2007); Wallis and Wickramasinghe (2004); Wesson (2011).
71. Nakamura (1986).
72. shCherbak and Makukov (2013).
73. Fry (2000).
74. Gusev and Schulze-Makuch (2004), p. 209.
75. Tlusty (2010).
76. For elaboration on this historical parallel, see Ćirković (2012), esp. Chapter 2.
77. The first notion of this kind was given by Scheffer (1994); for the modern-day hypothesis, see Gurzadyan (2005), Gurzadyan and Allahverdyan (2016), and Gurzadyan and Penrose (2016). Compare also Carrigan (2006). As a science-fiction motif, it had existed for quite some time prior; a masterful and dark example is given in Charles Stross's *Scratch Monkey*, which was originally written in 1993 (Stross [1993] 2011).
78. e.g. Calude (2002); Chaitin (2006).
79. Webb (2015), p. 122.

80. Lem [1971] (1999).
81. Sagan (1985).
82. For a fictional example on a smaller scale, see Reynolds (2008).
83. Hoyle (1983, 1994).
84. Davies (1978); Tipler (1982). For an extended discussion of this interesting early example of anthropic reasoning, see Ćirković (2004c).
85. Stapledon (1937); Baxter (2012).
86. Here we should also mention 'Zebra', from another science-fiction legend, Philip K. Dick. In his later years (1974–82), Dick experienced a series of mystical events and quasi-religious visions. In searching for a plausible explanation, he came to some rather wild speculations, in part inspired by Gnosticism, Zen, and the Neoplatonic theology of Johannes Scotus Eriugena. 'At one point, he attributed his experiences to Zebra, a hypothetical giant intelligence that remains invisible because it looks like the environment, as some insects do—but Zebra looks like the *whole* environment' (Wilson and Hill 1998, p. 154; emphasis in the original).
87. Lem (1986a), p. 288.
88. Lem [1971] (1999).
89. The name is, perhaps, an allusion to Paul Valéry's Mr Edmond Teste, a perfect intellectual (Valéry 1989). Monsieur Teste's motto is '*Que peut un homme*?', meaning 'Of what is man capable?' If we substitute 'an intelligent being' for 'man', we obtain the keystone question of 'The New Cosmogony'.
90. Lem [1971] (9), p. 208, emphasis in the original.
91. Ibid. p. 223.
92. Ibid. pp. 228–9. This cosmological optimism stands in sharp contrast to Lem's usual cognitive and moral pessimism as encountered in better-known works such as *His Master's Voice* (Lem [1968] 1999) and *Fiasco* (Lem 1987). Note, furthermore, how this is done through the mouth of the narrator's—an already older and distinguished scientist's—*teacher*, as if Lem wished to convey the impression that it is even a *conservative* thought to have in the context!
93. Compare, for instance, Carroll (2006). While the idea of fundamental 'constants' changing with cosmic time has been around since Dirac's 'large number hypothesis' in the 1930s, the real impetus to study such theories came from both observational and theoretical work in the 1990s; see Uzan (2003) and references therein. For widely distributed computing and cognition, see Hutchins (1995) and Clark and Chalmers (1998).
94. Olum (2004).
95. For criticisms, see Ho and Monton (2005) and Ćirković (2006).
96. Cf. Bostrom (2001, 2002).
97. Newman and Sagan (1981), p. 320.

98. Conway Morris (1998, 2003a, 2018); Plantinga (2011). It is, however, extremely doubtful that abandoning scientific naturalism in the context of SETI and Fermi's puzzle would make them happy—after all, both the **Simulation hypothesis** and the **New cosmogony** could be regarded as reductio ad absurdum of many traditional religious ontologies. Conway Morris seemingly preferred the 'rare Earth' hypothesis at the time of his 2003 book (Conway Morris 2003a), although his provocative 2018 paper is much more open to consideration of hypotheses in the solipsist category.
99. e.g. Hsu and Zee (2006); Beane, Davoudi, and Savage (2014).
100. Lebon (1985).
101. Wilcox (1956). See also my discussion of the Krell machine in the context of Kardashev's classification in Ćirković (2015).
102. If that is possible (i.e. if stronger forms of computationalism about minds hold true). Note that Lebon's proposal does not depend on that particular condition: artificial 'mirror brains' could be regarded as merely passive repositories, not having any life or cognition of their own.
103. Carruth (2013).

## CHAPTER 5

1. Fuentes (1976). I have used a modified translation from Prof. Michael Abeyta's ambitious monograph on *Terra Nostra* (Abeyta 2006).
2. Arendt [1963] (2007); Frayn (2006).
3. e.g. Midgley (1985); Fukuyama (2002).
4. Dijkstra (1989).
5. A highly indicative example is Gonzalez and Richards (2004).
6. Ward and Brownlee (2000).
7. For example, see the interviews in Impey (2010).
8. See, for instance, Chyba and Hand (2005); Sterelny (2005); Bounama, von Bloh, and Franck (2007); Prantzos (2008); Forgan and Rice (2010); Benner (2009); Vukotić (2010); Badescu (2011); and Vukotić et al. (2016).
9. Ward and Brownlee (2000), pp. 222–34.
10. For the general impact theory of Moon's origin, see, for instance, Canup and Asphaug (2001) and Halliday (2008).
11. Note that even this claim is spurious, since some impacts are arguably *less accidental* than others! In particular, a version of the impact hypothesis for the origin of Moon, put forward by Belbruno and Gott (2005), claims that the impactor, commonly called Theia, had formed in Earth's stable Lagrangian points, similar to Jupiter's Trojan asteroids, sometime before it drifted to the slow collision. If they are right, the impact was no random accident and the sequence of events leading to the formation of the Moon might not be as rare as it seems.

12. For planets without oceans, a conventional zero-altitude level could be defined, as was done for Mars.

13. This is what astronomers would dub 'a single complex biosphere per $L^*$ galaxy'. This is, obviously, an extremely small frequency, taking into account recent results which suggest that the number of planetary systems in the Galaxy (which is close enough to a typical $L^*$ galaxy) is on the order of $10^{10}$ or even $10^{11}$. See Lineweaver and Grether (2003), Cassan et al. (2012), and Petigura et al. (2013).

14. Basic physical and chemical preconditions have been fulfilled much earlier, according to Lineweaver's already discussed results.

15. And we do not observe them—in spite of recent strong surge in interest for looking; see J. T. Wright et al. (2014a, b), Griffith et al. (2015), and Zackrisson et al. (2015).

16. And, as indicated by interviews in Impey (2010), this is still a substantial majority.

17. Hanson (1998b); emphasis in the original.

18. In addition to Hanson, this applies to, for instance, Hart (1975), Bond (1982), Carter (1983, 1993, 2008), Martin and Bond (1983), and Ward (2005).

19. e.g. McMullin (1980); Kahane (2014); some interviewees in Impey (2010).

20. See, in particular, Fry (1995), Luisi (2006) and references therein.

21. Even smaller probabilities have been occasionally cited in the literature. Thus, Eigen (1992) cites the probability of random assembly of a polymer with a thousand nucleotides corresponding to a single gene as 1 part in $10^{602}$. This sort of 'superastronomical' number has led Hoyle and Wickramasinghe (1981, 1999) to invoke either an eternal universe—in contradiction with cosmology—or a creative agency. The (in)famous metaphor of the random assembly of a Boeing 747 out of components found in a junkyard, cited by Sir Fred Hoyle, nicely expresses this sort of desperation, which has, luckily enough, been overcome in the modern theories of abiogenesis. Both a review and a devastating critique of the creationist misuse of small probabilities is provided by Carrier (2004).

22. e.g. Yockey (1977); Hoyle (1983); Hoyle and Wickramasinghe (1999).

23. Pross (2012).

24. See, for example, Chaisson (1997, 2001, 2003), Pérez-Mercader (2002), and Bedau (2003).

25. Wesson (1990).

26. J. T. Wright et al. (2014a, b).

27. Chopra and Lineweaver (2016).

28. See also Doolittle (2014) and Janković and Ćirković (2016). In the latter reference, the term 'conceivability' is used to denote 'recipe-based' habitability.

29. Simpson (2017).

30. e.g. Volk (2002).

31. Heller and Armstrong (2014).

32. Cockell (2014).

33. Cf. figures 2 and 3 in Chopra and Lineweaver (2016).
34. Schwartzman and Middendorf (2011).
35. Cramer (1986). Cramer's 'The Alternate View' columns are quite interesting reading, although at time misleading and error prone (e.g. in the cited column, he claims that Luis Alvarez was awarded the 1968 Nobel Prize in physics for the discovery of the antiproton, while in fact it was Segrè and Chamberlain who discovered it and were awarded the prize in 1959). Most of the columns can be found at http://www.npl.washington.edu/av/ (last accessed 3 February 2018). See also Lem (1986b).
36. Ward and Brownlee (2000), especially pp. 171–5. One is left to speculate whether they were unacquainted with Cramer's ideas, or the source was deemed not serious and 'academic' enough.
37. Raup (1992); Schroeder (2002). Important elements for proper understanding of the underlying idea have been first put forward in the (in)famous criticism of early SETI projects by George Gaylord Simpson (1964). A detailed discussion of this particular solution is given in Ćirković (2005).
38. Schroeder (2002), p. 108; emphasis in the original.
39. Jaynes (1976), pp. 36–41. This brilliant but highly controversial book could be understood as a warning against naive views of cognitive phenomena—unfortunately, often present in SETI circles. Much less controversial post-behaviourism consensus among cognitive scientists leads to the same downgrading of consciousness; see, for example, Nørretranders (1999) and Koch (2004).
40. Schroeder (2002), p. 110; emphasis in the original.
41. Sober (1993), p. 22.
42. Dawkins (1989).
43. Wallace (1903). Ironically enough, Wallace also (in)famously insisted—to Darwin's dismay—that human mind shows so many non-adaptive features, like music appreciation, abstract mathematics, or spiritual communication, that it is a sure sign of an intervention by higher intelligence. Although precise tracing of adaptationism in astrobiology and SETI studies would be an interesting project in the history of science, untackled so far, it seems obvious that at least some of the early enthusiasm of the SETI founding fathers in the 1960s and 1970s was fuelled by their own overblown view of the adaptive value of intelligence. The criticism from Simpson (1964) and others was right on target in this respect (cf. Ćirković 2014).
44. Gould and Vrba (1982).
45. Schroeder (2002), pp. 178–9.
46. e.g. Bradbury and Vehrencamp (2011); Hutton et al. (2015).
47. Of course, once we accept the downgrading of the role of consciousness in the entire SETI business, we need to face the risk of inverse mistake as well: that truly conscious signals are mistaken for complex adaptive behaviours without any true 'self' behind them. This is the main topic of Schroeder's compatriot Peter Watts'

brilliant 'first contact' novel *Blindsight* (Watts 2006). The difficulty of detecting and recognizing locally adapted sophonts has been an occasional theme in science fiction for quite a long time; see Blish (1957).

48. Perhaps the fullest—and the most extreme—exposition of this pan-adaptationist creed can be found in Dennett (1995); for a strong criticism, see Ahouse (1998).
49. e.g. Russell (1983, 1995); Chela-Flores (2003); Rospars (2011).
50. For a detailed review of further problematic issues with this intriguing hypothesis, see Ćirković, Dragićević, and Berić-Bjedov (2005).
51. Lipunov (1997).
52. e.g. Darling (2001); Chyba and Hand (2005); Impey (2010); Forgan and Rice (2010); Ćirković (2012); Petigura et al. (2013).
53. Horner and Jones (2009), p. 75. Also see Horner and Jones (2008, 2012).
54. The opposite of the de re statement is called the de dicto statement (= one pertaining to the language used by humans to describe things). In the context under consideration, the de dicto alternative—the one dealing with words and their meaning only—would be that we are using the terms 'habitable' and 'Earth-like' interchangeably, as synonyms. Obviously, this is uninteresting, although the confusion among the two reigns, in particular in popular media descriptions of astrobiological work on habitability. On the de dicto reading, we could not talk about the habitability of Europa or Titan, obviously. So, adoption of the de dicto statement equating habitability and Earth-likeness would be a big step back for astrobiology.
55. Ćirković (2012), pp. 133–8.
56. Mackie (2006).
57. Franck et al. (2001, 2007); Forgan and Rice (2010).

## CHAPTER 6

1. Keys (1999); Diamond (2005). The already mentioned controversial theory of Julian Jaynes even ascribes the very *origin* of human consciousness and the modern-day mentality to large-scale stresses and convulsions accompanying catastrophic events (Jaynes 1976).
2. Lovecraft [1936] (2005). As was unfortunately frequent with Lovecraft, the novel was written in 1931, but only published—even that in the serialized pulp form in the pre-Campbellian *Astounding*—in 1936. 'Lovecraft boldly challenged that most entrenched dogma of art—that human beings should necessarily and exclusively be the centre of attention in every aesthetic creation—and his defiance of the 'humanocentric pose' is ineffably refreshing' (Joshi 2004, p. 652). Today, *At the Mountains of Madness* is one of the most popular works in the Lovecraftian canon, but its often uncanny relationship to astrobiology has not been acknowledged yet, in spite of the author being hailed as a 'literary Copernicus' (Leiber 2001).
3. Lovecraft [1936] (2005), p. 71.

4. See historical reviews in Raup (1991), Huggett (1997), and Palmer (2003).

5. For a finely written review of the Velikovskian controversy from a sceptical, although idiosyncratic point of view, see Bauer (1984).

6. Alvarez et al. (1980). In earlier literature (roughly before 2008), the boundary was denoted as the Cretaceous/Tertiary (or K/T).

7. The new paradigm is exemplified, for example, by Foote (2003), Benton (2003), and Erwin (2006).

8. Baker (1998), p. 180.

9. e.g. Raup and Valentine 1983; Maher and Stevenson (1988); Rampino and Self (1992); Ambrose (1998); Bostrom and Ćirković (2008).

10. Here, 'random' is used as synonymous with 'uncorrelated', without presuming anything about the underlying causal structure of such events. From the point of view of an advanced intelligence with great observing and computing power, such events might indeed be both deterministic and predictable—but they are *effectively* random from the point of view of the affected habitats and primitive minds like ours. There are no large-scale temporal or spatial correlations between them. (*Small-scale* correlations are to be expected in any case. For instance, if a supernova explodes close to a habitable planet, the pulse of ionizing radiation will arrive first, followed by accelerated cosmic-ray particles and possibly some massive ejecta as well—those are all potentially catastrophic processes, but they will be temporally and spatially correlated on scales small in comparison to both evolutionary timescales and GHZ spatial scales. Correlations induced by the Galactic gravitational potential—notably passages of planetary systems through the spiral arms or vertical oscillations of stars perpendicular to the Galactic plane—belong to this category as well (see Rampino and Stothers 1984; Matese and Whitman 1992; Matese and Whitmire 1996; Rampino 1997; Leitch and Vasisht 1998; Zank and Frisch 1999; Shaviv 2002a, b).) Note that the reason why we should expect our past historical experiences to be a poor guide for assessing the rate of catastrophes elsewhere is clear: we are subject to strong observation-selection effects biasing our conclusions (Ćirković, Sandberg, and Bostrom 2010).

11. Clube and Napier (1984, 1990); Hut et al. (1987); van den Bergh (1994); Chyba (1997); Firestone et al. (2007).

12. Rampino (2002); Mason, Pyle, and Oppenheimer (2004).

13. For a sampling of the huge and heterogeneous literature on the subject, see Krasovsky and Shklovsky (1957); Schindewolf (1962); Terry and Tucker (1968); Laster, Tucker, and Terry (1968); Ruderman (1974); Whitten et al. (1976); Hunt (1978); Thorsett (1995); Collar (1996); Dar (1997); Byakov, Stepanov, and Stepanova (1997, 2001); Dar, Laor, and Shaviv (1998); Leonard and Bonnell (1998); Sergi et al. (2001); Dar and De Rújula (2002); Scalo and Wheeler (2002); Karam (2002); Hartmann, Kretschmer, and Diehl (2002); Gehrels et al. (2003); Smith, Scalo, and Wheeler (2004); Thomas et al. (2005a, b); Thomas and Melott (2006); Galante and Horvath (2007); Thomas et al. (2008); Thomas (2009); Martín et al. (2009, 2010); Dartnell

(2011); Beech (2011b); Bougroura and Mimouni (2011); Melott and Thomas (2011); Svensmark (2012); Marinho, Paulucci, and Galante (2014); Li and Zhang (2015); and Gowanlock (2016).

14. Hoffman et al. (1998); Kirschvink et al. (2000); Pavlov et al. (2005).

15. Hence the Venusian scenario: very early habitability with possibly large oceans, cut short by the onset of the runaway greenhouse warming; see Rasoonl and de Bergh (1970) and Kasting (1988). While we have no reason to believe abiogenesis occurred on early Venus, the dominant hypothesis for the climate evolution of Venus clearly allows that, for somewhat different planetary and stellar parameters, the period of early habitability could have been prolonged. This would, in turn, mean that, on a Venus-analogue planet in the GHZ, life could emerge, take root, spread, and evolve—only to have biological evolution cut short by the boiling oceans soon afterwards (Caldeira and Kasting 1992; Franck et al. 2000).

16. A partial example in this sense is the idea of Abbas and Abbas (1998) that dark matter can, if gathered in clumps, trigger supervolcanic events in Earth's interior (see also Rampino 2015)—or even cause cancers in living beings (Abbas, Abbas, and Mohanty 1998; Freese and Savage 2012). These are wild cards in the sense of being distal causal mechanisms, not proximal ones.

17. Greaves et al. (2004). For general properties of the Tau Ceti system and its potential for habitability so far, see Cantrell, Henry, and White (2013) and Lawler et al. (2014). It seems, at the moment, that there are two planets in the circumstellar habitable zone of this nearby star, one of them being marginally habitable, similar to the case of Mars in our Solar System.

18. This need not *necessarily* be a bad thing; as discussed in Section 5.3, Ward and Brownlee (2000) admit that *some* impact stress can be rather beneficial for evolving observers. Here we consider the extreme tail of the severity distribution of impacts. For an introduction to the complex topic of the relationship between catastrophes and planetary habitability, see Hanslmeier (2009).

19. Annis (1999a); Ćirković and Vukotić (2008).

20. Annis (1999a). Prior to that, Thorsett (1995) was the first to point out that evidence indicating that the cosmological origin of GRBs has important astrobiological consequences *for our own Galaxy* (and, indeed, the life on Earth). For the modern-day view of GRB physics, see, for example, Piran (2000); Kumar and Piran (2000); Lloyd-Ronning, Fryer, and Ramirez-Ruiz (2002); Mészáros (2002); and Woosley and Bloom (2006).

21. Scalo and Wheeler (2002).

22. Lineweaver et al. (2004).

23. Clarke (1981); other conceivable astrobiological 'reset' mechanisms are discussed in Vukotić and Ćirković (2007, 2008). Fictional treatments of the dramatic idea of the Galactic Centre explosion are given in Hoyle and Hoyle (1973) and Egan (1997). Note that the Hoyles' novel was in part inspired by the then ongoing battle about the true nature of quasar red shifts. Against the majority view that those high red shifts

were of cosmological origin—thus indicating huge distances and luminosities involved—Sir Fred Hoyle's collaborators and friends like Geoffrey and Margarite Burbidge, Holton Arp, and others often invoked a picture in which quasars are nearby objects expelled with huge velocities from the Milky Way nucleus. This would necessitate highly energetic, explosive events to take place occasionally in the centre of our own stellar system. A modern fictional discussion of GRBs as playing the key role in accounting for Fermi's puzzle is given by Baxter (2002).

24. Dar, Laor, and Shaviv (1998); Shea and Smart (2000); Dermer and Holmes (2005); Griessmeier et al. (2005); Erlykin and Wolfendale (2010); Atri, Melott, and Karam (2014).
25. Thomas et al. (2005a); Thomas and Melott (2006); Thomas and Honeyman (2008).
26. Crutzen and Brühl (1996); Stozhkov (2003); Svensmark (2012).
27. Schindewolf (1962).
28. e.g. Simpson (1968).
29. Piran and Jimenez (2014), p. 1.
30. Melott et al. (2004).
31. e.g. Brakenridge (1981, 2011); Iyudin (2002); Benítez, Maíz-Apellániz, and Canelles (2002); Firestone (2014); Melott et al. (2015).
32. Note, however, that it does not seem to require fine-tuning to account for observations in other galaxies such as those of Annis (1999b) or Griffith et al. (2015), as long as the look-back time is sufficiently large.
33. Newman and Sagan (1981).
34. Some further predictions are discussed in Ćirković and Vukotić (2008).
35. Cf. von Hoerner (1978). An even bigger turnaround was made by Iosif Shklovsky, who, from roughly 1965 onwards, warned about the dangers of thermonuclear weapons as the chief reason why we have not received any signal of clearly artificial origin. Enrico Fermi himself, although no SETI researcher, has seemingly subscribed to such pessimism as well.
36. See reviews in Rees (2003) and in Bostrom and Ćirković (2008) and references therein; for the least widely known geoengineering threats, see also Ćirković and Cathcart (2004) and Baum, Maher, and Haqq-Misra (2013). Caldeira, Bala, and Cao (2013) offer a comprehensive introduction to the wide field of geoengineering.
37. This might remind some readers of the 'patchwork-quilt' solution to the paradox advocated by Webb (2002, 2015) in the concluding chapter of his popular survey. However, in contrast to Webb's construction, here we have the decomposition of a single explanatory hypothesis into individual causes, rather than the largely artificial union of many quite heterogeneous hypotheses.
38. See, for instance, Hardwig (1997) and Cholbi (2011). The whole fascinating subject matter of the bioethics of suicide has not yet tackled seriously the problems posed by contemporary and near-future advances in biotechnology and neuroscience.

39. 'Intentional' here, as elsewhere, means that there are *some* actors who intentionally take a particular action. It does not mean that there is any form of consensus, or even majority agreement on that. Considering the fact that the physical—and presumably social, although that is far less clear—capacity of individuals has been increasing exponentially through time in human history, and that the very definition of an advanced technological civilization requires that the overall capacities at the disposal of such a civilization are vast by human standards, it is clear that the issue of the internal distribution of such capacities comes to the fore.

40. For a gruesomely dramatic description of this possibility, see the portrayal of Quintan civilization in Lem's disturbing novel *Fiasco* (Lem 1987). While this fictional scenario includes elements of other solutions of Fermi's paradox, such as **Self-Destruction, Advanced Version**, the bulk of the obstacles to contact comes from totalitarian and xenophobic nature of both Quintan power blocs.

41. Orwell [1949] (1984), pp. 176–7.

42. Beyerchen (1992); Cornwell (2003).

43. Zubrin (1999), pp. 256–8. While Zubrin uses the 'Type 3' label for a large, technologically advanced civilization engaged in colonization of the Galaxy, his taxonomy is somewhat different from usual Kardashev's types, and his label would correspond to a 2.x civilization on the Kardashev scale.

44. Lem [1959] (1989).

45. e.g. Caplan (2008).

46. One can imagine, though, some contrived exceptions to this. A major plot device in Lem's *His Master's Voice* revolves around a new and particularly deadly type of nuclear weapon allegedly described in the alien message (Lem [1968] 1999). If the message were true and if many other young civilizations in this region of the Galaxy were to receive the same message, one might get an increased and spatially correlated rate of self-destruction on different planets. If destructive Dawkinsian memes are travelling through space in the form of messages of any kind—or superadvanced computer viruses!—this would fall under the **Deadly Probes** category discussed in Section 6.6.

47. Dick (1962).

48. Zamyatin [1924] (1993).

49. And this would have been visible from vast distances... if we happened to know something about the Leader's—his species, that is—morphology. A humorous take on a similar idea can be found in Charlies Stross's excellent short novel *Missile Gap* (Stross 2006).

50. Obviously, the lack of robustness against natural catastrophes makes them eliminable by **The Gigayear of Living Dangerously**.

51. Brin (1983); Hanson (1998a).

52. Papagiannis (1978); Forgan and Elvis (2011); Davies and Wagner (2013).

53. Adams and Laughlin (1997); Sandberg (2000).

54. Hanson (1998a).

55. Diamond (2005).

56. e.g. Tipler (1980, 1981a, b, 2003); Barrow and Tipler (1986).

57. Freitas (1980).

58. See Saberhagen (1992, 1998), Benford (1977, 1983, 1996), Reynolds (2000, 2002, 2003), and Schroeder (2002). The immense popularity of the *Mass Effect* trilogy of computer games, published in 2007–12, as well as its many spin-offs in different media, also contributed to the promotion of this scenario, since the major plot of these games revolve about 'the Reapers', which are just glorified—and for dramatic purposes, not very efficient—deadly von Neumann machines. The recent surge of interest in this hypothesis can be seen in Barlow (2013), Wiley (2014), Starling and Forgan (2014), Cooper (2014), and Kowald (2015), some details of which will be discussed below.

59. Brin (1983), pp. 296–7, 307. The other is a version of ecological devastation hypothesis, to be considered in Chapter 7.

60. Also, scenarios in which self-replicating machines are released in the environment and destroy their parent civilization could be regarded as a special case of the **Self-Destruction, Advanced Version**, as considered in Section 6.4.

61. The same pertains to the more contrived scenarios of, for instance, technology for creation of von Neumann probes, either nanotechnological or large, being feasible in principle but unfeasible in practice, requiring the amount of time for creation and spreading to be greater than the timescale for extinction of civilizations for other reasons or even the future age of the universe.

62. Kowald (2015).

63. A literary depiction of a similar problem has been given by Lem in *The Invincible* (Lem [1964] 1973). Although limited to a single planet, the story of a human crew of an interstellar battleship essentially powerless against unintelligent, simple, self-replicating 'beings' filling the entire planetary ecosystem is quite instructive in this regard. 'The race is not to the swift, nor the battle to the strong, neither yet bread to the wise, nor yet riches to men of understanding, nor yet favour to men of skill; but time and chance happeneth to them all' (Ecclesiastes 9:11, KJV).

64. e.g. Freitas and Merkle (2004); Wang et al. (2011).

65. This is where the crucial difference from simple biological self-replicating systems kicks in: self-replicating interstellar probes could, in principle, make decisions about the usage of resources, and thus exhaust them in a systematic and planned manner.

66. Freitas (1983); Freitas and Valdes (1980, 1985); Steel (1995); Forgan and Elvis (2011).

67. Stanislaw Lem calls it 'necroevolution' (Lem [1964] 1973), while Gregory Benford in his 'Galactic Center' novels dubs it simply 'machine civilization' (e.g. Benford 1983).

68. Some of the relevant results in the field of artificial life (frequently abbreviated as *A-Life*) have been summarized by Bedau (2003); Salzberg, Antony, and Sayama (2004); and Virgo et al. (2012).

69. The smallpox virus is adapted for no other host but humans and, in its *Variola major* variant, it infects and kills about a third of its victims so quickly and efficiently that its spread in a population is seriously limited. Its eradication from the natural environment in 1979 has been a seminal triumph of modern medicine (Fenner et al. 1988). On the other hand, the diphtheria toxin is produced by the bacterium only when the bacterium itself is infected with a bacteriophage.

70. Or at least until the physical eschatological processes bring about the dissolution of the Milky Way in about $10^{19}$ years (Adams and Laughlin 1997).

71. Bostrom (2006, 2014).

72. Brin (1983), pp. 296–7.

73. Parkinson (2005).

74. di Lampedusa [60] (2007), p. 40.

75. Tolkien (1983).

76. e.g. Kurzweil (2005).

77. Tsiolkovsky (1933); see also Lytkin, Finney, and Alepko (1995); Lipunov (1997).

78. e.g. Vinge (1986, 1992); Johansen and Sornette (2001); Kurzweil (2005). See also Tipler (1994) for an entirely different take on the same topic.

79. This is by no means surprising, considering the conclusion of many literary theorists that the 'quest for Transcendence' (with or without the capital 't') is a major attractor in the entire science-fiction narrative since the presumed 'Golden Age' (e.g. Broderick 2003). Without remotely pretending to be exhaustive, some of the fictional examples of various concepts of transcendence include Clarke (1953, 1968), Strugatsky and Strugatsky [1985] (1987), J. C. Wright (2002, 2003a, b), and Wilson (2003). A somewhat detached, satirical, or critical view of this particular science-fiction trope can be found in Schroeder (2005) and Stross (2005, 2009).

80. e.g. Chalmers (2010); Eden et al. (2012); Bostrom (2014).

81. Clarke (1953); see also Samuelson (1973). Subsequently, Clarke considered the topic of transcendence in a less overt way in his *Space Odyssey* series, starting from Clarke (1968).

82. A particularly intriguing aspect of Clarke's novel is his postulated fork in the evolutionary trajectories of intelligent species: one leading to the rational, scientific utopia of the Overlords, and the other to the mystical transcendence of the Overmind. While the first is repeatedly deemed inferior in some relevant sense (and dubbed an evolutionary dead end), one might speculate that this is just a dramatic vehicle for Clarke. In any case, this is the first speculative proposal—clearly inspired by Stapledon—to establish an 'evolutionary diagram' of advanced technological species; compare it with the fictional 'Ortega–Nilssen chart' and the corresponding fictional astrobiological theory in Lem (1987, pp. 91, 113, 123, 192–3).

83. Smart (2012).

84. While it seems a mandatory cliché to mention Clarke's Third Law ('any sufficiently advanced technology is indistinguishable from magic') at this point, an additional remark seems in order. Human intuitive understanding of technology—and, indeed, anything we deem 'artificial'—presupposes the capacity to clearly distinguish between the actor and the actor's environment; as we have seen when discussing the **New Cosmogony** (Section 4.7), this might not be the case even now, much less so in the advanced design space of posthuman/advanced extraterrestrial civilizations.

85. Gros (2005); Ćirković and Bradbury (2006).

86. **Transcension** might be in the better position here; Smart (2012) suggests several validation tests which could provide instrumental prediction without a deeper understanding of motivations and drives.

87. Consequently, it is impossible to state confidently whether the transcendence hypotheses in fact resolve **StrongFP**, that is, what additional assumptions are necessary for this rather vague concept to be a viable solution. On the other hand, the obvious—and rather dramatic—importance of this scenario for the future of humanity remains a strong motivation for further research.

88. Sorkin (2009). The 'too-big-to-fail' theory asserts that certain financial institutions are so large and so interconnected that their failure would be disastrous to the economy and so therefore they must be supported by the government when they face difficulties. Clearly, they arise only after some threshold of complexity in economic networks is reached.

## CHAPTER 7

1. Burroughs (1981), p. 153.
2. Ibid. p. 166.
3. Brill (1962).
4. e.g. Föllmer (1974).
5. Interesting variations on the topic have been given by Crawford (1993), Shechtman (2006), Kecskes (2009), and Brin (2014) and its subsequent discussion.
6. Banks (1987, 1996, 2008).
7. e.g. Zubrin (1999); Cockell (2008). The possibility of there being a diversity of lifestyles in space has been one of the major motivations for O'Neill's famous concept of space habitats (O'Neill 1977 and see references in Chapter 8). It is certainly no surprise that the concept of 'extraterrestrial liberty' has been one of the foremost science-fiction tropes; for an especially poignant example, see John C. Wright's *The Golden Oecumene* trilogy (J. C. Wright 2002, 2003a, b).
8. Boltzmann (1895), p. 414.
9. e.g. Freitas (1980); Fogg (1988); Tough (1998); Zubrin (1999); Vulpetti (1999); Rose and Wright (2004); Crawford (2009); Long (2012); Nicholson and Forgan (2013).

10. Barrow (1999).
11. As does, for instance, Seth Shostak, in an especially illuminating (though not for the intended reasons, presumably) 2006 piece, from which I borrowed the title of this section: http://www.space.com/2647-seti-barking-wrong-tree.html (last accessed 24 December 2017). It is a good, brief summary of about everything that is *wrong* with the orthodox SETI way of thinking; the short shrift which Shostak gives to Fermi's paradox is just one of the worrisome prejudices in the article.
12. The handle for this hypothesis is due to the hard-science-fiction author Hal Clement, who described several such worlds in his opus, notably Dhrawn in Clement (1971). Brin (1983) also ascribes the basic concept to Clement.
13. Laughlin, Bodenheimer, and Adams (1997).
14. See, for example, Ksanfomaliti (1986), Heath et al. (1999), Tarter et al. (2007), Driscoll and Barnes (2015), and Gale and Wandel (2017); for a discovery of a complex planetary system around a nearby extreme red dwarf, see Anglada-Escudé et al. (2014). Even this is topped by the discovery of a terrestrial planet in a habitable zone around Proxima Centauri, the closest star and the closest red dwarf, reported by the European Southern Observatory (Anglada-Escudé et al. 2016). For the general habitability of the closest star system, see Beech (2011a).
15. 'Terraforming' is in quotation marks here, since it should be understood as pertaining to any particular species of sophonts and not necessarily something like humans or the terrestrial planetary conditions. Better terms include 'ecopoiesis' or 'planetary engineering', although they are less used for historical reasons.
16. Schroeder (2002, 2014). In particular, in the latter novel, future humans use the low temperatures around brown dwarfs when undergoing a specific form of aestivation/hibernation which prolongs both personal and cultural timescales; see Section 7.4.
17. Liu et al. (2013).
18. e.g. Pigliucci (2010); Gray (2015). See also the response to—an admittedly small—part of Pigliucci's book dealing with SETI and Fermi's paradox in Ćirković (2012), esp. Chapter 7.
19. Landis (1998); Kinouchi (2001) Galera, Galanti, and Kinouchi (2018).
20. Bjørk (2007); Hart (1975); E. Jones (1976, 1981); Newman and Sagan (1981).
21. Hair and Hedman (2013), p. 51.
22. Golden, Ackley, and Lytle (1998); Solé and Valverde (2001).
23. Ćirković and Bradbury (2006). See also complementary considerations in Ćirković (2008).
24. e.g. Landauer (1961); Brillouin (1962).
25. Fixsen (2009); Ade et al. (2014).
26. Tryka et al. (1993). The somewhat surprising observation that Triton is colder than more distant objects like Pluto or Sedna is explained through Triton's extremely

high (icy) albedo. I neglect the cases where the term 'object' is taken in perhaps too liberal a sense, like CMB or indirectly detected black holes.

27. Of course, this is not the *radiation temperature*, which is simply a weighted average of the temperatures of stars producing it (and, since those are very bright, white-to-bluish stars, it is very high, measured in tens of thousands of kelvins)—it is important to avoid this confusion. Another proviso for the statement to be true is that the solid body under consideration has to be significantly larger than the mean wavelength of absorbed radiation, which is satisfied in all contexts of interest.

28. Cf. Dick (2003).

29. Egan (1997). Although it is not explicitly discussed in the novel, 'polises', which contain the uploaded minds in *Diaspora*, are supposed to be small objects, on the order of metres (Egan, private communication, 29 April 2017).

30. e.g. Learned, Pakvasa, and Zee (2009); Stancil et al. (2012).

31. Most strikingly, the idea of advanced technological civilization inhabiting the outer fringes of the Milky Way has been suggested—although without the thermodynamical rationale—by Vernon Vinge in *A Fire upon the Deep* (Vinge 1992). Vinge vividly envisages 'zone boundaries' separating dead and low-tech environments from the truly advanced societies inhabiting regions at the boundary of the disc and high above the Galactic plane. This is roughly analogous to the low-temperature regions outlined in Ćirković and Bradbury (2006) as comprising the most probable GTZ.

32. Zuckerman (1985).

33. Extraction of energy from a black hole is possible through the Penrose–Christodoulou process; see Christodoulou (1970), Penrose and Floyd (1971), and Wald (1974). For stellar uplifting as a means for prolonging the life of a star, as well as for providing rough material for other engineering purposes, see Criswell (1985) and Beech (2007).

34. Tipler (1986). See also note 15 to Chapter 2 above.

35. Beech (1990), pp. 185–6 (emphasis M. M. Ć.).

36. Spivey (2015).

37. Armstrong and Sandberg (2013).

38. Since Spivey suggests using antimatter propulsion for intergalactic migration, this could take a form of search for annihilation gamma rays, along the lines suggested by Harris (1986). At a glance, Spivey's hypothesis suffers from some astrophysical problems as well: a sizeable proportion of cluster galaxies are old, early type elliptical and S0 galaxies, with very weak star formation and metal-poor chemical composition. If advanced civilizations do not possess access to cheap nuclear transmutation ('femtotechnology', 'controlled nucleosynthesis'; cf. Adamenko, Selleri, and van der Merwe 2007), this might be a significant obstacle. On the other hand, if cheap transmutation is feasible, an advantage of metal-poor galaxies might be that the probability of locally evolved sophonts is suppressed as is the rate of risky stellar explosions.

39. de San (1981). See also Stephenson (1979).

40. O'Neill (1977).

41. e.g. Harris (1986, 1990, 2002). Of course, the search for propulsion signatures should be expanded and deepened in the future.

42. Sandberg, Armstrong, and Ćirković (2016). The literary reference applies to prehuman and superhuman beings from the literary opus of Howard P. Lovecraft and many subsequent authors (e.g. Joshi and Schultz 2001). While opinions wildly differ on the contentious issue of naturalism versus supernaturalism in Lovecraft's work, in recent criticism, the dominant idea is that the Great Old Ones are in fact advanced aliens, devoid of any anthropocentrism (e.g. Joshi 2004); hence, Fritz Leiber dubbed Lovecraft 'a literary Copernicus' (Leiber 2001). As discussed at the beginning of Chapter 6, Lovecraft [1936] (2005) contains a number of proto-astrobiological considerations, and the plot of that short, brilliant novel includes members of the ancient non-human race being unwittingly woken from many million years of aestivation. See also Lovecraft (1999, 2005).

43. Schroeder (2014).

44. Haqq-Misra and Baum (2009). For the 'city-state' model, see Ćirković and Bradbury (2006), Ćirković (2008), and Smart (2012).

45. von Hoerner (1975); Papagiannis (1984).

46. e.g. Dudley-Flores and Gangale (2012).

47. Criswell (1985); Beech (1990, 2007). I shall present a list (albeit not an exhaustive one) of astroengineering manifestations and artefacts in Chapter 8.

48. Nunn, Guy, and Bell (2014).

49. Ibid. p. 2.

50. For the obesity pandemic, see the poignant review Swinburn et al. (2011).

51. Zubrin (1999).

52. Gottlieb and Lima (2014). Somewhat ironically, this editorial comment in an extremely prestigious cardiology journal (*Journal of the American College of Cardiology*) uses Fermi's paradox as an example in purely epistemological sense. Gottlieb and Lima do not cite or endorse Nunn et al.'s theory, which appeared about the same time in *Nutrition and Metabolism* but, taking into account the frequency of coronary artery disease in the modern world and its link to obesity, the reasoning of Nunn et al. applies.

53. G. Miller (2006); Mansilla, Perkis, and Ebrahimi (2014).

54. e.g. Grene (1958); Gould (2002); Larson (2004).

55. Grene (1958), p. 112.

56. Schwartzman and Middendorf (2011).

57. As mentioned, Webb's popular and extraordinary diligent *Seventy-Five Solutions to the Fermi's Paradox* is more comprehensive, at the cost of much overlap between proposed hypotheses (Webb 2015).

## CHAPTER 8

1. According to the transcript of the news briefing of the US Department of Defense on 12 February 2002 (http://archive.defense.gov/Transcripts/Transcript.aspx?TranscriptID=2636, last accessed 2 November 2017). The totally misplaced mockery of the statement in the subsequent negatively biased press coverage just confirmed the simple but sad truth that, when it comes to logic and epistemology, most journalists are completely indistinguishable from monkeys hitting keyboard at random to occasionally—once in a googolplex or so years—produce a *Hamlet*.
2. As with so many other things related to the subject matter of this book, this Rumsfeldian moment was anticipated by Stanislaw Lem. In *Fiasco*, one of the rare comic situations in an otherwise grave and depressing narrative takes place when the commander of the human contact team sent to Quinta, after waking up from a hibernation-like state, asks the powerful and supremely rational ship computer about a message received from the mothership:

   > 'How long is the message?'
   > 'Three thousand, six hundred and sixty words. Should I give the text?'
   > 'Summarize it.'
   > 'I cannot summarize unknowns.'
   > 'How many unknowns?'
   > 'That, too, is an unknown.'...
   > 'Give the general picture.'
   > 'A picture of unclear things is unclear' (Lem 1987, pp. 143–4).

   This sort of absurdist humour is unavoidable when we consider how poorly human language is adapted for discussing the deep problems of this highly complex universe (cf. Nozick 1981; Hofstadter 2000).
3. Eco (2009).
4. For bioterrorism in relation to Fermi's paradox, see Cooper (2013). Nanotechnology risks are discussed on an introductory level in Phoenix and Treder (2008). For geo-engineering-related risks, see Ćirković and Cathcart (2004) and Baum, Maher, and Haqq-Misra (2013). The risk of creating negatively charged 'strangelets'—pieces of strange matter which would accrete more and more normal matter, ultimately converting our entire planet—in accelerators has been considered by Jaffe et al. (2000) for the Brookhaven collider and by the official CERN study of Ellis et al. (2008) for the Large Hadron Collider. The brilliant Czech novelist Karel Čapek described an extreme form of explosive named after the infamous Indonesian volcano Krakatoa in his eponymous 1922 novel *Krakatit* (Čapek [1922] 1975).
5. The corresponding narrative would be that of the total number of sites potentially harbouring sophonts; half is lost by missing the 'Gaian' window of opportunity (as per the **GAIAN WINDOW**); half of the rest comprises sophonts which self-destruct upon reaching the threshold of nuclear power (**STOP WORRYING AND LOVE THE BOMB**); the rest move to nearby brown dwarfs and have no interest in Main Sequence stars like the Sun (as per the **BROWN EMPIRE**), presumably refraining

from astroengineering and leaving other traces and manifestations as well. While there is no meaningful way of calculating the relevant probabilities so far, we could give such a scenario a weaker interpretation in terms of relative frequencies averaged over a cosmologically relevant spatio-temporal volume.

6. On the other hand, such a situation might lead to new insights—violations of symmetries leading to conservation of baryon and lepton numbers in particle physics have historically led to discovery of new quantum numbers and even new disciplines, like black-hole thermodynamics. See, for example, Kragh (1999).
7. Kitcher (1990). See also Weisberg and Muldoon (2009).
8. e.g. Glen (1994); Raup (1999).
9. Hanson (1998b).
10. Perhaps the 'latest' Late Great Filter is provided by PERMANENCE, although here we are dealing with the effect of diminishing returns: as discussed in Chapter 5, the suppression of detectability by the slow dysgenic decline as per this hypothesis might be simply too slow. We can legitimately raise the question, why don't we perceive vestiges of the last interstellar empire, their enclaves, or their artefacts? Besides, although it conforms to wording, PERMANENCE probably violates the *spirit* of Hanson's idea: a civilization lasting 50 or 100 million years after the invention of space flight should be counted as successful, not something 'filtered out'!
11. See an extended discussion in Ćirković (2015).
12. The Newtonian mechanics analogy also holds here, since the fact that Newtonian dynamics deals with *large* bodies, such as planets and satellites (as well as apples), enables it to ignore complexities and singularities which occur at very small spatial scales and which are dealt with later in the context of *quantum* mechanics.
13. Crutzen and Stoermer (2000); Ruddiman (2013). For the precursor history of similar ideas, which includes the overlap with Russian cosmists and early futurologists, see Young (2012) and Siddiqi (2008).
14. Zalasiewicz et al. (2014).
15. Hosek (2007), pp 138–9.
16. Calvino [1969] (2009), p. 276.
17. Crowe (1986); Raulin-Cerceau and Bilodeau (2012).
18. Sagan [1973] (2000), pp. 229–32. By the way, this widely read book (i.e. no obscure, hard-to-find *samizdat*) proves how false are sceptics' charges (e.g. see Gray 2015) that nobody was aware of Fermi's paradox before Hart (1975). For other early views on manifestations and artefacts, see Paprotny (1977), von Hoerner (1980), Freitas (1985a), Troitskij (1989), and Rubtsov (1991).
19. e.g. Gray and Marvel (2001).
20. See, for instance, Shklovsky and Sagan (1966), Sagan (1973a, b), and Michaud (2007).
21. Jackson (2015).
22. See DeBiase (2008), however, for a bold discussion of possibly destructive process.

23. Korpela, Sallmen, and Leystra Greene (2015).
24. Criswell (1985); Beech (1990, 1993, 2007, 2010).
25. Harris (1986, 1990, 2002). For the general issue of antimatter propulsion, see Long (2012) and references therein.
26. Paprotny (1977); Whitmire and Wright (1980); Borra (2012).
27. Valdes and Freitas (1986).
28. Lin, Gonzalez Abad, and Loeb (2014); Kuhn and Berdyugina (2015); Stevens, Forgan, and James (2016).
29. Guillochon and Loeb (2015); Lingam and Loeb (2017).
30. Kecskes (1998); Forgan and Elvis (2011).
31. Maccone (2011a); Gillon (2014).
32. Korycansky et al. (2001); McInnes (2002); Sanchez and McInnes (2012).
33. Learned et al. (2012); Chennamangalam et al. (2015). For obvious reasons, this type of activity has the strongest crossover with artefacts. Gregory Benford offers a fictional account of the manifestations of the Milky Way centre engineering in his Galactic Centre Saga (e.g. Benford 1977, 1983, 1996).
34. While there are no references for these, it might be useful to think in meta-terms here and consider the inverse question: *what manifestations are provably not possible—and are not listed above?* Cf. Barrow (1999).
35. e.g. Nicholls (2000).
36. Dyson (1960a, b, 1966); Suffern (1977); Badescu (1995); Inoue and Yokoo (2011); Osmanov (2016). Probably the most bizarre version—even for a field such as the exploratory engineering!—is the 'anti-Dyson sphere' suggested by Opatrný, Richterek, and Bakala (2016); in this case, the shell is constructed around a black hole, which serves as a reliable heat *sink*, instead of being a source.
37. Harrop and Schulze-Makuch (2010); Scheffer (2011); J. T. Wright et al. (2016). For the general issue of detectability of irregular-shaped transiting objects, see Arnold (2005); for a particular category of astroengineering artefact, the Class A stellar engine, see Forgan (2013).
38. These are special cases of the general category of O'Neill colonies (O'Neill 1977), shaped as rotated rings in order to create effective gravity on the inner side of the ring; see Figure 8.4. The best-known example from science-fiction literature is Larry Niven's Ringworld in the eponymous novel (Niven 1970) and its sequels. The concept was heavily popularized by the *Halo* series of video games, published 2001–16, where the artefacts are called 'installations'; in 2006 physicist and programmer Kevin Grazier wrote an essay entitled 'Halo Science 101', analysing the properties of such structures (http://www.gamasutra.com/view/feature/1462/halo_science_101.php, last accessed 1 March 2017). In Iain M. Banks' *Culture* novels, such artefacts are called 'orbitals' (e.g. Banks 1987, 2008), while Karl Schroeder uses the label 'coronels' in his impressive *Lady of Mazes* (Schroeder 2005).

39. Pearson (1975); Birch (1982, 1983); Bolonkin (2005); Fawcett, Laine, and Nugent (2006). Among many fictional descriptions, Reynolds (2001) is especially memorable. Lem's *Fiasco* contains a description of an orbital ring of ice, built in order to lower the ocean level on the planet below as a part of geoengineering mega-project, which could fit in this category as well (Lem 1987).

40. Birch (1982, 1983); Pearson, Oldson, and Levin (2006).

41. Birch (1991); Roy, Kennedy, and Fields (2009, 2014). For a dramatic fictional description, see Banks (2008).

42. Fogg (1989); Ćirković (2016).

43. Bradbury (1997); Sandberg (2000).

44. Niven (1974). An interesting application of classical field theory to Alderson discs is given at Jens Kleimann's web page (http://www.tp4.rub.de/~jk/science/gravity/chapt_alderson.html, last accessed 1 March 2017). Kardashev (1985) includes an amusing (and somewhat subversive!) cartoon. See also a poignant fictional description in Stross (2006), and a smaller version in the neo-noir movie *Dark City* (Proyas 1998).

45. e.g. Lior (2013). For the detection of such objects, see Loeb and Turner (2012). The concept of shielding swarms for protection against electromagnetic and/or cosmic-ray pulses from supernovas, magnetars, or GRBs has been introduced by Ćirković and Vukotić (2016).

46. E. Jones (1985a); Early (1989); Teodorani (2014).

47. Lacki (2015).

48. Lacki (2016).

49. Davies (2012); Carrigan (2012); Armstrong and Sandberg (2013); M. Jones (2015). For precursor views, see Sagan and Walker (1966), Sagan (1973a), Freeman and Lampton (1975), and Armitage (1976).

50. Of course, not *so* close as to preclude the habitability of the planet. Many such cases are possible—for instance, a planet circling a star in a wide binary where the other component is an old neutron star at, say, 10,000 AU seems habitable in principle, although the effects of starquakes and magnetic instabilities on habitability in such systems should be investigated. Options considered in Opatrný et al. (2016) then come into play.

51. Cameron (2009). This is true if such objects are allowed by the correct cosmogonic theory—which we do not know at present!

52. Artefacts whose purpose is unclear even within known contexts belong to this wild-card category as well. Such instances are unlikely to arise in observational astronomy, since local context is hard to firmly establish over interstellar distances, but are often encountered in science-fiction discourse where the context is postulated as a given or where the revelation of functional purpose is a part of the plot. Such is the alien acceleration structure (as well as the much larger 'living' structure) in Alastair Reynolds' *Pushing Ice*; the purpose of this structure is discovered only after the fact (Reynolds 2005b).

53. Considering, for example, the case in which the same civilization uses antimatter propulsion for transport between those systems. The already mentioned work of Harris (1986, 2002) indicates that such spacecraft could be potentially detectable with the advancement of gamma-ray astronomy. In contrast to many other forms of emission, including radio leakage from planetary surfaces or orbits around planets, detection signatures of antimatter–matter annihilation would be very difficult, if not impossible, to intentionally supress. So, even a hermit or a paranoid civilization could be more detectable than expected, due to its interstellar traffic.
54. Bradbury et al. (2011); Zackrisson et al. (2015).
55. Papagiannis (1978); Freitas and Valdes (1980, 1985); Arkhipov (1996, 1997); Kecskes (1998); Matloff and Martin (2005); Davies and Wagner (2013); see also Haqq-Misra and Kopparapu (2012).
56. The grounds have been put by Sagan and Walker (1966). For searches, see Slysh (1985); Jugaku, Noguchi, and Nishimura (1995); Tilgner and Heinrichsen (1998); Jugaku and Nishimura (2003); Timofeev, Kardashev, and Promyslov (2000); and Carrigan (2009).
57. Valdes and Freitas (1986).
58. Harris (1986, 2002).
59. Annis (1999b); Zackrisson et al. (2015).
60. J. T. Wright et al. (2014a, b); Griffith et al. (2015).
61. Arnold (2005); J. T. Wright et al. (2016).
62. Again, a good recent example is Gray (2015).
63. Unless, as proposed by Olson (2015), intelligent beings begin to modify cosmological distribution of matter. This intriguing study can be regarded as the first rudimentary blueprint for the activity of Lem's Players!
64. J. T. Wright et al. (2014a, b); Griffith et al. (2015); J. T. Wright et al. (2016); Zackrisson et al. (2016).
65. e.g. Hart (1995), p. 224: 'Most intelligent races should see no other civilizations in their galaxy; indeed, they should see no others in the entire portion of the universe (including perhaps $10^{22}$ stars) which they are able to observe with their telescopes.'
66. Bounama, von Bloh, and Franck (2007); Franck, von Bloh, and Bounama (2007).
67. e.g. Ward (2005).
68. Suthar and McKay (2012). Ward and Brownlee (2000) noticeably squirm around this issue whenever it appears (e.g. pp. 29–31), putting larger emphasis on the *temporal* structure of habitability than the spatial one. Nowhere in their narrative is there any suggestion that the Milky Way is unique or atypical; see, however, Dayal et al. (2015) for a different view.
69. There are, admittedly, multiple subtler issues at work here. For galaxies more distant than some threshold value, the time delay becomes comparable or larger than the timescale for GHZ expansion (cf. Section 2.3); this, in turn, means that the GHZ at

larger distances is too narrow to expect even a single complex biosphere ('Gaia') in any observed galaxy, although that galaxy is expected to host multiple Gaias unobservable at present. While this threshold value could be computed, knowing the look-back time in the new standard cosmological model and the rate of chemical enrichment driving the GHZ expansion, it would not be realistically useful: if the reasoning behind the **Gaian window** is anywhere near correct, we are dealing with a whole series of temporal windows of opportunity and unrepeatable events, not just one or a few well-defined timescales. On the other hand, in contrast to **Permanence**, and in even sharper contrast to those solutions which posit multiple simultaneous civilizations during a large fraction of the galactic age (such as **Interstellar Containment** or **Persistence**), it could be argued that the **Gaian window** enables easier ascent up Kardashev's ladder. For those rare civilizations which finally emerge, the likelihood of encountering any resistance to appropriating resources of their home galaxy is vanishingly small.

70. Bostrom (2014).
71. Dyson (1979).
72. Note that this applies even if the physiology in question is explicitly *human* physiology. A research physiologist may regard the system she investigates—a pathogen and the specific tissue it attacks—as isolated from herself and the rest of the universe (abstracting the specific laboratory apparatus she needs). On some high philosophical level, she might be aware that her findings influence the welfare and perhaps even existence of many human beings, possibly even herself. But the parameters of the problem and her 'questions posed to the Nature' will not change as a consequence of such abstract awareness. In particular, the difficulty of the problem does not depend upon whether the disease under scrutiny is extremely rare or as common as flu.
73. Bostrom (2002).
74. In Ćirković (2015), I have outlined a few scenarios 'defeating' Kardashev's classification, in the sense that they do not fall easily into any particular Kardashev type.
75. Many minor religious traditions, including ancient Gnosticism, as well as some fascinating modern thinkers such as Emil Cioran and Philip K. Dick, do talk in vague terms about the 'evil Demiurge' creating the material world around us. In the context of scientific rationalism, especially if we accept naturalism, this view does not make much sense.
76. Which may or may not be a good thing. As H. P. Lovecraft aptly indicated in his fiction, the **Great Old Ones** might not like being awakened by upstarts!
77. See Rescher (1985) and Kukla (2008, 2010) for dissenting opinions. Ernan McMullin, no great friend of SETI studies himself (e.g. McMullin 1980), has successfully—at least in the humble view of this author—demolished this whole 'different science' narrative in his review of Regis's anthology (McMullin 1989, pp. 101–2). For another, more ambiguous view of this issue, see Hamming (1998).
78. Brin (1983); Kowald (2015).

79. Armstrong and Sandberg (2013). A possible objection here is that intergalactic voyage requires completely new level of reliability over long timescales, which might lead to excessive failure rate and weakening of the 'infection'.
80. After being warned, perhaps, by extraterrestrial analogues of Hanson (1998a)!
81. Newton-Smith (1980); Price (1996).
82. The qualifier in parenthesis indicates that other, extremely low-probability possibilities exist. For instance, all other sophonts could destroy themselves just in time for the Posthuman Dominion to arise without unnecessary delays and unpleasantries such as interstellar wars. Analyses indicate this would require vigorous use of the Infinite Improbability Drive (Adams 1979).
83. Kahane (2014).

## CHAPTER 9

1. And, importantly from the point of view of the song's longevity and cultural impact, it was topping the charts a few weeks later, during the original Woodstock Festival.
2. e.g. Grene (1958); Bowler (1992).
3. Dick (2003); Davies (2010); Bradbury et al. (2011); Carrigan (2012); Flores Martinez (2014); Lacki (2016).
4. Matthew 22:14 (KJV; see http://biblehub.com/kjv/, last accessed 28 December 2017). Ironically, it is proponents of 'intelligent' design and other latter-day creationist nonsense who often charge 'established' science with being overly elitist, narrow-minded, and selective.
5. Davies (2010), p. 205.
6. Hofstadter (2000), especially pp. 713–17. See also de Smit and Lenstra (2003).
7. Ibid. pp. 716–17.
8. A simple example here is a version of the teleological meaning: 'the blemish serves as a placeholder for Escher's signature', which could then lead us into a discourse on labelling, authorship, individuality in graphic arts, and so on. Taking into account the results of de Smit and Lenstra (2003), we can think about a number of other meanings connected to the Droste effect (the fractal replication of an image inside itself).
9. e.g. Zubrin (1999); Levinson (2003); Cockell (2007); Nunn et al. (2014); Klee (2017). Most of the respected authors condemn this trend, of course.
10. Hoyle (1983), p. 251.
11. Simpson (1964).
12. Kundera (2007), p. 37; emphasis in the original.
13. Remember the 'Order of the Dolphin', formed at the very first SETI conference in Green Bank in 1961; see, for example, Grinspoon (2003) and Michaud (2007).
14. The precursor to this (and still the very best of the sorry bunch) is Arendt [1963] (2007). For subsequent samples, see Westfahl (1997), deGroot (2006), and Klee (2017).

15. Ćirković (2012).
16. Des Marais and Walter (1999); Des Marais et al. (2008).
17. One may add other items here, in accordance with either Freud's immodest claim or the equally immodest claims of the founding fathers of the field of artificial intelligence. Under the umbrella term 'cosmological revolution', I refer to the period spanning roughly 1929–65, from the discovery of the Hubble expansion to the discovery of the primordial microwave background. The prescient study of Mazlish (1967) postulated that the rise of information technology would overcome the 'fourth discontinuity' between man and machine artefacts. So, after each triumph of Copernicanism, we can adapt to a new reality: 'In an important sense, it can be contended, once man is able to accept this situation, he is in harmony with the rest of existence' (p. 3).
18. Kuiper and Morris (1977). A reprint edition of *The Cyclops Report* is Oliver et al. [1971] (1996).
19. Bold voices rising above that despicable anti-cosmic consensus are, for instance, Zubrin (1999), Levinson (2003), Gilster (2004), and deGrasse Tyson (2012).
20. Reynolds (2005a).
21. A fine example of Copernicanism motivating concrete SETI study is the recent work of Heller and Pudritz (2016), where the question *who can observe us via transit observations?* has been studied. While there can be no guarantees—or indeed indications—that there are actually inhabited planetary systems in Earth's transit zone (which is that part of space from which an observer could detect Earth's transits across the Sun), it is still reasonable to investigate this inverse problem since, if there are such sophonts, mutual SETI discovery will make subsequent communications easier. Of course, the vantage point of technologically advanced sophonts might not correspond to their home system—or even any planetary system at all. But that is not the end of the story—for a further 'turn of the screw' in the eternal Copernican/anthropocentric tug of war, see Kipping and Teachey (2016): we still may not detect anything anomalous, since the transits could be stealthy! The same could be applied to the transits of artificial objects as studied in Arnold (2005) and J. T. Wright et al. (2016).
22. This is a more general formulation, since we are talking about 'civilizations' or even 'species' as a matter of ingrained habit, while those terms might lose their meaning in the general context and for extremely high-complexity regions of the landscape. Not only might postbiological evolution suppress or mute these anthropocentric concepts, but we might develop more radical solutions, similar to the ones discussed in connection with the **New cosmogony** or **Transcendence/Transcension hypothesis.**
23. Stapledon (1930); Blish (1957).
24. Davies (2010), p. 161.
25. Dick (2003); Parkinson (2005); Ćirković and Bradbury (2006).
26. J. C. Wright (2011), p. 228.

27. Barrow (1999).
28. One could speculate further that the dichotomy between form and function will disappear as well at a sufficiently high level of complexity. For a beautiful fictional discussion of this issue, see Robson (2005). Alternatively, the dichotomy could be erased intentionally, as described in Schroeder's *Ventus* and *Lady of Mazes* (Schroeder 2001, 2005).
29. See, for instance, Cockell (2007).
30. Nazaretyan (2005); Stewart (2010, 2012); Last (2017).
31. e.g. Bedau (2003); Virgo et al. (2012).
32. Last (2017), p. 42.
33. e.g. Fukuyama (2002). Fukuyama dubs the concept of postbiological humanity 'the most dangerous idea' humanity faces in this century.
34. National Intelligence Council (2008).
35. e.g. Chaouchi (2010); Swan (2012).
36. Swan (2013). 'There is a paradox that on one hand humans are becoming increasingly dependent upon technology for everything including interacting with the outside world, while on the other hand technology is providing a richer, more detailed, controllable, and personal relationship with the world' (p. 96).
37. Hutchins (1995).
38. Lem [1964] 2013, pp. 68–70.
39. Of course, as discussed in Chapter 4, our universe might be a simulation. Even there, we can distinguish between 'on-level' subuniverses created in (say) a particle physics lab, and 'next-simulation-level' universes created on our own (albeit second-level-simulated!) supercomputers. After all, our particle physics mini-universe will be tightly constrained by our observed physics, in contrast to the simulated universe where, for example, one could have three or thirteen low-energy interactions, instead of the familiar four (gravitation, the weak force, electromagnetism, and the strong/nuclear force).
40. Fogg (1989); Ćirković (2016).
41. Farhi and Guth (1987); Farhi, Guth, and Guven (1990); Linde (1992); Holt (2004); Ansoldi and Guendelman (2007).
42. Harrison (1995).
43. At least such was my impression when, being a junior grad-school student at the State University of New York at Stony Brook when Harrison's paper was published, we discussed it at a journal club seminar!
44. Gardner (2004, 2007); see also Stewart (2010, 2012); Flores Martinez (2014).
45. Stewart argues that it is exactly the relative strength of an evolutionary perspective on future studies which favours universe creation for future humans as well (Stewart 2010, 2012).

46. Hsu and Zee (2006); Beane, Davoudi, and Savage (2014).
47. http://www.darwinproject.ac.uk/letter/DCP-LETT-2690.xml, last accessed 3 June 2017.
48. Simpson (1964); Barrow and Tipler (1986); Mayr (1993); Conway Morris (2003a, b); Lineweaver (2008). While some of Simpson's and the Simpson-inspired criticism is no longer valid (cf. Ćirković 2014), the anthropocentric charge still stands. Only very recently (cf. references cited in Section 8.2) has this dominance of anthropocentrism in SETI studies weakened.
49. Barrow and Tipler (1986); see also Mayr (1993).
50. e.g. Sagan and Walker (1966); Sagan and Salpeter (1976); Kardashev (1964, 1985, 1986, 1997).
51. An example is provided in a review by Shostak (2003): 'While many authors... have suggested novel wavelengths that might be monitored... none have proved compelling to the SETI community' (p. 113). Shostak should be credited, though, for rather shyly considering alternative ideas like the postbiological evolution and artificial intelligence at all; the orthodoxy mostly ignores them entirely.
52. Hart (1975).
53. Hart (1979). It is important to understand that Hart's pessimistic conclusion about very narrow circumstellar habitable zones is obsolete today, and for clearly established reasons (Levenson 2015). One might further speculate that Hart's subsequent engagement with fringe politics of a rather repugnant kind—racism/'white nationalism'/'saving Western civilization'—did not help with taking his astrobiological views seriously (Swain and Nieli 2003). It might be speculated as to the degree to which his extreme anthropocentrism in astrobiology and his extreme social and political views were related.
54. Bates (1978). The NASA ADS indicates that there were exactly *two citations* (!) of this study in the entire 1978–2016 period—in spite of the prominence of the author, who in 1978 was awarded peerage. See also Bates (1974).
55. The best-known and most egregious example is Drake (1976).
56. See the fine review of Korpela (2012) and references therein.
57. Bracewell (1975); Sagan and Salpeter (1976); Oliver et al. [1971] (1996); Kardashev (1964).
58. Kardashev (1985), p. 497.
59. Sagan (1994, 1995).

# REFERENCES


Aaronovitch, D. 2009, *Voodoo Histories: The Role of the Conspiracy Theory in Shaping Modern History* (Jonathan Cape, London).

Abbas, S. and Abbas, A. 1998, 'Volcanogenic Dark Matter and Mass Extinctions', *Astropart. Phys.* **8**, 317–20.

Abbas, S., Abbas, A., and Mohanty, S. 1998, 'Double Mass Extinctions and the Volcanogenic Dark Matter Scenario' (preprint, http://xxx.lanl.gov/abs/astro-ph/9805142, last accessed 1 April 2016).

Abeyta, M. 2006, *Fuentes, Terra Nostra, and the Reconfiguration of Latin American Culture* (University of Missouri Press, Columbia).

Aczel, A. D. 1998, *Probability 1: Why There Must Be Intelligent Life in the Universe* (Harcourt Brace, New York).

Adamenko, S. V., Selleri, F., and van der Merwe, A. (eds.). 2007, *Controlled Nucleosynthesis: Breakthroughs in Experiment and Theory* (Springer, Dordrecht).

Adams, D. 1979, *The Hitchhiker's Guide to the Galaxy* (Pan Books, London).

Adams, F. C. and Laughlin, G. 1997, 'A Dying Universe: The Long-Term Fate and Evolution of Astrophysical Objects', *Rev. Mod. Phys.* **69**, 337–72.

Adams, F. C., Hollenbach, D., Laughlin, G., and Gorti, U. 2004, 'Photoevaporation of Circumstellar Disks Due to External Far-Ultraviolet Radiation in Stellar Aggregates', *Astrophys. J.* **611**, 360–79.

Ade, P. A. R. et al. (Planck Collaboration) 2014, 'Planck 2013 Results. XVI. Cosmological Parameters', *Astron. Astrophys.* **571**, article id. A16 (66 pp.).

Ahouse, J. C. 1998, 'The Tragedy of a Priori Selectionism: Dennett and Gould on Adaptationism', *Biol. Philos.* **13**, 359–91.

Allègre, C. J., Manhès, G., and Gopel, C. 1995, 'The Age of the Earth', *Geochim. Cosmochim. Acta* **59**, 1445–56.

Almár, I. 1992, 'Analogies between Olbers' Paradox and the Fermi Paradox', *Acta Astronaut.* **26**, 253–6.

Almár, I. and Tarter, J. 2011, 'The Discovery of ETI as a High-Consequence, Low-Probability Event', *Acta Astronaut.* **68**, 358–61.

Alvarez, L. W., Alvarez, W., Asaro, F., and Michel, H. V. 1980, 'Extraterrestrial Cause for the Cretaceous-Tertiary Boundary Extinction', *Science* **208**, 1095–108.

Ambartsumian, V. A., Kardashev, N. S., and Troitskii, V. S. (eds.). 1986, *Problem of the Search for Life in the Universe* (Izdatel'stvo Nauka, Moscow).

Ambrose, S. H. 1998, 'Late Pleistocene Human Population Bottlenecks, Volcanic Winter, and Differentiation of Modern Humans', *J. Hum. Evol.* **34**, 623–51.

Anglada-Escudé, G. et al. 2014, 'Two Planets around Kapteyn's Star: A Cold and a Temperate Super-Earth Orbiting the Nearest Halo Red Dwarf', *Mon. Notices Royal Astron. Soc.* **443**, L89–L93.

Anglada-Escudé, G. et al. 2016, 'A Terrestrial Planet Candidate in a Temperate Orbit around Proxima Centauri', *Nature* **536**, 437–40.

Annis, J. 1999a, 'An Astrophysical Explanation for the Great Silence', *J. Brit. Interplanet. Soc.* **52**, 19–22.

Annis, J. 1999b, 'Placing a Limit on Star-Fed Kardashev Type III Civilisations', *J. Brit. Interplanet. Soc.* **52**, 33–6.

Ansoldi, S. and Guendelman, E. I. 2007, 'Solitons as Key Parts to Produce a Universe in the Laboratory', *Found. Phys.* **37**, 712–22.

Arendt, H. [1963] 2007, 'The Conquest of Space and the Stature of Man', *New Atlantis* **18**, 43–55.

Arkhipov, A. V. 1996, 'On the Possibility of Extraterrestrial-Artefact Finds on the Earth', *Observatory* **116**, 175–6.

Arkhipov, A. V. 1997, 'Extraterrestrial Technogenic Component of the Meteoroid Flux', *Astrophys. Space Sci.* **252**, 67–71.

Armitage, J. 1976, 'The Prospect of Astro-Palaeontology', *J. Brit. Interplanet. Soc.* **30**, 466–9.

Armstrong, S. and Sandberg, A. 2013, 'Eternity in Six Hours: Intergalactic Spreading of Intelligent Life and Sharpening the Fermi Paradox', *Acta Astronaut.* **89**, 1–13.

Arnold, L. 2013, 'Transmitting Signals over Interstellar Distances: Three Approaches Compared in the Context of the Drake Equation', *Int. J. Astrobiol.* **12**, 212–17.

Arnold, L. F. A. 2005, 'Transit Light-Curve Signatures of Artificial Objects', *Astrophys. J.* **627**, 534–9.

Asghari, N. et al. 2004, 'Stability of Terrestrial Planets in the Habitable Zone of Gl 777 A, HD 72659, Gl 614, 47 Uma and HD 4208', *Astron. Astrophys.* **426**, 353–65.

Ashkenazi, M. 2017, *What We Know About Extraterrestrial Intelligence: Foundations of Xenology* (Springer, Cham).

Ashworth, S. 2014, 'A Parameter Space as an Improved Tool for Investigating Extraterrestrial Intelligence', *J. Brit. Interplanet. Soc.* **67**, 224–31.

ATLAS Collaboration 2013, 'Evidence for the Spin-0 Nature of the Higgs Boson using ATLAS Data', *Phys. Lett. B* **726**, 120–44.

Atri, D., Melott, A. L., and Karam, A. 2014, 'Biological Radiation Dose from Secondary Particles in a Milky Way Gamma-Ray Burst', *Int. J. Astrobiol.* **13**, 224–8.

Bada, J. L. 2004, 'How Life Began on Earth: A Status Report', *Earth Planet. Sci. Lett.* **226**, 1–15.

Badescu, V. 1995, 'On the Radius of Dyson's Sphere', *Acta Astronaut.* **36**, 135–8.

Badescu, V. 2011, 'Free-Floating Planets as Potential Seats for Aqueous and Non-Aqueous Life', *Icarus* **216**, 485–91.

Badescu, V. and Cathcart, R. B. 2000, 'Stellar Engines for Kardashev's Type II Civilisations', *J. Brit. Interplanet. Soc.* **53**, 297–306.

Badescu, V., Cathcart, R. B., and Schuiling, R. D. (eds.). 2006, *Macro-Engineering: A Challenge for the Future* (Springer, New York).

Bailer-Jones, C. A. L. 2009, 'The Evidence for and against Astronomical Impacts on Climate Change and Mass Extinctions: A Review', *Int. J. Astrobiol.* **8**, 213–39.

Baker, V. R. 1998, 'Catastrophism and Uniformitarianism: Logical Roots and Current Relevance in Geology', *Geol. Soc., Lond., Spec. Pub.* **143**, 171–82.

Balashov, Yu. 1994, 'Uniformitarianism in Cosmology: Background and Philosophical Implications of the Steady-State Theory', *Stud. Hist. Philos. Sci. A* **25**, 933–58.

Balázs, B. A. 1986, 'Galactic Position of Our Sun and the Optimization of Search Strategies for Detecting Extraterrestrial Civilizations', *Acta Astronaut.* **13**, 123–6.

Balázs, B. A. 2000, 'SETI and the Galactic Belt of Intelligent Life', in G. Lemarchand and K. Meech (eds.) *Bioastronomy 99: A New Era in the Search for Life* (ASP Conference Series, San Francisco), 441–4.

Ball, J. A. 1973, 'The Zoo Hypothesis', *Icarus* **19**, 347–9.

Banks, I. M. 1987, *Consider Phlebas* (Orbit, London).

Banks, I. M. 1996, *Excession* (Orbit, London).

Banks, I. M. 2008, *Matter* (Orbit, London).

Barkun, M. 2003, *A Culture of Conspiracy: Apocalyptic Visions in Contemporary America* (University of California Press, Berkeley).

Barlow, M. T. 2013, 'Galactic Exploration by Directed Self-Replicating Probes, and Its Implications for the Fermi Paradox', *Int. J. Astrobiol.* **12**, 63–8.

Barnes, L. A. 2012, 'The Fine-Tuning of the Universe for Intelligent Life', *Publ. Astron. Soc. Aust.* **29**, 529–64.

Barrow, J. D. 1999, *Impossibility: The Limits of Science and the Science of Limits* (Oxford University Press, Oxford).

Barrow, J. D., Conway Morris, S., Freeland, S. J., and Harper, C. L., Jr. 2008, *Fitness of the Cosmos for Life: Biochemistry and Fine-Tuning* (Cambridge University Press, Cambridge).

Barrow, J. D. and Tipler, F. J. 1986, *The Anthropic Cosmological Principle* (Oxford University Press, New York).

Basalla, G. 2006, *Intelligent Life in the Universe* (Oxford University Press, Oxford).

Bates, D. R. 1974, 'CETI: Put Not Your Trust in Beacons', *Nature* **252**, 432–3.

Bates, D. R. 1978, 'On Making Radio Contact with Extraterrestrial Civilizations', *Astrophys. Space Sci.* **55**, 7–13.

Bauer, H. H. 1984, *Beyond Velikovsky: The History of a Public Controversy* (University of Illinois Press, Urbana).

Baum, S. D. 2010, 'Is Humanity Doomed? Insights from Astrobiology', *Sustainability* **2**, 591–603.

Baum, S. D., Haqq-Misra, J. D., and Domagal-Goldman, S. D. 2011, 'Would Contact with Extraterrestrials Benefit or Harm Humanity? A Scenario Analysis', *Acta Astronaut.* **68**, 2114–29.

Baum, S. D., Maher, T. M. Jr, and Haqq-Misra, J. 2013, 'Double Catastrophe: Intermittent Stratospheric Geoengineering Induced by Societal Collapse', *Environ. Syst. Decis.* **33**, 168–80.

Baxter, S. 2000, 'The Planetarium Hypothesis: A Resolution of the Fermi Paradox', *J. Br. Interplan. Soc.* **54**, 210–16.

Baxter, S. 2002, *Manifold: Space* (Del Rey, New York).

Baxter, S. 2012, 'Where Was Everybody? Olaf Stapledon and the Fermi Paradox', *J. Br. Interplan. Soc.* **65**, 7–12.

Beane, S. R., Davoudi, Z., and Savage, M. J. 2014, 'Constraints on the Universe as a Numerical Simulation', *Eur. Phys. J. A* **50**, article id. 148 (9 pp.).

Beatty, J. 2006, 'Replaying Life's Tape', *J. Philos.* **103**, 336–62.

Beck, L. W. 1972, 'Extraterrestrial Intelligent Life', *P. Am. Philos. Soc.* **45**, 5–21.

Bedau, M. A. 2003, 'Artificial Life: Organization, Adaptation and Complexity from the Bottom up', *Trends Cogn. Sci.* **7**, 505–12.

Beech, M. 1990, 'Blue Stragglers as Indicators of Extraterrestrial Civilisations?' *Earth Moon Planets* **49**, 177–86.

Beech, M. 1993, 'Aspects of an Astroengineering Option', *J. Br. Interplan. Soc* **46**, 317–23.

Beech, M. 2007, *Rejuvenating the Sun and Avoiding Other Global Catastrophes* (Springer, New York).

Beech, M. 2010, 'A Dark Sun Rising: It's a Solar Wrap', *J. Brit. Interplanet. Soc.* **63**, 104–7.

Beech, M. 2011a, 'Exploring α Centauri: From Planets, to a Cometary Cloud, and Impact Flares on Proxima', *Observatory* **131**, 212–24.

Beech, M. 2011b, 'The Past, Present and Future Supernova Threat to Earth's Biosphere', *Astrophys. Space Sci.* **336**, 287–302.

Beer, M. E., King, A. R., Livio, M., and Pringle, J. E. 2004, 'How Special is the Solar System?' *Mon. Notices Royal Astron. Soc.* **354**, 763–8.

Behroozi, P. and Peeples, M. S. 2015, 'On the History and Future of Cosmic Planet Formation', *Mon. Notices Royal Astron. Soc.* **454**, 1811–17.

Beisbart, C. and Jung, T. 2006, 'Privileged, Typical, or Not Even That? Our Place in the World According to the Copernican and the Cosmological Principles', *J Gen Philos Sci* **37**, 225–56.

Belbruno, E. and Gott, J. R., III. 2005, 'Where Did the Moon Come from?' *Astron. J.* **129**, 1724–45.

Bell, E. A., Boehnke, P., Harrison, T. M., and Mao, W. L. 2015, 'Potentially Biogenic Carbon Preserved in a 4.1 Billion-Year-Old Zircon', *P Natl. Acad. Sci. USA* **112**, 14518–21.

Beltzer, T. 2000, '*Last Year at Marienbad*: An Intertextual Meditation', *Senses of Cinema*, vol. **10** (online article, http://sensesofcinema.com/2000/10/marienbad/, last accessed 15 September 2015).

Benford, G. 1977, *In the Ocean of Night* (The Dial Press/James Wade, New York).

Benford, G. 1983, *Across the Sea of Suns* (Simon & Schuster, New York).

Benford, G. 1996, *Sailing Bright Eternity* (Bantam Spectra, New York).

Benítez, N., Maíz-Apellániz, J., and Canelles, M. 2002, Evidence for Nearby Supernova Explosions', *Phys. Rev. Lett.* **88**, 081101.

Benner, S. 2009, *Life, the Universe... and the Scientific Method* (The FfAME Press, Gainesville).

Benner, S. A. 2010, 'Defining Life', *Astrobiology* **10**, 1021–30.

Bennett, J. O. and Shostak, S. 2011, *Life in the Universe* (3rd edition, Benjamin Cummings, San Francisco).

Bennett, C. L. et al. 2013, 'Nine-Year Wilkinson Microwave Anisotropy Probe (WMAP) Observations: Final Maps and Results', *Astrophys. J. Suppl.* **208**, article id. 20 (54 pp.).

Benton, M. J. 2003, *When Life Nearly Died: The Greatest Mass Extinction of All Time* (Thames and Hudson, London).

Beyerchen, A. 1992, 'What We Now Know About Nazism and Science', *Social Research* **59**, 615–41.

Binney, J. and Merrifield, M. 1998, *Galactic Astronomy* (Princeton University Press, Princeton).

Binney, J. and Tremaine, S. 2008, *Galactic Dynamics* (2nd edition, Princeton University Press, Princeton).

Bioy Casares, A. [1940] 1996, *La invención de Morel* (Penguin Books, London).

Birch, P. 1982, 'Orbital Ring Systems and Jacob's Ladders: I', *J. Brit. Interplanet. Soc.* **35**, 475–97.

Birch, P. 1983, 'Orbital Ring Systems and Jacob's Ladders: III', *J. Brit. Interplanet. Soc.* **36**, 231–8.

Birch, P. 1990, 'The Peer Hypothesis' (unpublished manuscript, http://buildengineer.com/www.paulbirch.net/PeerHypothesis.pdf, last accessed 15 September 2015).

Birch, P. 1991, 'Supramundane Planets', *J. Brit. Interplanet. Soc.* **44**, 169–82.

Bjørk, R. 2007, 'Exploring the Galaxy Using Space Probes', *Int. J. Astrobiol.* **6**, 89–93.

Blanton, M. R. et al. 2003, 'The Galaxy Luminosity Function and Luminosity Density at Redshift $z = 0.1$', *Astrophys. J.* **592**, 819–38.

Blish, J. 1957, *The Seedling Stars* (Gnome Press, New York).

Bolonkin, A. A. 2005, *Non-Rocket Space Launch and Flight* (Elsevier, London).

Boltzmann, L. 1895, 'On Certain Questions of the Theory of Gases', *Nature* **51,** 413–14.

Bond, A. 1982, 'On the Improbability of Intelligent Extraterrestrials', *J. Brit. Interplanet. Soc.* **35**, 195–207.

Bond, H. E., Nelan, E. P., VandenBerg, D. A., Schaefer, G. H., and Harmer, D. 2013, 'HD 140283: A Star in the Solar Neighborhood that Formed Shortly after the Big Bang', *Astrophys. J. Lett.* **765**, article id. L12 (5 pp.).

Bondi, H. 1961, *Cosmology* (2nd edition, Cambridge University Press, London).

Bondi, H. and Gold, T. 1948, 'The Steady-State Theory of the Expanding Universe', *Mon. Not. R. Astr. Soc.* **108**, 252–70.

Borra, E. F. 2012, 'Searching for Extraterrestrial Intelligence Signals in Astronomical Spectra, Including Existing Data', *Astron. J.* **144**, article id. 181 (7 pp.).

Bostrom, N. 2001, 'The Doomsday Argument, Adam & Eve, $UN^{++}$ and Quantum Joe', *Synthese* **127**, 359–87.

Bostrom, N. 2002, *Anthropic Bias: Observation Selection Effects in Science and Philosophy* (Routledge, New York).

Bostrom, N. 2003a, 'Are You Living in a Computer Simulation?', *Philos. Q.* **53**, 243–55.

Bostrom, N. 2003b, 'Astronomical Waste: The Opportunity Cost of Delayed Technological Development', *Utilitas* **5**, 308–14.

Bostrom, N. 2006, 'What is a Singleton?', *Linguist. Philos. Invest.* **5**, 48–54.

Bostrom, N. 2009, 'The Future of Humanity', in J.-K. B. Olsen, E. Selinger, and S. Riis (eds.) *New Waves in Philosophy of Technology* (Palgrave McMillan, New York), 186–216.

Bostrom, N. 2014, *Superintelligence: Paths, Dangers, Strategies* (Oxford University Press, Oxford).

Bostrom, N. and Ćirković, M. M. (eds.). 2008, *Global Catastrophic Risks* (Oxford University Press, Oxford).

Boudry, M., Blancke, S., and Braeckman, J. 2010, 'How Not to Attack Intelligent Design Creationism: Philosophical Misconceptions About Methodological Naturalism', *Found. Sci.* **15**, 227–44.

Bougroura, H. and Mimouni, J. 2011, 'Threshold Distance of Supernova Explosion Emitting Neutrino Flux Dangerous to Life on Earth', *Afr. Rev. Phys.* **6**, article id. 0022 (5 pp.).

Bounama, C., von Bloh, W., and Franck, S. 2007, 'How Rare Is Complex Life in the Milky Way?' *Astrobiology* **7**, 745–55.

Bouvier, A. and Wadhwa, M. 2010, 'The Age of the Solar System Redefined by the Oldest Pb-Pb Age of a Meteoritic Inclusion', *Nat. Geosci.* **3**, 637–41.

Bovy, J. et al. 2014, 'The APOGEE Red-clump Catalog: Precise Distances, Velocities, and High-resolution Elemental Abundances over a Large Area of the Milky Way's Disk', *Astrophys. J.* **790**, article id. 127 (21 pp.).

Bowler, P. J. 1992, *The Eclipse of Darwinism* (The Johns Hopkins University Press, Baltimore).

Bowyer, S., Zeitlin, G., Tarter, J., Lampton, M., and Welch, W. J. 1983, 'The Berkeley Parasitic SETI Program', *Icarus* **53**, 147–55.

Boyajian, T. S. et al. 2016, 'Planet Hunters X. KIC 8462852: Where's the Flux?' *Mon. Not. R. Astr. Soc.* **457**, 3988–4004.

Bracewell, R. N. 1970, 'Communications from Superior Galactic Communities', *Nature* **186**, 670–1.

Bracewell, R. N. 1975, *The Galactic Club: Intelligent Life in Outer Space* (W. H. Freeman, San Francisco).

Bradbury, R. J. 1997, 'Matrioshka Brains' (unpublished manuscript, http://web.archive.org/web/20100115101633/http://www.aeiveos.com/~bradbury/MatrioshkaBrains/MatrioshkaBrainsPaper.html, last accessed 15 September 2015).

Bradbury, R. J., Ćirković, M. M., and Dvorsky, G. 2011, 'Dysonian Approach to SETI: A Fruitful Middle Ground?' *J. Brit. Interplanet. Soc.* **64**, 156–65.

Bradbury, J. W. and Vehrencamp, S. L. 2011, *Principles of Animal Communication* (Sinauer Associates Inc., Sunderland, MA).

Brakenridge, G. R. 1981, 'Terrestrial Paleoenvironmental Effects of a Late Quaternary-Age Supernova', *Icarus* **46**, 81–93.

Brakenridge, G. R. 2011, 'Core-Collapse Supernovae and the Younger Dryas/Terminal Rancholabrean Extinctions', *Icarus* **215**, 101–6.

Brill, R. H. 1962, 'A Note on the Scientist's Definition of Glass', *J. Glass Stud.* **4**, 127–38.

Brillouin, L. 1962, *Science and Information Theory* (Academic Press, New York).

Brin, G. D. 1983, 'The "Great Silence": The Controversy Concerning Extraterrestrial Intelligence', *Q. J. Roy. Astron. Soc.* **24**, 283–309.

Brin, G. D. 2014, 'SETI, METI and the Paradox of Extraterrestrial Life: Is There a Libertarian Perspective?' (online article, http://www.cato-unbound.org/2014/12/01/david-brin/seti-meti-paradox-extraterrestrial-life-there-libertarian-perspective, last accessed 15 August 2015).

Broderick, D. 2003, 'New Wave and Backwash: 1960–1980', in E. James and F. Mendelsohn (eds.) *The Cambridge Companion to Science Fiction* (Cambridge University Press, Cambridge), 48–63.

Brown, J. R. 1993, *The Laboratory of the Mind: Thought Experiments in the Natural Sciences* (Routledge, London).

Brown, W. R., Geller, M. J., Kenyon, S. J., and Kurtz, M. J. 2005, 'Discovery of an Unbound Hypervelocity Star in the Milky Way Halo', *Astrophys. J.* **622**, L33–6.

Brown, W. R., Geller, M. J., Kenyon, S. J., and Kurtz, M. J. 2007, 'Hypervelocity Stars. III. The Space Density and Ejection History of Main-Sequence Stars from the Galactic Center', *Astrophys. J.* **671**, 1708–16.

Browne, D. 2004, 'Do Dolphins Know Their Own Minds?' *Biol. Philos.* **19**, 633–53.

Buccino, A. P., Lemarchand, G. A., and Mauas, P. J. D. 2006, 'Ultraviolet Constraints around the Circumstellar Habitable Zones', *Icarus* **183**, 491–503.

Burchell, M. J. 2006, 'W(h)ither the Drake Equation?' *Int. J. Astrobiol.* **5**, 243–50.

Burroughs, W. S. 1981, *Cities of the Red Night* (Viking Press, New York).

Butterfield, H. [1949] 1997, *The Origins of Modern Science* (Free Press, New York).

Byakov, V. M., Stepanov, S. V., and Stepanova, O. P. 1997, 'Quasi-Regular Staying of Solar System in Supernova Remnants and Natural Earth History', *Radiat. Phys. Chem.* **49**, 299–305.

Byakov, V. M., Stepanov, S. V., and Stepanova, O. P. 2001, 'Role of Ionizing Radiation in the Natural History of Earth', *Radiat. Phys. Chem.* **60**, 297–301.

Caldeira, K., Bala, G., and Cao, L. 2013, 'The Science of Geoengineering', *Annu. Rev. Earth Planet. Sci.* **41**, 231–56.

Caldeira, K. and Kasting, J. F. 1992, 'The Life Span of the Biosphere Revisited', *Nature* **360**, 721–3.

Caldwell, R. R. and Stebbins, A. 2008, 'A Test of the Copernican Principle', *Phys. Rev. Lett.* **100**, article id. 191302 (4 pp.).

Calude, C. S. 2002, *Information and Randomness: An Algorithmic Perspective* (Springer, Berlin).

Calvino, I. [1969] 2009, *The Complete Cosmicomics* (translated by M. McLaughlin, T. Parks, and W. Weaver; Penguin Books, London).

Cameron, J. (director). 2009, *Avatar* (DVD, Lightstorm Entertainment, Dune Entertainment, Ingenious Media/20th Century Fox).

Campbell, J. B. 2006, 'Archaeology and Direct Imaging of Exoplanets', in C. Aime and F. Vakili (eds.) *Direct Imaging of Exoplanets: Science and Techniques*, Proceedings of the IAU Colloquium No. 200 (Cambridge University Press, Cambridge), 247–50.

Cantrell, J. R., Henry, T. J., and White, R. J. 2013 'The Solar Neighborhood XXIX: The Habitable Real Estate of Our Nearest Stellar Neighbors', *Astron. J.* **146**, article id. 99 (20 pp.).

Canup, R. and Asphaug, E. 2001, 'Origin of the Moon in a Giant Impact near the End of the Earth's Formation', *Nature* **412**, 708–12.

Čapek, K. [1922] 1975, *Krakatit* (Rourke Publishing, Vero Beach, FL).

Caplan, B. 2008, 'The Totalitarian Threat', in N. Bostrom and M. M. Ćirković (eds.) *Global Catastrophic Risks* (Oxford University Press, Oxford), 498–513.

Carlotto, M. J. 2007, 'Detecting Patterns of a Technological Intelligence in Remotely Sensed Imagery', *J. Brit. Interplanet. Soc.* **60**, 28–39.

Carrier, R. C. 2004, 'The Argument from Biogenesis: Probabilities against a Natural Origin of Life', *Biol. Philos.* **19**, 739–64.

Carrigan, R. A., Jr. 2006, 'Do Potential SETI Signals Need To Be Decontaminated?', *Acta Astronaut.* **58**, 112–17.

Carrigan, R. A., Jr. 2009, 'IRAS-Based Whole-Sky Upper Limit on Dyson Spheres', *Astrophys. J.* **698**, 2075–86.

Carrigan, R. A., Jr. 2012, 'Is Interstellar Archeology Possible?' *Acta Astronaut.* **78**, 121–6.

Carroll, S. M. 2006, 'Is Our Universe Natural?' *Nature* **440**, 1132–6.

Carruth, S. (director). 2013, *Upstream Color* (DVD, Independent Pictures/Metrodome Dist.).

Carter, B. 1983, 'The Anthropic Principle and its Implications for Biological Evolution', *Philos. Trans. R. Soc. London A* **310**, 347–63.

Carter, B. 1993, 'The Anthropic Selection Principle and the Ultra-Darwinian Synthesis', in F. Bertola and U. Curi (eds.) *The Anthropic Principle, Proceedings of the Second Venice Conference on Cosmology and Philosophy* (Cambridge University Press, Cambridge), 33–66.

Carter, B. 2008, 'Five- or Six-Step Scenario for Evolution?' *Int. J. Astrobiol.* **7**, 177–82.

Carter, B. 2012, 'Hominid Evolution: Genetics versus Memetics', *Int. J. Astrobiol.* **11**, 3–13.

Cartin, D. 2014, 'Quantifying the Fermi Paradox in the Local Solar Neighbourhood', *J. Brit. Interplanet. Soc.* **67**, 119–26.

Cassan, A. et al. 2012, 'One or More Bound Planets per Milky Way Star from Microlensing Observations', *Nature* **481**, 167–9.

Casti, J. L. 1989, *Paradigms Lost: Images of Man in the Mirror of Science* (William Morrow and Comp., New York).

Cavicchioli, R. 2002, 'Extremophiles and the Search for Extraterrestrial Life', *Astrobiology* **2**, 281–92.

Chaisson, E. J. 1997, 'The Rise of Information in an Evolutionary Universe', *World Futures* **50**, 447–55.

Chaisson, E. J. 2001, *Cosmic Evolution: The Rise of Complexity in Nature* (Harvard University Press, Cambridge, MA).

Chaisson, E. J. 2003, 'A Unifying Concept for Astrobiology', *Int. J. Astrobiol.* **2**, 91–101.

Chaitin, G. 2006, 'From Leibniz to Ω: Epistemology as Information Theory', *Collapse* **1**, 27–51.

Chalmers, D. J. 2010, 'The Singularity: A Philosophical Analysis', *J. Conscious. Stud.* **17**, 7–65.

Chaouchi, H. 2010, *The Internet of Things* (Wiley-ISTE, London).

Chela-Flores, J. 2003, 'Testing Evolutionary Convergence on Europa', *Int. J. Astrobiol.* **2**, 307–12.

Chennamangalam, J., Siemion, A. P. V., Lorimer, D. R., and Werthimer, D. 2015, 'Jumping the Energetics Queue: Modulation of Pulsar Signals by Extraterrestrial Civilizations', *New Astron.* **34**, 245–9.

Chernavskii, D. S. 2000, 'The Origin of Life and Thinking from the Viewpoint of Modern Physics', *Physics-Uspekhi* **43**, 151–76.

Cholbi, M. 2011, *Suicide: The Philosophical Dimensions* (Broadview, Peterborough).

Chon-Torres, O. 2017, 'Astrobioethics', *Int. J. Astrobiol.*, **17**, 51–6.

Chopra, A. and Lineweaver, C. H. 2016, 'The Case for a Gaian Bottleneck: The Biology of Habitability', *Astrobiology* **16**, 7–22.

Christodoulou, D. 1970, 'Reversible and Irreversible Transformations in Black-Hole Physics', *Phys. Rev. Lett.* **25**, 1596–7.

Chyba, C. F. 1997, 'Catastrophic Impacts and the Drake Equation', in C. B. Cosmovici, S. Bowyer, and D. Werthimer (eds.) *Astronomical and Biochemical Origins and the Search for Life in the Universe* (Editrice Compositori, Bologna), pp. 157–64.

Chyba, C. F. and Hand, K. 2005, 'Astrobiology: The Study of the Living Universe', *Annu. Rev. Astron. Astrophys.* **43**, 31–74.

Ćirković, M. M. 2003, 'The Thermodynamical Arrow of Time: Reinterpreting the Boltzmann-Schuetz Argument', *Found. Phys.* **33**, 467–90.

Ćirković, M. M. 2004a, 'Earths: Rare in Time, Not Space?' *J. Brit. Interplanet. Soc.* **57**, 53–9.

Ćirković, M. M. 2004b, 'On the Temporal Aspect of the Drake Equation and SETI, *Astrobiology* **4**, 225–31.

Ćirković, M. M. 2004c, 'The Anthropic Principle and the Duration of the Cosmological Past', *Astronomical and Astrophysical Transactions* **23**, 567–97.

Ćirković, M. M. 2005, '"Permanence" - An Adaptationist Solution to Fermi's Paradox?' *J. Brit. Interplanet. Soc.* **58**, 62–70.

Ćirković, M. M. 2006, 'Too Early? On the Apparent Conflict of Astrobiology and Cosmology', *Biol. Philos.* **21**, 369–79.

Ćirković, M. M. 2008, 'Against the Empire', *J. Brit. Interplanet. Soc.* **61**, 246–54.

Ćirković, M. M. 2009, 'Fermi's Paradox: The Last Challenge for Copernicanism?' *Serb. Astron. J.* **178**, 1–20.

Ćirković, M. M. 2012, *The Astrobiological Landscape: Philosophical Foundations of the Study of Cosmic Life* (Cambridge University Press, Cambridge).

Ćirković, M. M. 2013, 'Who Are the SETI Skeptics?' *Acta Astronaut.* **89**, 38–45.

Ćirković, M.M. 2014, 'Evolutionary Contingency and SETI Revisited', *Biol. Philos.* **29**, 539–57.

Ćirković, M. M. 2015, 'Kardashev's Classification at 50+: A Fine Vehicle with Room for Improvement', *Serb. Astron. J.* **191**, 1–15.

Ćirković, M. M. 2016, 'Stellified Planets and Brown Dwarfs as novel Dysonian SETI Signatures', *J. Brit. Interplanet. Soc.* **69**, 92–6.

Ćirković, M. M. and Bradbury, R. J. 2006, 'Galactic Gradients, Postbiological Evolution and the Apparent Failure of SETI', *New Ast.* **11**, 628–39.

Ćirković, M. M. and Cathcart, R. B. 2004, 'Geo-Engineering Gone Awry: A New Partial Solution of Fermi's Paradox', *J. Brit. Interplanet. Soc.* **57**, 209–15.

Ćirković, M. M., Dragićević, I., and Berić-Bjedov, T. 2005, 'Adaptationism Fails to Resolve Fermi's Paradox', *Serb. Astron. J.* **170**, 89–100.

Ćirković, M. M. and Lanzetta, K. M. 2000, 'On the Small-Scale Clustering of Lyα Forest Clouds', *Mon. Notices Royal Astron. Soc.* **315**, 473–8.

Ćirković, M. M., Sandberg, A., and Bostrom, N. 2010, 'Anthropic Shadow: Observation Selection Effects and Human Extinction Risks', *Risk Anal.* **30**, 1495–506.

Ćirković M. M. and Vukotić, B. 2008, 'Astrobiological Phase Transition: Towards Resolution of Fermi's Paradox', *Orig. Life Evol. Biospheres* **38**, 535–47.

Ćirković, M. M. and Vukotić, B. 2016, 'Long-Term Prospects: Mitigation of Supernova and Gamma-Ray Burst Threat to Intelligent Beings,' *Acta Astronautica*, **129**, 438–46.

Ćirković, M. M., Vukotić, B., and Dragićević, I. 2008, 'Galactic "Punctuated Equilibrium": How to Undermine Carter's Anthropic Argument in Astrobiology', *Astrobiology* **9**, 491–501.

Clark, A. and Chalmers, D. 1998, 'The Extended Mind', *Analysis* **58**, 7–19.

Clarke, A. C. 1953, *Childhood's End* (Ballantine Books, New York).

Clarke, A. C. 1968, *2001: A Space Odyssey* (Hutchinson, London).

Clarke, J. N. 1981, 'Extraterrestrial Intelligence and Galactic Nuclear Activity', *Icarus* **46**, 94–6.

Cleland, C. E. 2012, 'Life without Definitions', *Synthese* **185**, 125–44.

Cleland, C. E. and Chyba, C. F. 2002, 'Defining "Life"', *Orig. Life Evol. Biosph.* **32**, 387–93.

Cleland, C. E. and Copley, S. D. 2005, 'The Possibility of Alternative Microbial Life on Earth', *Int. J. Astrobiol.* **4**, 165–73.

Clement, H. 1971, *Star Light* (Ballantine Books, New York).

Clube, S. V. M. and Napier, W. M. 1984, 'The Microstructure of Terrestrial Catastrophism', *Mon. Not. R. Astron. Soc.* **211**, 953–68.

Clube, S. V. M. and Napier, W. M. 1990, *The Cosmic Winter* (Basil Blackwell, Oxford).

Cocconi, G. and Morrison, P. 1959, 'Searching for Interstellar Communications', *Nature* **184**, 844–6.

Cockell, C. S. 2007, *Space on Earth: Saving Our World by Seeking Others* (Macmillan, New York).

Cockell, C. S. 2008, 'An Essay on Extraterrestrial Liberty', *J. Brit. Interplanet. Soc.* **61**, 255–75.

Cockell, C. S. 2014, 'Habitable Worlds with No Signs of Life', *Phil. Trans. R. Soc. A* **372**, article id. 20130082 (15 pp.).

Cohen, J. and Stewart, I. 2002, *What Does a Martian Look Like? The Science of Extraterrestrial Life* (John Wiley & Sons, Hoboken, NJ).

Colavito, J. 2005, *The Cult of Alien Gods: H. P. Lovecraft and Extraterrestrial Pop Culture* (Prometheus Books, Amherst).

Collar, J. I. 1996, 'Biological Effects of Stellar Collapse Neutrinos', *Phys. Rev. Lett.* **76**, 999–1002.

Condon, E. U. and Gillmor, D. S. 1969, *Final Report of the Scientific Study of Unidentified Flying Objects* (Dutton, New York).

Conway Morris, S. 1998, *The Crucible of Creation* (Oxford University Press, Oxford).

Conway Morris, S. 2003a, *Life's Solution: Inevitable Humans in a Lonely Universe* (Cambridge University Press, Cambridge).

Conway Morris, S. 2003b, 'The Navigation of Biological Hyperspace', *Int. J. Astrobiol.* **2**, 149–52.

Conway Morris, S. 2018, 'Three Explanations for Extraterrestrials: Sensible, Unlikely, Mad', *Int. J. Astrobiol.* **First View**, http://doi.org/10.1017/S1473550416000379.

Cooper, M. W. 1991, *Behold A Pale Horse* (Light Technology Publishing, Flagstaff).

Cooper, J. 2013, 'Bioterrorism and the Fermi Paradox', *Int. J. Astrobiol.* **12**, 144–8.

Cooper, K. 2014, 'The Interstellar Ethics of Self-Replicating Probes', *J. Brit. Interplanet. Soc.* **67**, 258–60.

Cornwell, J. 2003, *Hitler's Scientists: Science, War and the Devil's Pact* (Penguin Books, London).

Coulter, G. R., Klein, M. J., Backus, P. R., and Rummel, J. D. 1994, 'Searching for Intelligent Life in the Universe: NASA's High Resolution Microwave Survey', *Adv. Space Biol. Med.* **4**, 189–224.

Cramer, J. G. 1986, 'The Pump of Evolution', *Analog Science Fiction & Fact Magazine*, January issue, 124–7.

Crane, L. 1994, 'Possible Implications of the Quantum Theory of Gravity' (preprint, http://arxiv.org/abs/hep-th/9402104, last accessed 15 December 2017).

Cranford, J. L. 2015, *Astrobiological Neurosystems: Rise and Fall of Intelligent Life Forms in the Universe* (Springer, New York).

Crawford, I. A. 1993, 'Space, World Government, and "The End of History"', *J. Brit. Interplanet. Soc.* **46**, 415–20.

Crawford, I. A. 2009, 'The Astronomical, Astrobiological and Planetary Science Case for Interstellar Spaceflight', *J. Brit. Interplanet. Soc.* **62**, 415–21.

Crawford, L. 2013, 'Freak Observers and the Simulation Argument', *Ratio* **26**, 250–64.

Crick, F. H. C. and Orgel, L. E. 1973, 'Directed Panspermia', *Icarus* **19**, 341–6.

Criswell, D. 1985, 'Solar System Industrialization: Implications for Interstellar Migration', in B. Finney and E. Jones, *Interstellar Migration and the Human Experience* (University of California Press, Oakland), 50–87.

Cronin, J. W. 2004, *Fermi Remembered.* (University of Chicago Press, Chicago).

Croswell, K. 1988, 'Does Barnard's Star have Planets', *Astronomy* **16** (March issue), 15–17.

Crowe, M. J. 1986, *The Extraterrestrial Life Debate 1750–1900* (Cambridge University Press, Cambridge).

Crowe, M. J. 1997, 'A History of the Extraterrestrial Life Debate', *Zygon* **32**, 147–62.

Crutzen, P. J. and Brühl, C. 1996, 'Mass Extinctions and Supernova Explosions', *Proc. Natl. Acad. Sci. USA* **93**, 1582–4.

Crutzen, P. J. and Stoermer, E. F. 2000, 'The "Anthropocene"', *Global Change Newslett.* **41**, 17–18.

Csicsery-Ronay, I., Jr. 2013, 'The Summa and the Fiction', *Science Fiction Stud.* **40**, 451–62.

Dar, A. 1997, 'Life Extinctions by Neutron Star Mergers', in Y. Giraud-Heraud and J. Tran Thanh Van (eds.) *Very High Energy Phenomena in the Universe; Morion Workshop* (Editions Frontieres, Paris), 379–86.

Dar, A. and De Rújula, A. 2002, 'The Threat to Life from Eta Carinae and Gamma-Ray Bursts', in A. Morselli and P. Picozza (eds.) *Astrophysics and Gamma Ray Physics in Space*, Frascati Physics Series Vol. 24 (Laboratori Nazionali di Frascati, Frascati), 513–23.

Dar, A., Laor, A., and Shaviv, N. J. 1998, 'Life Extinctions by Cosmic Ray Jets', *Phys. Rev. Lett.* **80**, 5813–16.

Darling, D. 2001, *Life Everywhere: The Maverick Science of Astrobiology* (Basic Books, New York).

Dartnell, L. R. 2011, 'Ionizing Radiation and Life', *Astrobiology* **11**, 551–82.

Davies, P. C. W. 1978, 'Cosmic Heresy?' *Nature* **273**, 336–7.

Davies, P. C. W. 1995, *Are We Alone? Philosophical Implications of the Discovery of Extraterrestrial Life* (Basic Books, New York).

Davies, P. C. W. 2010, *The Eerie Silence: Renewing Our Search for Alien Intelligence* (Houghton Mifflin Harcourt, Boston).

Davies, P. C. W. 2012, 'Footprints of Alien Technology', *Acta Astronaut.* **73**, 250–7.

Davies, P. C. W., Benner, S. A., Cleland, C. E., Lineweaver, C. H., McKay, C. P., and Wolfe-Simon, F. 2009, 'Signatures of a Shadow Biosphere', *Astrobiology* **9**, 241–9.

Davies, P. C. W. and Wagner, R. V. 2013, 'Searching for Alien Artifacts on the Moon', *Acta Astronaut.* **89**, 261–5.

Dawkins, R. 1989, *The Selfish Gene* (Oxford University Press, Oxford).

Dayal, P., Cockell, C., Rice, K., and Mazumdar, A. 2015, 'The quest for cradles of life: using the fundamental metallicity relation to hunt for the most habitable type of galaxy,' Astrophys. J. Lett. 810, article id. L2 (5pp).

Deardorff, J. W. 1986, 'Possible Extraterrestrial Strategy for Earth', *Q. J. Roy. Astron. Soc.* **27**, 94–101.

Deardorff, J. W. 1987, 'Examination of the Embargo Hypothesis as an Explanation for the Great Silence', *J. Brit. Interplanet. Soc.* **40**, 373–9.

Deardorff, J., Haisch, B., Maccabee, B., and Puthof, H. E. 2005, 'Inflation-Theory Implications for Extraterrestrial Visitation', *J. Brit. Interplanet. Soc.* **58**, 43–50.

DeBiase, R. L. 2008, 'Effects of Collisions upon a Partial Dyson Sphere', *J. Brit. Interplanet. Soc.* **61**, 386–94.

deGrasse Tyson, N. 2012, *Space Chronicles: Facing the Ultimate Frontier* (W. W. Norton, New York).

De Groot, G. 2006, *Dark Side of the Moon: The Magnificent Madness of the American Lunar Quest* (NYU Press, New York).

del Peloso, E. F., da Silva, L., and Arany-Prado, L. I. 2005, 'The Age of the Galactic Thin Disk from Th/Eu Nucleocosmochronology. II. Chronological Analysis', *Astron. Astrophys.* **434**, 301–8.

Dennett, D. 1995, *Darwin's Dangerous Idea: Evolution and the Meanings of Life* (Simon & Schuster, New York).

Dermer, C. D. and Holmes, J. M. 2005, 'Cosmic Rays from Gamma-Ray Bursts in the Galaxy', *Astrophys. J.* **628**, L21–4.

de San, M. G. 1981, 'The Ultimate Destiny of an Intelligent Species: Everlasting Nomadic Life in the Galaxy', *J. Brit. Interplanet. Soc.* **34**, 219–37.

de Smit, B. and Lenstra, H. W. 2003, 'The Mathematical Structure of Escher's Print Gallery', *Notices Amer. Math. Soc* **50**, 446–51.

de Sousa António, M. R. and Schulze-Makuch, D. 2011, 'The Power of Social Structure: How We Became an Intelligent Lineage', *Int. J. Astrobiol.* **10**, 15–23.

Des Marais, D. J. and Walter, M. R. 1999, 'Astrobiology: Exploring the Origins, Evolution, and Distribution of Life in the Universe', *Annu. Rev. Ecol. Syst.* **30**, 397–420.

Des Marais, D. J. et al. 2008, 'The NASA Astrobiology Roadmap', *Astrobiology* **8**, 715–30.

di Lampedusa, G. T. [1960] 2007, *The Leopard: A Novel* (translated by A. Colquhuon; Pantheon Books, New York).

Di Scala, S. M. 2005, 'Science and Fascism: The Case of Enrico Fermi', *Politics Religion Ideol.* **6**, 199–211.

Diamond, J. 1999, 'To Whom It May Concern', *The New York Times Magazine*, 5 December 1999, 68–9.

Diamond, J. 2005, *Collapse: How Societies Choose to Fail or Succeed* (Viking Press, New York).

Dick, P. K. 1962, *The Man in the High Castle* (Putnam, New York).

Dick, S. J. 1996, *The Biological Universe: The Twentieth-Century Extraterrestrial Life Debate and the Limits of Science* (Cambridge University Press, Cambridge).

Dick, S. J. 1997, 'The Biophysical Cosmology: The Place of Bioastronomy in the History of Science', in C. B. Cosmovici, S. Bowyer, and D. Werthimer (eds.) *Astronomical and Biochemical Origins and the Search for Life in the Universe*, Proceedings of the IAU Colloquium No. 161 (Editrice Compositori, Bologna), 785–8.

Dick, S. J. 2003, 'Cultural Evolution, the Postbiological Universe and SETI', *Int. J. Astrobiol.* **2**, 65–74.

Dick, S. J. and Lupisella, M. L. (eds.). 2010, *Cosmos and Culture: Cultural Evolution in a Cosmic Context* (NASA, Washington, DC).

Dijkstra, E. W. 1989, 'By Way of Introduction' (unpublished manuscript EWD1041, http://www.cs.utexas.edu/users/EWD/transcriptions/EWD10xx/EWD1041.html, last accessed 20 October 2016).

Dittmer, J. N. 2007, 'Colonialism and Place Creation in *Mars Pathfinder* Media Coverage', *Geogr. Rev.* **97**, 112–30.

Dodd, M. S. et al. 2017, 'Evidence for Early Life in Earth's Oldest Hydrothermal Vent Precipitates', *Nature* **543**, 60–4.

Doolittle, W. F. 2014, 'Natural Selection through Survival Alone, and the Possibility of Gaia', *Biol. Philos.* **29**, 415–23.

Drake, F. 1976, 'On Hands and Knees in Search of Elysium', *MIT Technol. Rev.* **78** (June), 22–9.

Drake, F. 2011, 'The Search for Extra-Terrestrial Intelligence', *Phil. Trans. R. Soc. A* **369**, 633–43.

Drake, F. D. 1962, *Intelligent Life in Space* (Macmillan, New York).

Driscoll, P. E. and Barnes, R. 2015, 'Tidal Heating of Earth-like Exoplanets around M Stars: Thermal, Magnetic, and Orbital Evolutions', *Astrobiology* **15**, 739–60.

Dudley-Flores, M. and Gangale, T. 2012, 'Forecasting the Political Economy of the Inner Solar System', *Astropolitics* **10**, 183–233.

Dunér, D., Parthemore, J., Persson, E., and Holmberg, G. 2013, *The History and Philosophy of Astrobiology: Perspectives on Extraterrestrial Life and the Human Mind* (Cambridge Scholars Publishing, Newcastle upon Tyne).

Duric, N. and Field, L. 2003, 'On the Detectability of Intelligent Civilizations in the Galaxy', *Serb. Astron. J.* **167**, 1–10.

Dyson, F. J. 1960a, 'Search for Artificial Stellar Sources of Infrared Radiation', *Science* **131**, 1667–8.

Dyson, F. J. 1960b, 'Reply', *Science* **132**, 252–3.

Dyson, F. J. 1966, 'The Search for Extraterrestrial Technology', in R. E. Marshak, ed., *Perspectives in Modern Physics* (Interscience Publishers, New York), 641–55.

Dyson, F. J. 1979, 'Time without End: Physics and Biology in an Open Universe', *Rev. Mod. Phys.* **51**, 447–60.

Early, J. T. 1989, 'Space-Based Solar Shield to Offset Greenhouse Effect', *J. Brit. Interplanet. Soc.* **42**, 567–9.

Eco, U. 2009, *The Infinity of Lists: An Illustrated Essay* (Rizzoli, New York).

Eden, A., Moor, J. H., Søraker, J. H., and Steinhart, E. (eds.). 2012, *Singularity Hypotheses: A Scientific and Philosophical Assessment* (Springer, Berlin).

Egan, G. 1997, *Diaspora* (Orion/Millennium, London).

Egan, G. 2002, *Schild's Ladder* (HarperCollins, New York).

Egan, G. 2008a, *Incandescence* (Gollancz, London).

Egan, G. 2008b, 'Riding the Crocodile', in *Dark Integers and Other Stories* (Subterranean Press, Burton), 51–102.

Ehrenfreund, P. et al. (eds.). 2004, *Astrobiology: Future Perspectives* (Kluwer, Dordrecht).

Eigen, M. 1992, *Steps towards Life* (Oxford University Press, Oxford).

Eldredge, N. and Gould, S. J. 1972, 'Punctuated Equilibria: An Alternative to Phyletic Gradualism', in T. M. Schopf (ed.) *Models in Paleobiology* (Freeman Cooper, San Francisco), 82–115.

Ellis, J. et al. (LHC Safety Assessment Group) 2008, 'Review of the Safety of LHC Collisions' (report, http://lsag.web.cern.ch/lsag/LSAG-Report.pdf, last accessed 3 February 2018).

Epstein, R. J. and Zhao, Y. 2009, 'The Threat That Dare Not Speak Its Name', *Perspect. Biol. Med.* **52**, 116–25.

Erlykin, A. D. and Wolfendale, A. W. 2010, 'Long Term Time Variability of Cosmic Rays and Possible Relevance to the Development of Life on Earth', *Surv. Geophys.* **31**, 383–98.

Erwin, D. H. 2006, *Extinction: How Life on Earth Nearly Ended 250 Million Year Ago* (Princeton University Press, Princeton).

Farhi, E. and Guth, A. H. 1987, 'An Obstacle to Creating a Universe in the laboratory', *Phys. Lett. B* **183**, 149–55.

Farhi, E., Guth, A. H., and Guven, J. 1990, 'Is it Possible To Create a Universe In The Laboratory by Quantum Tunneling?' *Nucl. Phys. B* **339**, 417–90.

Fawcett, B., Laine, M., and Nugent, T., Jr. 2006, *Liftport: Opening Space to Everyone* (Meisha Merlin Publishing, Decatur).

Feng, F. and Bailer-Jones, C. A. L. 2013, 'Assessing the Influence of the Solar Orbit on Terrestrial Biodiversity', *Astrophys. J.* **768**, article id. 152 (21 pp.).

Fenner, F., Henderson, D. A., Arita, I., Ježek, Z., and Ladnyi, I. D. 1988, *Smallpox and Its Eradication* (World Health Organization, Geneva).

Ferguson, N. (ed.). 1999, *Virtual History: Alternatives and Counterfactuals* (Basic Books, New York).

Ferrando, F. 2013, 'Posthumanism, Transhumanism, Antihumanism, Metahumanism, and New Materialisms: Differences and Relations', *Existenz* **8**, 26–32.

Feynman, R. P., Leighton, R. B., and Sands, M. 1964, *The Feynman Lectures on Physics* (Addison-Wesley, Reading, MA).

Filipović, M. D., Horner, J., Crawford, E. J., Tothill, N. F. H., and White, G. L. 2013, 'Mass Extinction and the Structure of the Milky Way', *Serb. Astron. J.* **187**, 43–52.

Finney, B. 1990, 'The Impact of Contact', *Acta Astronaut.* **21**, 117–21.

Firestone, R. B. 2014, 'Observation of 23 Supernovae That Exploded <300 pc from Earth during the Past 300 kyr', *Astrophys. J.* **789**, article id. 29 (11 pp.).

Firestone, R. B. et al. 2007, 'Evidence for an Extraterrestrial Impact 12,900 Years Ago That Contributed to the Megafaunal Extinctions and the Younger Dryas Cooling', *Proc. Natl. Acad. Sci. USA* **104**, 16016–21.

Fixsen, D. J. 2009, 'The Temperature of the Cosmic Microwave Background', *Astrophys. J.* **707**, 916–20.

Flores Martinez, C. L. 2014, 'SETI in the Light of Cosmic Convergent Evolution', *Acta Astronaut.* **104**, 341–9.

Fogg, M. J. 1987, 'Temporal Aspects of the Interaction among the First Galactic Civilizations: The "Interdict Hypothesis"', *Icarus* **69**, 370–84.

Fogg, M. J. 1988, 'The Feasibility of Intergalactic Colonisation and its Relevance to SETI', *J. Brit. Interplanet. Soc.* **41**, 491–6.

Fogg, M. J. 1989, 'Stellifying Jupiter: A First Step to Terraforming the Galilean Satellites', *J. Brit. Interplanet. Soc.* **42**, 587–92.

Föllmer, H. 1974, 'Random Economies with Many Interacting Agents', *J. Math. Econ.* **1**, 51–62.

Foote, M. 2003, 'Origination and Extinction through the Phanerozoic: A New Approach', *J. Geol,* **111**, 125–48.

Forgan, D., Dayal, P., Cockell, C., and Libeskind, N. 2017, 'Evaluating Galactic Habitability Using High-Resolution Cosmological Simulations of Galaxy Formation', *Int. J. Astrobiol.* **16**, 60–73.

Forgan, D. H. 2011, 'Spatio-Temporal Constraints on the Zoo Hypothesis, and the Breakdown of Total Hegemony', *Int. J. Astrobiol.* **10**, 341–7.

Forgan, D. H. 2013, 'On the Possibility of Detecting Class A Stellar Engines Using Exoplanet Transit Curves', *J. Brit. Interplanet. Soc.* **66**, 144–54.

Forgan, D. H. and Elvis, M. 2011, 'Extrasolar Asteroid Mining as Forensic Evidence for Extraterrestrial Intelligence', *Int. J. Astrobiol.* **10**, 307–13.

Forgan, D. H. and Rice, K. 2010, 'Numerical Testing of the Rare Earth Hypothesis Using Monte Carlo Realization Techniques', *Int. J. Astrobiol.* **9**, 73–80.

Forward, R. L. 1980, *The Dragon Egg* (Ballantine Books, New York).

Forward, R. L. 1986, 'Feasibility of Interstellar Travel: A Review', *Acta Astronaut.* **14**, 243–52.

Franck, S., Block, A., von Bloh, W., Bounama, C., and Schellnhuber, H.-J. 2001, 'Planetary Habitability: Estimating the Number of Gaias in the Milky Way', in P. Ehrenfreund, O. Angerer, and B. Battrick (eds.) *Proceedings of the First European Workshop on Exo- and Astro-Biology*, ESA SP-496 (ESA Publications Division, Noordwijk), 73–8.

Franck, S., Block, A., von Bloh, W., Bounama, C., and Schellnhuber, H.-J., and Svirezhev, Y. 2000, 'Reduction of Biosphere Life Span as a Consequence of Geodynamics', *Tellus* **52B**, 94–107.

Franck, S., von Bloh, W., and Bounama, C. 2007, 'Maximum Number of Habitable Planets at the Time of Earth's Origin: New Hints for Panspermia and the Mediocrity Principle', *Int. J. Astrobiol.* **6**, 153–7.

Frank, A. and Sullivan, W. 2014, 'Sustainability and the Astrobiological Perspective: Framing Human Futures in a Planetary Context', *Anthropocene* **5**, 32–41.

Frayn, M. 2006, *The Human Touch: Our Part in the Creation of a Universe* (Faber and Faber, London).

Freeman, J. and Lampton, M. 1975, 'Interstellar Archaeology and the Prevalence of Intelligence', *Icarus* **25**, 368–9.

Freese, K. and Savage, C. 2012, 'Dark Matter Collisions with the Human Body', *Phys. Lett. B* **717**, 25–8.

Freitas, R. A. and Merkle, R. C. 2004, *Kinematic Self-Replicating Machines* (Landes Bioscience, Georgetown, TX).

Freitas, R. A., Jr. 1980, 'A Self-Reproducing Interstellar Probe', *J. Brit. Interplanet. Soc.* **33**, 251–64.

Freitas, R. A., Jr. 1983, 'Extraterrestrial Intelligence in the Solar System: Resolving the Fermi Paradox', *J. Brit. Interplanet. Soc.* **36**, 496–500.

Freitas, R. A., Jr. 1985a, 'Observable Characteristics of Extraterrestrial Technological Civilizations', *J. Brit. Interplanet. Soc.* **38**, 106–12.

Freitas, R. A., Jr. 1985b, 'There Is No Fermi Paradox', *Icarus* **62**, 518–20.

Freitas, R. A., Jr and Valdes, F. 1980, 'A Search for Natural or Artificial Objects Located at the Earth-Moon Libration Points', *Icarus* **42**, 442–7.

Freitas, R. A., Jr and Valdes, F. 1985, 'The Search for Extraterrestrial Artifacts (SETA)', *Acta Astronaut.* **12**, 1027–34.

Freudenthal, H. 1960, *Lincos: Design of a Language for Cosmic Intercourse* (North-Holland Pub. Co., Amsterdam).

Friedman, M. 1974, 'Explanation and Scientific Understanding', *J. Philos.* **71**, 5–19.

Friedman, S. 2002, 'UFOs: Challenge to SETI Specialists' (online article, http://www.stantonfriedman.com/index.php?ptp=articles&fdt=2002.05.13, last accessed 20 February 2018).

Fry, I. 1995, 'Are the Different Hypotheses on the Emergence of Life as Different as They Seem?' *Biol. Philos.* **10**, 389–417.

Fry, I. 2000, *The Emergence of Life on Earth: A Historical and Scientific Overview* (Rutgers University Press, New Brunswick).

Fry, I. 2011, 'The Role of Natural Selection in the Origin of Life', *Orig. Life Evol. Biosph.* **41**, 3–16.

Fry, I. 2012, 'Is Science Metaphysically Neutral?' *Stud. Hist. Philos. Biol. Biomed, Sci,* **43**, 665–73.

Fuentes, C. 1976, *Terra Nostra* (translated by M. S. Peden; Farrar, Straus and Giroux, New York).

Fukuyama, F. 2002, *Our Posthuman Future: Consequences of the Biotechnology Revolution* (Farrar, Straus and Giroux, New York).

Fuller, R. B. [1969] 2008, *Operating Manual for Spaceship Earth* (Lars Müller Publishers, Zürich).

Galantai, Z. 2004, 'Long Futures and Type IV Civilizations', *Periodica Polytechnica Ser. Soc. Man. Sci.* **12**, 83–9.

Galante, D. and Horvath, J. E. 2007, 'Biological Effects of Gamma-Ray Bursts: Distances for Severe Damage on the Biota', *Int. J. Astrobiol.* **6**, 19–26.

Gale, J. and Wandel, A. 2017, 'The Potential of Planets Orbiting Red Dwarf Stars To Support Oxygenic Photosynthesis and Complex Life', *Int. J. Astrobiol.* **16**, 1–9.

Galera, E., Galanti, G. R., and Kinouchi, O. 2018, 'Invasion Percolation Solves Fermi Paradox but Challenges SETI Projects,' *Int. J. Astrobiol.*, in press.

Garber, S. J. 1999, 'Searching for Good Science: The Cancellation of NASA's SETI Program', *J. Brit. Interplanet. Soc.* **52**, 3–12.

Gardner, J. N. 2004, 'The Physical Constants as Biosignature: An Anthropic Retrodiction of the Selfish Biocosm Hypothesis', *Int. J. Astrobiol.* **3**, 229–36.

Gardner, J. N. 2007, *The Intelligent Universe: AI, ET, and the Emerging Mind of the Cosmos* (New Page Books, Pompton Plains, NJ).

Gardner, M. 1979, 'Mathematical Games: A Pride of Problems, Including One That Is Virtually Impossible', *Sci. Am.* **241**(December issue), 22–30.

Gehrels, N., Laird, C. M., Jackman, C. H., Cannizzo, J. K., Mattson, B. J., and Chen, W. 2003, 'Ozone Depletion from Nearby Supernovae', *Astrophys. J.* **585**, 1169–76.

Gehrels, N., Ramirez-Ruiz, E., and Fox, D. B. 2009, 'Gamma-Ray Bursts in the Swift Era', *Ann. Rev, Astron. Astrophys,* **47**, 567–617.

Gerhard, O. 2002, 'Mass Distribution in Our Galaxy', *Space Sci. Rev.* **100**, 129–38.

Gies, D. R. and Helsel, J. W. 2005, 'Ice Age Epochs and the Sun's Path through the Galaxy', *Astrophys. J.* **626**, 844–8.

Gillispie, C. C. 2000, *Pierre-Simon Laplace, 1749–1827: A Life in Exact Science* (Princeton University Press, Princeton).

Gillon, M. 2014, 'A Novel SETI Strategy Targeting the Solar Focal Regions of the Most Nearby Stars', *Acta Astronaut.* **94**, 629–33.

Gillon, M. et al. 2017, 'Seven Temperate Terrestrial Planets around the Nearby Ultracool Dwarf Star TRAPPIST-1', *Nature* **542**, 456–60.

Gilster, P. 2004, *Centauri Dreams: Imagining and Planning Interstellar Exploration* (Copernicus Books, New York).

Gindilis, L. M. et al. 1969, *Extraterrestrial Civilizations: Problems of Interstellar Communication* (Nauka, Moscow, in Russian).

Glade, N., Ballet, P., and Bastien, O. 2012, 'A Stochastic Process Approach of the Drake Equation Parameters', *Int. J. Astrobiol.* **11**, 103–8.

Glaz, A. 2014, 'Rorschach, We Have a Problem! The Linguistics of First Contact in Watts's Blindsight and Lem's His Master's Voice', *Sci. Fict. Stud.* **41**, 364–91.

Gleiser, M. 2010, 'Drake Equation for the Multiverse: From the String Landscape to Complex Life', *Int. J. Mod. Phys. A D* **19**, 1299–308.

Glen, W. (ed.). 1994, *The Mass-Extinction Debates: How Science Works in a Crisis* (Stanford University Press, Stanford).

Golden, K. M., Ackley, S. F., and Lytle, V. I. 1998, 'The Percolation Phase Transition in Sea Ice', *Science* **282**, 2238–41.

Goncharov, G. N. and Orlov, V. V. 2003, 'Global Repeating Events in the History of the Earth and the Motion of the Sun in the Galaxy', *Astron. Rep.* **47**, 925–33.

Gonzalez, G. 2005, 'Habitable Zones in the Universe', *Orig. Life Evol. Biospheres* **35**, 555–606.

Gonzalez, G., Brownlee, D., and Ward, P. 2001, 'The Galactic Habitable Zone: Galactic Chemical Evolution', *Icarus* **152**, 185–200.

Gonzalez, G. and Richards, J. 2004, *The Privileged Planet: How Our Place in the Cosmos Is Designed for Discovery* (Regnery Publishing, Washington, DC).

Gottlieb, I. and Lima, R. S. L. 2014, 'The Fermi Paradox and Coronary Artery Disease', *J. Am. Coll. Cardiol.* **64**, 693–5.

Gould, S. J. 1987, *The Flamingo's Smile: Reflections in Natural History* (W. W. Norton, New York).

Gould, S. J. 1989, *Wonderful Life: The Burgess Shale and the Nature of History* (W. W. Norton, New York).

Gould, S. J. 1996, *Full House: The Spread of Excellence from Plato to Darwin* (Three Rivers Press, New York).

Gould, S. J. 2002, *The Structure of Evolutionary Theory* (Belknap Press, Cambridge, MA).

Gould, S. J. 2003, *The Hedgehog, the Fox, and the Magister's Pox: Mending and Minding the Misconceived Gap between Science and the Humanities* (Vintage Books, London).

Gould, S. J. and E. S. Vrba 1982, 'Exaptation: A Missing Term in the Science of Form', *Paleobiology* **8**, 4–15.

Gowanlock, M. G. 2016, 'Astrobiological Effects of Gamma-Ray Bursts in the Milky Way Galaxy', *Astrophys. J.* **832**, article id. 38 (12 pp.).

Gray, R. H. 2015, 'The Fermi Paradox Is Neither Fermi's Nor a Paradox', *Astrobiology* **15**, 195–9.

Gray, R. H. and Marvel, K. B. 2001, 'A VLA Search for the Ohio State "Wow"', *Astrophys. J.* **546**, 1171–7.

Gray, R. H. and Mooley, K. 2017, 'A VLA Search for Radio Signals from M31 and M33', *Astron. J.* **153**, article id. 110 (12 pp.).

Greaves, J. S., Wyatt, M. C., Holland, W. S., and Dent, W. R. F. 2004, 'The Debris Disc around $\tau$ Ceti: A Massive Analogue to the Kuiper Belt', *Mon. Notices Royal Astron. Soc.* **351**, L54–8.

Grene, M. 1958, 'Two Evolutionary Theories (I)', *Brit. J. Phil. Sci.* **9**, 110–27.

Griessmeier, J.-M., Stadelmann, A., Motschmann, U., Belisheva, N. K., Lammer, H., and Biernat, H. K. 2005, 'Cosmic Ray Impact on Extrasolar Earth-Like Planets in Close-In Habitable Zones', *Astrobiology* **5**, 587–603.

Griffith, R. L., Wright, J. T., Maldonado, J., Povich, M. S., Sigurdsson, S., and Mullan, B. 2015, 'The Ĝ Infrared Search for Extraterrestrial Civilizations with Large Energy

Supplies. III. The Reddest Extended Sources in WISE', *Astrophys. J. Suppl. Ser.* **217**, article id. 25 (34 pp.).

Grinspoon, D. 2003, *Lonely Planets: The Natural Philosophy of Alien Life* (HarperCollins, New York).

Gros, C. 2005, 'Expanding Advanced Civilizations in the Universe', *J. Brit. Interplanet. Soc.* **58**, 108–10.

Gudmundsson, E. H. and Björnsson, G. 2002, 'Dark Energy and the Observable Universe', *Astrophys. J.* **565**, 1–16.

Guillochon, J. and Loeb, A. 2015, 'SETI via Leakage from Light Sails in Exoplanetary Systems', *Astrophys. J. Lett.* **811**, article id. L20 (6 pp.).

Gurzadyan, V. G. 2005, 'Kolmogorov Complexity, String Information, Panspermia and the Fermi Paradox', *Observatory* **125**, 352–5.

Gurzadyan, A. V. and Allahverdyan, A. E. 2016, 'Non-Random Structures in Universal Compression and the Fermi Paradox', *Eur. Phys. J. Plus* **131**, article id. 26 (6 pp.).

Gurzadyan, V. G. and Penrose, R. 2016, 'CCC and the Fermi Paradox', *Eur. Phys. J. Plus* **131**, article id. 11 (5 pp.).

Gusev, V. A. and Schulze-Makuch, D. 2004, 'Genetic Code: Lucky Chance or Fundamental Law of Nature?' *Physics of Life Reviews* **1**, 202–29.

Gustafsson, B. 1998, 'Is the Sun a Sun-like Star?' *Space Sci. Rev.* **85**, 419–28.

Haines, G. K. 1999, 'CIA's Role in the Study of UFOs, 1947–90: A Die-Hard Issue', *Intel. Nat. Sec.* **14**, 26–48.

Hair, T. W. and Hedman, A. D. 2013, 'Spatial Dispersion of Interstellar Civilizations: A Probabilistic Site Percolation Model in Three Dimensions', *Int. J. Astrobiol.* **12**, 45–52.

Haldane, J. B. S. [1927] 2017, *Possible Worlds and Other Essays* (Taylor & Francis, Milton Park).

Halley, J. W. 2012, *How Likely is Extraterrestrial Life?* (Springer, Dordrecht).

Halliday, A. N. 2008, 'A Young Moon-Forming Giant Impact at 70–110 Million Years Accompanied by Late-Stage Mixing, Core Formation and Degassing of the Earth'. *Phil. Trans. R. Soc. A* **366**, 4163–81.

Hamming, R. W. 1998, 'Mathematics on a Distant Planet', *Am. Math. Mon.* **105**, 640–50.

Hanslmeier, A. 2009, *Habitability and Cosmic Catastrophes* (Springer, Berlin).

Hanson, R. 1998a, 'Burning the Cosmic Commons: Evolutionary Strategies for Interstellar Colonization' (preprint, http://hanson.gmu.edu/filluniv.pdf, last accessed 15 September 2015).

Hanson, R. 1998b, 'The Great Filter: Are We Almost Past It?' (preprint, http://hanson.gmu.edu/greatfilter.html, last accessed 15 September 2015).

Haqq-Misra, J. D. and Baum, S. D. 2009, 'The Sustainability Solution to the Fermi Paradox', *J. Brit. Interplanet. Soc.* **62**, 47–51.

Haqq-Misra, J., Busch, M. W., Som, S. M., and Baum, S. D. 2013, 'The Benefits and Harm of Transmitting into Space', *Space Policy* **29**, 40–8.

Haqq-Misra, J. and Kopparapu, R. K. 2012, 'On the Likelihood of Non-Terrestrial Artifacts in the Solar System' *Acta Astronaut.* **72**, 15–20.

Hardwig, J. 1997, 'Is There a Duty to Die?' *Hastings Cent. Rep.* **27**, 34–42.

Harris, M. J. 1986, 'On the Detectability of Antimatter Propulsion Space-Craft', *Astrophys. Space Sci.* **123**, 297–303.

Harris, M. J. 1990, 'A Search for Linear Alignments of Gamma Ray Burst Sources', *J. Brit. Interplanet. Soc.* **43**, 551–5.

Harris, M. J. 2002, 'Limits from CGRO/EGRET Data on the Use of Antimatter as a Power Source by Extraterrestrial Civilizations', *J. Brit. Interplanet. Soc.* **55**, 383–93.

Harrison, E. R. 1995, 'The Natural Selection of Universes Containing Intelligent Life', *Q. J. Roy. Astron. Soc.* **36**, 193–203.

Harrop, B. L. and Schulze-Makuch, D. 2010, 'The Solar Wind Power Satellite as an Alternative to a Traditional Dyson Sphere and Its Implications for Remote Detection', *Int. J. Astrobiol.* **9**, 89–99.

Hart, M. H. 1975, 'Explanation for the Absence of Extraterrestrials on Earth', *Q. J. Roy. Astron. Soc.* **16**, 128–35.

Hart, M. H. 1979, 'Habitable Zones about Main Sequence Stars', *Icarus* **37**, 351–7.

Hart, M. H. 1995, 'Atmospheric Evolution, the Drake Equation and DNA: Sparse Life in an Infinite Universe', in B. Zuckerman and M. H. Hart (eds.) *Extraterrestrials. Where Are They?* (Cambridge University Press, Cambridge), 215–25.

Hartle, J. B. and Srednicki, M. 2007, 'Are We Typical?' *Phys. Rev. D* **75**, article id. 123523 (6 pp.).

Hartmann, D. H., Kretschmer, K., and Diehl, R. 2002, 'Disturbance Ecology from Nearby Supernovae', in *Proceedings of the 11th Workshop on Nuclear Astrophysics, Ringberg Castle, 2/11-16, 2002* (preprint, http://arxiv.org/abs/astro-ph/0205110, last accessed 31 October 2017).

Hatcher, W. S. 1982, *The Logical Foundations of Mathematics: Foundations and Philosophy of Science and Technology Series* (Pergamon Press, Oxford).

Havel, I. M. 1999, 'Living in Conceivable Worlds', *Found. Sci.* **3**, 375–94.

Hawass, Z. et al. 2010, 'Ancestry and Pathology in King Tutankhamun's Family', *JAMA* **303**, 638–47.

Hawking, S. W. and Ellis, G. F. R. 1973, *The Large Scale Structure of Space-Time* (Cambridge University Press, Cambridge).

Hawks, J. D. and Wolpoff, M. H. 2001, 'The Accretion Model of Neandertal Evolution', *Evolution* **55**, 1474–85.

Hawthorn, G. 1991, *Plausible Worlds: Possibility and Understanding in History and the Social Sciences* (Cambridge University Press, Cambridge).

Heath, M. J., Doyle, L. R., Joshi, M. M., and Haberle, R. M. 1999, 'Habitability of Planets around Red Dwarf Stars', *Orig. Life Evol. Biospheres* **29**, 405–24.

Heffernan, W. C. 1978, 'The Singularity of our Inhabited World: William Whewell and A. R. Wallace in Dissent', *J. Hist. Ideas* **39**, 81–100.

Heller, R. and Armstrong, J. 2014, 'Superhabitable Worlds', *Astrobiology* **14**, 50–66.

Heller R. and Pudritz R. E. 2016, 'The Search for Extraterrestrial Intelligence in Earth's Solar Transit Zone', *Astrobiology* **16**, 259–70.

Hengeveld, R. 2011, 'Definitions of Life Are Not Only Unnecessary, But They Can Do Harm to Understanding', *Found. Sci.* **16**, 323–5.

Herbison-Evans, D. 1977, 'Extraterrestrials on Earth', *Q. J. Roy. Astron. Soc.* **18**, 511–13.

Herwig, H. H. 1999, '*Geopolitik*: Haushofer, Hitler and Lebensraum', *J. Strategic Stud.* **22**, 218–41.

Ho, B. and Monton, B. 2005, 'Anthropic Reasoning Does Not Conflict with Observation', *Analysis* **65**, 42–5.

Hoffman, P. F., Kaufman, A. J., Halverson, G. P., and Schrag, D. P. 1998, 'A Neoproterozoic Snowball Earth', *Science* **281**, 1342–6.

Hofstadter, D. R. 2000, *Gödel, Escher, Bach: An Eternal Golden Braid* (20th-anniversary edition, Penguin Books, London).

Hofstadter, R. 2008, *The Paranoid Style in American Politics, and Other Essays* (Vintage Books, New York).

Holt, J. 2004, 'The Big Lab Experiment: Was Our Universe Created by Design?', *Slate*, 19 May 2004 (online article, http://www.slate.com/articles/arts/egghead/2004/05/the_big_lab_experiment.html, last accessed 15 September 2015).

Horneck, G. and Rettberg, P. (eds.). 2007, *Complete Course in Astrobiology* (Wiley-VCH, Weinheim).

Horner, J. and Jones, B. W. 2008, 'Jupiter: Friend or Foe? I: The Asteroids', *Int. J. Astrobiol.* **7**, 251–61.

Horner, J. and Jones, B. W. 2009, 'Jupiter: Friend or Foe? II: The Centaurs', *Int. J. Astrobiol.* **8**, 75–80.

Horner, J. and Jones, B. W. 2012, 'Jupiter: Friend or Foe? IV: The Influence of Orbital Eccentricity and Inclination', *Int. J. Astrobiol.* **11**, 147–56.

Horvat, M., Nakić, A., and Otočan, I. 2012, 'Impact of Technological Synchronicity on Prospects for SETI', *Int. J. Astrobiol.* **11**, 51–9.

Hosek, W. R. 2007, 'Economics and the Fermi Paradox', *J. Brit. Interplanet. Soc.* **60**, 137–41.

Hoyle, F. 1948, 'A New Model for the Expanding Universe', *Mon. Notices Royal Astron. Soc.* **108**, 372–82.

Hoyle, F. 1957, *The Black Cloud* (William Heinemann, London).

Hoyle, F. 1983, *The Intelligent Universe* (Michael Joseph, London).

Hoyle, F. 1994, *Home Is Where the Wind Blows: Chapters from a Cosmologist's Life* (University Science Books, Mill Valley).

Hoyle, F. [1964] 2005, *Of Men and Galaxies* (Prometheus Books, Amherst).

Hoyle, F. and Hoyle, G. 1973, *The Inferno* (William Heinemann, London).

Hoyle, F. and Wickramasinghe, N. 1981, *Evolution from Space* (J. M. Dent, London).

Hoyle, F. and Wickramasinghe, N. 1999, 'The Universe and Life: Deductions from The Weak Anthropic Principle', *Astrophys. Space Sci.* **268**, 89–102.

Hsu, S. and Zee, A. 2006, 'Message in the Sky', *Mod. Phys. Lett. A* **21**, 1495–500.

Huggenberger, S. 2008, 'The Size and Complexity of Dolphin Brains: A Paradox?' *J. Marine Biol. Assoc. UK* **88**, 1103–8.

Huggett, R. J. 1997, *Catastrophism: Asteroids, Comets, and Other Dynamic Events in Earth History* (Verso, London).

Hughes, J. 2004, *Citizen Cyborg: Why Democratic Societies Must Respond to the Redesigned Human of the Future* (Basic Books, New York).

Hunt, G. E. 1978, 'Possible Climatic and Biological Impact of Nearby Supernovae', *Nature* **271**, 430–1.

Hut, P. et al. 1987, 'Comet Showers as a Cause of Mass Extinctions', *Nature* **329**, 118–26.

Hutchins, E. 1995, *Cognition in the Wild* (MIT Press, Cambridge, MA).

Hutton, P., Seymoure, B. M., McGraw, K. J., Ligon, R. A., and Simpson, R. K. 2015, 'Dynamic Color Communication', *Curr. Opin. Behav. Sci.* **6**, 41–9.

Hynek, J. A. 1972, *The UFO Experience: A Scientific Inquiry* (Henry Regnery, Chicago).

Impey, C. (ed.). 2010, *Talking about Life: Conversations on Astrobiology* (Cambridge University Press, Cambridge).

Inoue, M. and Yokoo, H. 2011, 'Type III Dyson Sphere of Highly Advanced Civilisations around a Super Massive Black Hole', *J. Brit. Interplanet. Soc.* **64**, 58–62.

Iyudin, A. F. 2002, 'Terrestrial Impact of the Galactic Historical SNe', *J. Atmospheric Sol-Terr. Phys.* **64**, 669–76.

Jackson, A. A. 2015, 'Black Hole Beacon: Gravitational Lensing', *J. Brit. Interplanet. Soc.* **68**, 342–6.

Jacobson, M. 2014, 'The End of UFOs', *Daily Intelligencer*, Aug 9 2014 (online article, http://nymag.com/daily/intelligencer/2014/08/end-of-ufos.html, last accessed 15 January 2018).

Jaffe, R. L., Busza, W., Wilczek, F., and Sandweiss, J. 2000, 'Review of Speculative "Disaster Scenarios" at RHIC', *Rev. Mod. Phys.* **72**, 1125–40.

Jagers op Akkerhuis, G. A. J. M. 2010, 'Towards a Hierarchical Definition of Life, the Organism, and Death', *Found. Sci.* **15**, 245–62.

Jaime, L. G., Aguilar, L., and Pichardo, B. 2014, 'Habitable Zones with Stable Orbits for Planets around Binary Systems', *Mon. Notices Royal Astron. Soc.* **443**, 260–74.

Janković, S. and Ćirković, M. M. 2016, 'Evolvability Is an Evolved Ability: The Coding Concept as the Arch-Unit of Natural Selection', *Orig. Life Evol. Biospheres* **46**, 67–79.

Jaynes, J. 1976, *The Origin of Consciousness in the Breakdown of the Bicameral Mind* (Houghton Mifflin, Boston).

Johansen, A. and Sornette, D. 2001, 'Finite-Time Singularity in the Dynamics of the World Population, Economic and Financial Indices', *Physica A* **294**, 465–502.

Johnson, J. L. and Li, H. 2012, 'The First Planets: The Critical Metallicity for Planet Formation', *Astrophys. J.* **751**, article id. 81 (11 pp.).

Jones, E. M. 1976, 'Colonization of the Galaxy', *Icarus* **28**, 421–2.

Jones, E. M. 1981, 'Discrete Calculations of Interstellar Migration and Settlement', *Icarus* **46**, 328–36.

Jones, E. M. 1985a, 'A Manned Interstellar Vessel Using Microwave Propulsion: A Dysonship', *J. Brit. Interplanet. Soc.* **38**, 270–3.

Jones, E. M. 1985b, '"Where Is Everybody?" An Account of Fermi's Question' (preprint, http://lib-www.lanl.gov/la-pubs/00318938.pdf, last accessed 15 September 2015).

Jones, M. 2015, 'Reconsidering Macro-Artefacts in SETI Searches', *Acta Astronaut.* **116**, 161–5.

Joshi, S. T. 2004, *H. P. Lovecraft: A Life* (Necronomicon Press, Warwick).

Joshi S. T. and Schultz, D. E. 2001, *An H. P. Lovecraft Encyclopedia* (Hippocampus, New York).

Jugaku, J. and Nishimura, S. 2003. 'A Search for Dyson Spheres Around Late-Type Stars in the Solar Neighborhood', in R. Norris and F. Stootman (eds.) *Bioastronomy 2002: Life Among the Stars*, Proceedings of the 213th IAU Symposium (ASP Conference Series, San Francisco), 437–8.

Jugaku, J., Noguchi, K., and Nishimura, S. 1995, 'A Search for Dyson Spheres around Late-Type Stars in the Solar Neighborhood', in G. Seth Shostak, ed., *Progress in the Search for Extraterrestrial Life* (ASP Conference Series, San Francisco), 381–5.

Jung, C. G. 1959, *Flying Saucers: A Modern Myth of Things Seen in the Sky* (Routledge, London).

Kahane, G. 2014, 'Our Cosmic Insignificance', *Nous* **48**, 745–72.

Karam, P.A. 2002, 'Gamma and Neutrino Radiation Dose from Gamma Ray Bursts and Nearby Supernovae', *Health Phys.* **82**, 491–9.

Kardashev, N. S. 1964, 'Transmission of Information by Extraterrestrial Civilizations', *Sov. Astron.* **8**, 217–20.

Kardashev, N. S. 1985, 'On the Inevitability and Possible Forms of Supercivilizations', in M. D. Papagiannis, ed., *The Search for Extraterrestrial Life: Recent Developments*, Proceedings of the 112th IAU Symposium (Springer-Verlag, Dordrecht), 497–504.

Kardashev, N. S. 1986, 'On the Inevitability and Possible Forms of Supercivilizations', in *Problem of the Search for Life in the Universe* (Izdatel'stvo Nauka, Moscow), 25–30 (in Russian).

Kardashev, N. S. 1997, 'Cosmology and Civilizations', *Astrophys. Space Sci.* **252**, 25–40.

Kardashev, N. S. and Strelnitskij, V. S. 1988, 'Supercivilizations as Possible Products of the Progressive Evolution of Matter', in G. Marx (ed.) *Bioastronomy: The Next Steps* (Kluwer, Dordrecht), 295–302.

Kasting, J. F. 1988, 'Runaway and Moist Greenhouse Atmospheres and the Evolution of Earth and Venus', *Icarus* **74**, 472–94.

Kecskes, C. 1998, 'The Possibility of Finding Traces of Extraterrestrial Intelligence on Asteroids', *J. Brit. Interplanet. Soc.* **51**,175–9.

Kecskes, C. 2009, 'Evolution and Detectability of Advanced Civilizations', *J. Brit. Interplanet. Soc.* **62**, 316–19.

Kent, A. 2011, 'Too Damned Quiet?' (preprint, http://arxiv.org/abs/1104.0624, last accessed 15 February 2018).

Keys, D. 1999, *Catastrophe: An Investigation into the Origins of the Modern World* (Random House, New York).

Kinclová, L. 2015, 'Legitimacy of the "Humanitarian Military Intervention": An Empirical Assessment', *Peace Econ., Peace Sci. and Public Policy* **21**, 111–52.

King, S. 2009, *Under the Dome* (Scribner, New York).

Kinouchi, O. 2001, 'Persistence Solves Fermi Paradox but Challenges SETI Projects' (preprint, http://arxiv.org/abs/cond-mat/0112137, last accessed 15 February 2018).

Kipping, D. M. and Teachey, A. 2016, 'A Cloaking Device for Transiting Planets', *Mon. Notices Royal Astron. Soc.* **459**, 1233–41.

Kirschvink, J. L. et al. 2000, 'Paleoproterozoic Snowball Earth: Extreme Climatic and Geochemical Global Change and Its Biological Consequences', *Proc. Natl. Acad. Sci. USA* **97**, 1400–5.

Kitcher, P. 1981, 'Explanatory Unification', *Philos. Sci.* **48**, 507–31.

Kitcher, P. 1990, 'The Division of Cognitive Labor', *J. Philos.* **87**, 5–22.

Klass, P. J. 1974, *UFOs Explained* (Random House, New York).

Klee, R. 2017, 'Human Expunction', *Int. J. Astrobiol.* 16, 379–88.

Koch, C. 2004, *The Quest for Consciousness: A Neurobiological Approach* (Roberts & Company, Englewood, Colorado).

Koertge, N. (ed.). 1998, *A House Built on Sand: Exposing Postmodernist Myths about Science* (Oxford University Press, Oxford).

Kopparapu, R. K., Raymond, S. N., and Barnes, R. 2009, 'Stability of Additional Planets in and around the Habitable Zone of the HD 47186 Planetary System', *Astrophys. J. Lett.* **695**, L181-4.

Korhonen, J. M. 2013, 'MAD with Aliens? Interstellar Deterrence and Its Implications', *Acta Astronaut.* **86**, 201–10.

Korpela, E. J. 2012, 'SETI@home, BOINC, and Volunteer Distributed Computing', *Annu. Rev. Earth Planet. Sci.* **40**, 69–87.

Korpela, E. J., Sallmen, S. M., and Leystra Greene, D. 2015, 'Modeling Indications of Technology in Planetary Transit Light Curves: Dark-Side Illumination', *Astrophys. J.* **809**, article id. 139 (13 pp.).

Korycansky, D.G., Laughlin, G., and Adams, F. C. 2001, 'Astronomical Engineering: A Strategy For Modifying Planetary Orbits', *Astrophys. Space Sci.* **275**, 349–66.

Kosso, P. 2006, 'Introduction: The Epistemology of Archaeology', in G. G. Fagan (ed.) *Archaeological Fantasies: How Pseudoarchaeology Misrepresents the Past and Misleads the Public* (Routledge, London), 3--S22.

Kowald, A. 2015, 'Why Is There No von Neumann Probe on Ceres? Error-Catastrophe Can Explain the Fermi–Hart Paradox', *J. Brit. Interplanet. Soc.* **68**, 383–8.

Kragh, H. 1996, *Cosmology and Controversy* (Princeton University Press, Princeton).

Kragh, H. 1999, *Quantum Generations: A History of Physics in the Twentieth Century* (Princeton University Press, Princeton).

Kragh, H. 2004, *Matter and Spirit in the Universe: Scientific and Religious Preludes to Modern Cosmology* (Imperial College Press, London).

Kragh, H. 2007, *Conceptions of Cosmos from Myth to the Accelerating Universe: A History of Cosmology* (Oxford University Press, Oxford).

Krasovsky, V. I. and Shklovsky, I. S. 1957, 'Supernova Explosions and Their Possible Effect on the Evolution of Life on the Earth', *Dokl. Ac. Sci. USSR* **116**, 197–9.

Krissansen-Totton, J. et al. 2016, 'Is the Pale Blue Dot Unique? Optimized Photometric Bands for Identifying Earth-like Exoplanets', *Astrophys. J.* **817**, article id. 31 (20 pp.).

Ksanfomaliti, L. V. 1986, 'Problem of Habitation on Planetary Systems of Red Dwarf Stars', *J. Brit. Interplanet. Soc.* **39**, 416–17.

Kubrick, S. (director) [1968] 2006, *2001: A Space Odyssey* (DVD, Warner Home Video).

Kuhn, T. 1957, *The Copernican Revolution* (Harvard University Press, Cambridge).

Kuhn, J. R. and Berdyugina, S. V. 2015, 'Global Warming as a Detectable Thermodynamic Marker of Earth-Like Extrasolar Civilizations: The Case for a Telescope Like Colossus', *Int. J. Astrobiol.* **14**, 401–10.

Kuiper, T. B. H. and Morris, M. 1977, 'Searching for Extraterrestrial Civilizations', *Science* **196**, 616–21.

Kukla, A. 2001, 'SETI: On the Prospects and Pursuitworthiness of the Search for Extraterrestrial Intelligence', *Stud. Hist. Phil. Sci.* **32**, 31–67.

Kukla, A. 2008, 'The One World, One Science Argument', *Brit. J. Phil. Sci.* **59**, 73–88.

Kukla, A. 2010, *Extraterrestrials: A Philosophical Perspective* (Lexington Books, Lanham, Maryland).

Kumar, P. and Piran, T. 2000, 'Energetics and Luminosity Function of Gamma-Ray Bursts', *Astrophys. J.* **535**, 152–7.

Kundera, M. 2007, *The Curtain: An Essay in Seven Parts* (Faber and Faber, London).

Kurzweil, R. 1999, *The Age of Spiritual Machines: When Computers Exceed Human Intelligence* (Viking Press, New York).

Kurzweil, R. 2005, *The Singularity Is Near: When Humans Transcend Biology* (Viking Press, New York).

Kutschera, U. 2012, 'Wallace Pioneered Astrobiology Too', *Nature* **489**, 208.

Lachmann, M., Newman, M. E. J., and Moore, C. 2004, 'The Physical Limits of Communication or Why Any Sufficiently Advanced Technology Is Indistinguishable from Noise', *Am. J. Phys.* **72**, 1290–3.

Lacki, B. C. 2015, 'SETI at Planck Energy: When Particle Physicists Become Cosmic Engineers' (e-print, http://arxiv.org/abs/1503.01509, last accessed 15 February 2018).

Lacki, B. C. 2016, 'Type III Societies (Apparently) Do Not Exist' (e-print, http://arxiv.org/abs/1604.07844, last accessed 15 February 2018).

Lahav, N., Nir, S., and Elitzur, A. C. 2001, 'The Emergence of Life on Earth', *Prog. Biophys. Mol. Biol.* **75**, 75–120.

Lamb, D. 1997, 'Communication with Extraterrestrial Intelligence: SETI and Scientific Methodology', in D. Ginev and R. S. Cohen (eds.) *Issues and Images in the Philosophy of Science* (Kluwer, Dordrecht), 223–51.

Lamb, D. 2001, *The Search for Extraterrestrial Intelligence: A Philosophical Inquiry* (Routledge, New York).

Lampton, M. 2013, 'Information-Driven Societies and Fermi's Paradox', *Int. J. Astrobiol.* **12**, 312–13.

Landauer, R. 1961, 'Irreversibility and Heat Generation in the Computing Process', *IBM J. Res. Dev.* **5**, 183–91.

Landis, G. A. 1998, 'The Fermi Paradox: An Approach Based on Percolation Theory', *J. Brit. Interplanet. Soc.* **51**, 163–6.

Larson, E. J. 2004, *Evolution: The Remarkable History of a Scientific Theory* (Modern Library, New York).

Last, C. 2017, 'Big Historical Foundations for Deep Future Speculations: Cosmic Evolution, Atechnogenesis, and Technocultural Civilization', *Found. Sci.* **22**, 39–124.

Laster, H., Tucker, W. H., and Terry, K. D. 1968, 'Cosmic Rays from Nearby Supernovae: Biological Effects', *Science* **160**, 1138–9.

Laughlin, G. and Adams, F. C. 2000, 'The Frozen Earth: Binary Scattering Events and the Fate of the Solar System', *Icarus* **145**, 614–27.

Laughlin, G., Bodenheimer, P., and Adams, F. C. 1997, 'The End of the Main Sequence', *Astrophys. J.* **482**, 420–32.

Lawler, S. M. et al. 2014, 'The Debris Disc of Solar Analogue τ Ceti: Herschel Observations and Dynamical Simulations of the Proposed Multiplanet System', *Mon. Notices Royal Astron. Soc.* **444**, 2665–75.

Learned, J. G., Kudritzki, R.-P., Pakvasa, S., and Zee, A. 2012, 'The Cepheid Galactic Internet', *Contemp. Phys.* **53**, 113–18.

Learned, J. G., Pakvasa, S., and Zee, A. 2009, 'Galactic Neutrino Communication', *Phys. Lett. B* **671**, 15–19.

Lebon, B. 1985, 'A Rational Goal for Mankind: Progenitive Conception', *J. Brit. Interplanet. Soc.* **38**, 262–4.

Leiber, F., Jr. 2001, 'A Literary Copernicus', in D. Schweitzer, ed., *Discovering H. P. Lovecraft* (Wildside Press, Holicong), 7–16.

Leitch, E. M. and Vasisht, G. 1998, 'Mass Extinctions and the Sun's Encounters with Spiral Arms', *New Astron.* **3**, 51–6.

Lem, S. [1964] 1973, *The Invincible* (Sidgwick and Jackson, London).

Lem, S. 1984, *Microworlds: Writings on Science Fiction and Fantasy* (Harcourt Brace, Orlando).

Lem, S. 1986a, 'On Stapledon's *Last and First Men*', *Sci. Fict. Stud.* **13**, 272–91.

Lem, S. 1986b, *One Human Minute* (Harcourt Brace, San Diego).

Lem, S. 1987, *Fiasco* (Harcourt Brace, San Diego).

Lem, S. [1959] 1989, *Eden* (Harcourt Brace, San Diego).

Lem, S. [1971] 1999, 'The New Cosmogony', in *A Perfect Vacuum*, tr. M. Kandel (Northwestern University Press, Evanston), 197–227.

Lem, S. [1968] 1999, *His Master's Voice* (Northwestern University Press, Evanston).

Lem, S. [1964] 2013, *Summa Technologiae* (translated by J. Zylinska; University of Minnesota Press, Minneapolis).

Lemarchand, G. A. and Lomberg, J. 2009, 'Universal Cognitive Maps and the Search for Intelligent Life in the Universe', *Leonardo* **42**, 396–402.

Leonard, P. J. T. and Bonnell, J. T. 1998, 'Gamma-Ray Bursts of Doom', *Sky Telescope* **95**, 28–34.

Leslie, J. 1996, *The End of the World: The Ethics and Science of Human Extinction* (Routledge, London).

Levenson, B. P. 2015, 'Why Hart Found Narrow Ecospheres: A Minor Science Mystery Solved', *Astrobiology* **15**, 327–30.

Levinson, P. 2003, *Realspace: The Fate of Physical Presence in the Digital Age, On and Off Planet* (Routledge, New York).

Li, Y. and Zhang, B. 2015, 'Can Life Survive Gamma-Ray Bursts in the High-Redshift Universe?' *Astrophys. J.* **810**, article id. 41 (7 pp.).

Liddle, A. 2015, *An Introduction to Modern Cosmology* (3rd edition, John Wiley & Sons, Chichester).

Lin, H. W., Gonzalez Abad, G., and Loeb, A. 2014, 'Detecting Industrial Pollution in the Atmospheres of Earth-Like Exoplanets', *Astrophys. J. Lett.* **792**, article id. L7 (4 pp.).

Linde, A. D. 1988, 'Life after Inflation', *Phys. Lett. B* **211**, 29–31.

Linde, A. D. 1990, *Inflation and Quantum Cosmology* (Academic Press, San Diego).

Linde, A. D. 1992, 'Stochastic Approach to Tunneling and Baby Universe Formation', *Nuclear Physics B* **372**, 421–42.

Lineweaver, C. H. 1998, 'The Cosmic Microwave Background and Observational Convergence in the Omega_m-Omega_Lambda Plane', *Astrophys. J.* **505**, L69–L73.

Lineweaver, C. H. 2001, 'An Estimate of the Age Distribution of Terrestrial Planets in the Universe: Quantifying Metallicity as a Selection Effect', *Icarus* **151**, 307–13.

Lineweaver, C. H. 2008, 'Paleontological Tests: Human-Like Intelligence Is Not a Convergent Feature of Evolution', in J. Seckbach and M. Walsh (eds.) *From Fossils to Astrobiology* (Springer, Dordrecht), 353–68.

Lineweaver, C. H. and Chopra, A. 2012, 'The Habitability of Our Earth and Other Earths: Astrophysical, Geochemical, Geophysical, and Biological Limits on Planet Habitability', *Annu. Rev. Earth Planet. Sci.* **40**, 597–623.

Lineweaver, C. H. and Davis, T. M. 2002, 'Does the Rapid Appearance of Life on Earth Suggest that Life Is Common in the Universe?' *Astrobiology* **2**, 293–304.

Lineweaver, C. H., Fenner, Y., and Gibson, B. K. 2004, 'The Galactic Habitable Zone and the Age Distribution of Complex Life in the Milky Way', *Science* **303**, 59–62.

Lineweaver, C. H. and Grether, D. 2003, 'What Fraction of Sun-like Stars Have Planets?' *Astrophys. J.* **598**, 1350–60.

Lingam, M. and Loeb, A. 2017, 'Fast Radio Bursts from Extragalactic Light Sails', *Astrophys. J. Lett.* **837**, article id. L23 (5 pp.).

Lior, N. 2013, 'Mirrors in the Sky: Status, Sustainability, and Some Supporting Materials Experiments', *Renewable and Sustainable Energy Reviews* **18**, 401–15.

Lipunov, V. M. 1997, 'On the Problem of the Super Reason in Astrophysics', *Astrophys. Space Sci.* **252**, 73–81.

Lisse, C. M., Sitko, M. L., and Marengo, M. 2015, 'IRTF/SPeX Observations of the Unusual Kepler Light Curve System KIC8462852', *Astrophys. J. Lett.* **815**, article id. L27 (4 pp.).

Liu, W. M. and Chaboyer, B. 2000, 'The Relative Age of the Thin and Thick Galactic Disks', *Astrophys. J.* **544**, 818–29.

Liu, M. C. et al. 2013, 'The Extremely Red, Young L Dwarf PSO J318.5338-22.8603: A Free-Floating Planetary-Mass Analog to Directly Imaged Young Gas-Giant Planets', *Astrophys. J. Lett.* **777**, article id. L20 (7 pp.).

Livio, M. 1999, 'How Rare Are Extraterrestrial Civilizations, and When Did They Emerge?' *Astrophys. J.* **511**, 429–31.

Lloyd-Ronning, N. M., Fryer, C. L., and Ramirez-Ruiz, E. 2002, 'Cosmological Aspects of Gamma-Ray Bursts: Luminosity Evolution and an Estimate of the Star Formation Rate at High Redshifts', *Astrophys. J.* **574**, 554–65.

Loeb, A. 2014, 'The Habitable Epoch of the Early Universe', *Int. J. Astrobiol.* **13**, 337–9.

Loeb, A. and Turner, E. L. 2012, 'Detection Technique for Artificially Illuminated Objects in the Outer Solar System and Beyond', *Astrobiology* **12**, 290–4.

Long, K. F. 2012, *Deep Space Propulsion: A Roadmap to Interstellar Flight* (Springer, New York).

Lovecraft, H. P. 1999, *The H. P. Lovecraft Omnibus I: At the Mountains of Madness* (HarperCollins, London).

Lovecraft, H. P. [1936] 2005, *At the Mountains of Madness* (The Modern Library, New York).

Loveday, J., Peterson, B. A., Efstathiou, G., and Maddox, S. J. 1992, 'The Stromlo-APM Redshift Survey. I. The Luminosity Function and Space Density of Galaxies', *Astrophys. J.* **390**, 338–44.

Lucas, G. 2012, *Understanding the Archaeological Record* (Cambridge University Press, Cambridge).

Luisi, P. L. 2006, *The Emergence of Life: From Chemical Origins to Synthetic Biology* (Cambridge University Press, Cambridge).

Lyons, T. D. 2003, 'Explaining the Success of a Scientific Theory', *Philos. Sci.* **70**, 891–901.

Lyons, T. D. 2005, 'Towards a Purely Axiological Scientific Realism', *Erkenntnis* **63**, 167–204.

Lytkin, V., Finney, B., and Alepko, L. 1995, 'Tsiolkovsky, Russian Cosmism and Extraterrestrial Intelligence', *Q. J. Roy. Astron. Soc.* **36**, 369–76.

Maccone, C. 2010, 'The Statistical Fermi Paradox', *J. Brit. Interplanet. Soc.* **63**, 222–39.

Maccone, C. 2011a, 'A New Belt beyond Kuiper's: A Belt of Focal Spheres between 550 and 17,000 AU for SETI and Science', *Acta Astronaut.* **69**, 939–48.

Maccone, C. 2011b, 'A Mathematical Model for Evolution and SETI', *Orig. Life Evol. Biospheres* **41**, 609–19.

Mackie, P. 2006, *How Things Might Have Been: Individuals, Kinds, and Essential Properties* (Oxford University Press, Oxford).

Maddison, S. T., Kawata, D., and Gibson, B. K. 2002, 'Galactic Cannibalism: The Origin of the Magellanic Stream', *Astrophys. Space Sci.* **281**, 421–2.

Magalhães, J. P. de 2016, 'A Direct Communication Proposal To Test the Zoo Hypothesis', *Space Policy* **38**, 22–6.

Maher, K. A. and Stevenson, D. J. 1988, 'Impact Frustration of the Origin of Life', *Nature* **331**, 612–14.

Mansilla, W. A., Perkis, A., and Ebrahimi, T. 2014, 'Exploring the Impact of Food Craving and Pleasure Technologies on Aesthetic Experience in Digital Media', *Int. J. Hum. Comput. Interact.* **30**, 192–205.

Marengo, M., Hulsebus, A., and Willis, S. 2015, 'KIC 8462852: The Infrared Flux', *Astrophys. J.* **814**, article id. L15 (5 pp.).

Marinho, F., Paulucci, L., and Galante, D. 2014, 'Propagation and Energy Deposition of Cosmic Rays' Muons on Terrestrial Environments', *Int. J. Astrobiol.* **13**, 319–23.

Markley, R. 2005, *Dying Planet: Mars in Science and the Imagination* (Duke University Press, Durham).

Marochnik, L. S. 1983, 'On the Origin of the Solar System and the Exceptional Position of the Sun in the Galaxy', *Astrophys. Space Sci.* **89**, 61–75.

Marriner, N., Morhange, C., and Skrimshire, S. 2010, 'Geoscience Meets the Four Horsemen? Tracking the Rise of Neocatastrophism', *Glob. Planet. Change* **74**, 43–8.

Martin, A. R. and Bond, A. 1983, 'Is Mankind Unique?", *J. Brit. Interplanet. Soc.* **36**, 223–5.

Martin, O., Cardenas, R., Guimarais, M., Peñate, L., Horvath, J., and Galante, D. 2010, 'Effects of Gamma Ray Bursts in Earth's Atmosphere', *Astrophys. Space Sci.* **326**, 61–7.

Martín, O., Galante, D., Cárdenas, R., and Horvath, J. E. 2009, 'Short-Term Effects of Gamma Ray Bursts on Earth', *Astrophys. Space Sci.* **321**, 161–7.

Mash, R. 1993, 'Big Numbers and Induction in the Case for Extraterrestrial Intelligence', *Philos. Sci.* **60**, 204–22.

Mason, B. G., Pyle, D. M., and Oppenheimer, C. 2004, 'The Size and Frequency of the Largest Explosive Eruptions on Earth', *Bull. Volcanol.* **66**, 735–48.

Mason, J. W. (ed.). 2008, *Exoplanets: Detection, Formation, Properties, Habitability* (Springer Praxis Books, Chichester).

Matese, J. J. and Whitman, P. G. 1992, 'A Model of the Galactic Tidal Interaction with the Oort Comet Cloud', *Cel. Mech. and Dyn. Astron.* **54**, 13–35.

Matese, J. J. and Whitmire, D. 1996, 'Tidal Imprint of Distant Galactic Matter on the Oort Comet Cloud', *Astrophys. J. Lett.* **472**, L41–3.

Matheny, J. G. 2007, 'Reducing the Risk of Human Extinction', *Risk Anal.* **27**, 1335–44.

Matloff, G. L. and Martin, A. R. 2005, 'Suggested Targets for an Infrared Search for Artificial Kuiper Belt Objects', *J. Brit. Interplanet. Soc.* **58**, 51–61.

Maudlin, T. 2007, *The Metaphysics Within Physics* (Oxford University Press, Oxford).

Mautner, M. and Matloff, G. L. 1979, 'A Technical and Ethical Evaluation of Seeding Nearby Solar Systems', *J. Brit. Interplanet. Soc.* **32**, 419–23.

Mautner, M. N. 2004, *Seeding the Universe with Life: Securing Our Cosmological Future* (Legacy Books, Christchurch).

Mautner, M. N. 2005, 'Life in the Cosmological Future: Resources, Biomass and Populations', *J. Brit. Interplanet. Soc.* **58**, 167–80.

Mayr, E. 1993, 'The Search for Intelligence', *Science* **259**, 1522–3.

Mazlish, B. 1967, 'The Fourth Discontinuity', *Technol. Cult.* **8**, 1–15.

McConnell, B. 2001, *Beyond Contact: A Guide to SETI and Communicating with Alien Civilizations* (O'Reilly, Sebastopol, CA).

McInnes, C. R. 2002, 'Astronomical Engineering Revisited: Planetary Orbit Modification Using Solar Radiation Pressure', *Astrophys. Space Sci.* **282**, 765–72.

McKay, C. P. and Marinova, M. M. 2001, 'The Physics, Biology, and Environmental Ethics of Making Mars Habitable', *Astrobiology* **1**, 90–104.

McMullin, E. 1980, 'Persons in the Universe', *Zygon* **15**, 69–89.

McMullin, E. 1989, 'Having Fun with ET', *Biol. Philos.* **4**, 97–105.

Melott, A. L. and Thomas, B. C. 2011, 'Astrophysical Ionizing Radiation and Earth: A Brief Review and Census of Intermittent Intense Sources', *Astrobiology* **11**, 343–61.

Melott, A. L. et al. 2004, 'Did a Gamma-Ray Burst Initiate the late Ordovician Mass Extinction?' *Int. J. Astrobiol.* **3**, 55–61.

Melott, A. L., Usoskin, I. G., Kovaltsov, G. A., and Laird, C. M. 2015, 'Has the Earth Been Exposed to Numerous Supernovae within the Last 300 kyr?' *Int. J. Astrobiol.* **14**, 375–8.

Mészáros, P. 2002, 'Theories of Gamma-Ray Bursts', *Ann. Rev. Astron. Astrophys.* **40**, 137–69.

Michael, G. 2011, 'Extraterrestrial Aliens: Friends, Foes, or Just Curious?' *Skeptic* **16**, 46–53.

Michaud, M. A. G. 2007, *Contact with Alien Civilizations: Our Hopes and Fears about Encountering Extraterrestrials* (Copernicus Books, New York).

Midgley, M. 1985, *Evolution as a Religion: Strange Hopes and Stranger Fears* (Routledge, London).

Miletić, T. 2015, 'Extraterrestrial Artificial Intelligences and Humanity's Cosmic Future: Answering the Fermi Paradox through the Construction of a Bracewell-Von Neumann AGI', *J. Evol. Technol.* **25**, 56–73.

Miller, A. 2016, 'Realism', in E. N. Zalta (ed.) *The Stanford Encyclopedia of Philosophy* (Winter 2016 edition; online article http://plato.stanford.edu/archives/win2016/entries/realism/, last accessed 29 December 2017).

Miller, G. 2006, 'Runaway Consumerism Explains the Fermi Paradox' (online article, http://edge.org/response-detail/11475, last accessed 1 March 2016).

Mitton, S. 2005, *Conflict in the Cosmos: Fred Hoyle's Life in Science* (Joseph Henry Press, Washington, DC).

Mojzsis, S. J., Arrhenius, G., McKeegan, K. D., Harrison, T. M., Nutman, A. P., and Friend, C. R. L. 1996, 'Evidence for Life on Earth before 3,800 Million Years Ago', *Nature* **384**, 55–9.

Molière, J.-B. P. [1670] 2008, *The Bourgeois Gentleman* (translated by P. D. Jones; Project Gutenberg, Urbana; e-text, http://www.gutenberg.org/ebooks/2992, last accessed 15 September 2015).

Musso, P. 2012, 'The Problem of Active SETI: An overview', *Acta Astronautica*, **78**, 43–54.

Nakamura, H. 1986, 'SV40 DNA: A Message from $\epsilon$ Eri', *Acta Astronaut.* **13**, 573–8.

Napier, W. M. 2004, 'A Mechanism for Interstellar Panspermia', *Mon. Not. R. Astron. Soc.* **348**, 46–51.

Napier, W. M. 2007, 'Pollination of Exoplanets by Nebulae', *Int. J. Astrobiol.* **6**, 223–8.

National Intelligence Council. 2008, 'Disruptive Technologies Global Trends 2025. Six Technologies with Potential Impacts on US Interests out to 2025' (report, http://www.fas.org/irp/nic/disruptive.pdf, last accessed 30 June 2016).

Nazaretyan, A. P. 2005, 'Big (Universal) History Paradigm: Versions and Approaches', *Soc. Evol. Hist.* **4**, 61–86.

Neal, M. 2014, 'Preparing for Extraterrestrial Contact', *Risk Manage.* **16**, 63–87.

Nemiroff, R. J. 1994, 'A Century of Gamma Ray Burst Models', *Comment. Astrophys.* **17**, 189–205.

Newman, W. I. and Sagan, C. 1981, 'Galactic Civilizations: Population Dynamics and Interstellar Diffusion', *Icarus* **46**, 293–327.

Newton-Smith, W. H. 1980, *The Structure of Time* (Routledge & Kegan Paul, London).

Nicholls, P. 2000, 'Big Dumb Objects and Cosmic Enigmas: The Love Affair between Space Fiction and the Transcendental', in G. Westfahl, ed., *Space and Beyond. The Frontier Theme in Science Fiction* (Praeger, Westport), 11–23.

Nicholson, A. and Forgan, D. 2013, 'Slingshot Dynamics for Self-Replicating Probes and the Effect on Exploration Timescales', *Int. J. Astrobiol.* **12**, 337–44.

Niven, L. 1970, *Ringworld* (Ballantine Books, New York).

Niven, L. 1974, *A Hole in Space* (Ballantine Books, New York).

Noble, M., Musielak, Z. E., Cuntz, M. 2002, 'Orbital Stability of Terrestrial Planets inside the Habitable Zones of Extrasolar Planetary Systems', *Astrophys. J.* **572**, 1024–30.

Nørretranders, T. 1999, *The User Illusion: Cutting Consciousness Down to Size* (Penguin Books, New York).

North, J. 1965, *The Measure of the Universe: A History of Modern Cosmology* (Oxford University Press, London).

North, J. 1994, *The Fontana History of Astronomy and Cosmology* (Fontana Press, London).

Norton, J. D. 2004, 'Why Thought Experiments Do Not Transcend Empiricism', in C. Hitchcock, ed., *Contemporary Debates in the Philosophy of Science* (Blackwell, Oxford), 44–66.

Nozick, R. 1981, *Philosophical Explanations* (Harvard University Press, Cambridge, MA).

Nunn, A. V. W., Guy, G. W., and Bell, J. D. 2014, 'The Intelligence Paradox; Will ET Get the Metabolic Syndrome? Lessons from and for Earth', *Nutr. Metab.* **11**, article id. 34 (13 pp.).

Oliver, B. M. et al. [1971] 1996, *Project Cyclops, Second Printing* (SETI League and SETI Institute, Little Ferry, NJ).

Olson, S. J. 2015, 'Homogeneous Cosmology with Aggressively Expanding Civilizations', *Class. Quantum Grav.* **32**, article id. 215025 (24 pp.).

Olum, K. 2004, 'Conflict between Anthropic Reasoning and Observation', *Analysis* **64**, 1–8.

O'Neill, G. K. 1977, *The High Frontier: Human Colonies in Space*. (William Morrow & Company, New York).

Opatrný, T., Richterek, L., and Bakala, P. 2016, 'Life under a Black Sun', *Am. J. Phys.* **85**, 14–22.

Orwell, G. [1949] 1984, *1984* (Plume, New York).

Osmanov, Z. 2016, 'On the Search for Artificial Dyson-Like Structures around Pulsars', *Int. J. Astrobiol.* **15**, 127–32.

Osterbrock, D. E. 2001, *Walter Baade: A Life in Astrophysics* (Princeton University Press, Princeton).

Overholt, A. C., Melott, A. L., and Pohl, M. 2009, 'Testing the Link between Terrestrial Climate Change and Galactic Spiral Arm Transit', *Astrophys. J.* **705**, L101–L103.

Page, D. N. 2008, 'Typicality Derived', *Phys. Rev. D* **78**, article id. 023514 (7 pp.).

Palmer, T. 2003, *Perilous Planet Earth: Catastrophes and Catastrophism through the Ages* (Cambridge University Press, Cambridge).

Papagiannis, M. D. 1978, 'Are We All Alone, or Could They Be in the Asteroid Belt?' *Q. J. Roy. Astron. Soc.* **19**, 277–81.

Papagiannis, M. D. 1984, 'Natural Selection of Stellar Civilizations by the Limits of Growth', *Q. J. Roy. Astron. Soc.* **25**, 309–18.

Paprotny, Z. 1977, 'Nonradio Methods of SETI', *PoAn* **10**, 39–67 (in Polish).

Parkinson, B. 2005, 'The Carbon or Silicon Colonization of the Universe?' *J. Brit. Interplanet. Soc.* **58**, 111–16.

Partington, J. S. 2003, *Building Cosmopolis: The Political Thought of H. G. Wells* (Ashgate, London).

Pavlov, A. A., Toon, O. B., Pavlov, A. K., Bally, J., and Pollard, D. 2005, 'Passing through a Giant Molecular Cloud: "Snowball" Glaciations Produced by Interstellar Dust', *Geophys. Res. Lett.* **32**, L03705[4].

Peacock, J. A. 1999, *Cosmological Physics* (Cambridge University Press, Cambridge).

Pearson, J. 1975, 'The Orbital Tower: A Spacecraft Launcher Using the Earth's Rotational Energy', *Acta Astronaut.* **2**, 785–99.

Pearson, J., Oldson, J., and Levin, E. 2006, 'Earth Rings for Planetary Environment Control', *Acta Astronaut.* **58**, 44–57.

Peebles, P. J. E., Page, L. A., Jr, and Partridge, R. B. (eds.). 2009, *Finding the Big Bang* (Cambridge University Press, Cambridge).

Peña-Cabrera, G. V. Y. and Durand-Manterola, H. J. 2004, 'Possible Biotic Distribution in our Galaxy', *Adv. Space Res.* **33**, 114–17.

Penrose, R. 1989, *The Emperor's New Mind* (Oxford University Press, Oxford).

Penrose, R. and Floyd, R. M. 1971, 'Extraction of Rotational Energy from a Black Hole', *Nature* **229**, 177–9.

Pérez-Mercader, J. 2002, 'Scaling Phenomena and the Emergence of Complexity in Astrobiology', in G. Horneck and C. Baumstark-Khan (eds.) *Astrobiology: The Quest for the Conditions of Life* (Springer-Verlag, Berlin), 337–60.

Perogamvros, L. 2013, 'Consciousness and the Invention of Morel', *Front. Hum. Neurosci.* **7**, 61.

Petigura, E. A., Howard, A. W., and Marcy, G. W. 2013, 'Prevalence of Earth-Size Planets Orbiting Sun-Like Stars', *Proc. Natl. Acad. Sci. USA* **110**, 19273–8.

Phillips, J. 2012, 'Storytelling in Earth Sciences: The Eight Basic Plots', *Earth-Sci. Rev.* **115**, 153–62.

Phoenix, C. and Treder, M. 2008, 'Nanotechnology as Global Catastrophic Risk', in N. Bostrom, N. and M. M. Ćirković (eds.) *Global Catastrophic Risks* (Oxford University Press, Oxford), 481–503.

Pigliucci, M. 2008, 'Is Evolvability Evolvable?' *Nat. Rev. Genet.* **9**, 75–82.

Pigliucci, M. 2010, *Nonsense on Stilts: How to Tell Science from Bunk* (University of Chicago Press, Chicago).

Piran, T. 2000, 'Gamma-Ray Bursts: A Puzzle Being Resolved', *Phys. Rep.* **333/334**, 529–53.

Piran, T. and Jimenez, R. 2014, 'Possible Role of Gamma Ray Bursts on Life Extinction in the Universe', *Phys. Rev. Lett.* **113**, article id. 231102 (6 pp.).

Plantinga, A. 2011, *Where the Conflict Really Lies: Science, Religion, and Naturalism* (Oxford University Press, New York).

Popper, K. 1972, *Objective Knowledge: An Evolutionary Approach* (Oxford University Press, Oxford).

Prantzos, N. 2008, 'On the "Galactic Habitable Zone"', *Space Sci. Rev.* **135**, 313–22.

Price, H. 1996, *Time's Arrow and Archimedes' Point* (Oxford University Press, Oxford).

Price, H. 2007, 'Quining Naturalism', *J. Philos.* **104**, 375–402.

Pross, A. 2012, *What is Life? How Chemistry Becomes Biology* (Oxford University Press, Oxford).

Proyas, A. (director). 1998, *Dark City* (DVD, Mystery Clock Cinema/New Line Cinema).

Puccetti, R. 1969, *Persons: A Study of Possible Moral Agents in the Universe* (Herder & Herder, New York).

Rahmati, A., Pawlik, A. H., Raičević, M., and Schaye, J. 2013, 'On the Evolution of the H I Column Density Distribution in Cosmological Simulations', *Mon. Notices Royal Astron. Soc.* **430**, 2427–45.

Rampino, M. R. 1997, 'The Galactic Theory of Mass Extinctions: An Update', *Cel. Mech. and Dyn. Astron.* **69**, 49–58.

Rampino, M. R. 2002, 'Supereruptions as a Threat to Civilizations on Earth-Like Planets', *Icarus* **156**, 562–9.

Rampino, M. R. 2015, 'Disc Dark Matter in the Galaxy and Potential Cycles of Extraterrestrial Impacts, Mass Extinctions and Geological Events', *Mon. Notices Royal Astron. Soc.* **448**, 1816–20.

Rampino, M. R. and Self, S. 1992, 'Volcanic Winter and Accelerated Glaciation Following the Toba Super-Eruption', *Nature* **359**, 50–2.

Rampino, M. R. and Stothers, R. B. 1984, 'Terrestrial Mass Extinctions, Cometary Impacts and the Sun's Motion Perpendicular to the Galactic Plane', *Nature* **308**, 709–12.

Rasoonl, S. I. and de Bergh, C. 1970, 'The Runaway Greenhouse Effect and the Accumulation of $CO_2$ in the Atmosphere of Venus', *Nature* **226**, 1037–9.

Rauch, M. 1998, 'The Lyman Alpha Forest in the Spectra of QSOs', *Ann. Rev. Astron. Astrophys.* **36**, 267–316.

Raulin-Cerceau, F. 2010, 'What Possible Life Forms Could Exist on Other Planets: A Historical Overview', *Orig. Life Evol. Biospheres* **40**, 195–202.

Raulin-Cerceau, F. and Bilodeau, B. 2012, 'A Comparison between the 19th Century Early Proposals and the 20th–21st Century Realized Projects Intended To Contact Other Planets', *Acta Astronaut.* **78**, 72–9.

Raup, D. M. 1991, *Extinction: Bad Genes or Bad Luck?* (W. W. Norton, New York).

Raup, D. M. 1992, 'Nonconscious Intelligence in the Universe', *Acta Astronaut.* **26**, 257–61.

Raup, D. M. 1999, *The Nemesis Affair: A Story of the Death of Dinosaurs and the Ways of Science* (2nd edition, W. W. Norton, New York).

Raup, D. M. and Valentine, J. W. 1983, 'Multiple Origins of Life', *Proc. Natl. Acad. Sci. USA* **80**, 2981–4.

Rees, M. J. 2003, *Our Final Century? Will the Human Race Survive the Twenty-First Century?* (William Heinemann, London).

Rescher, N. 1985, 'Extraterrestrial Science', in E. Regis, Jr (ed.) *Extraterrestrials, Science and Alien Intelligence* (Cambridge University Press, Cambridge), 83–116.

Reynolds, A. 2000, *Revelation Space* (Gollancz, London).

Reynolds, A. 2001, *Chasm City* (Gollancz, London).

Reynolds, A. 2002, *Redemption Ark* (Gollancz, London).

Reynolds, A. 2003, *Absolution Gap* (Gollancz, London).

Reynolds. A. 2004, *Century Rain* (Gollancz, London).

Reynolds, A. 2005a, 'Feeling Rejected', *Nature* **437**, 788.

Reynolds, A. 2005b, *Pushing Ice* (Gollancz, London).

Reynolds, A. 2008, *House of Suns* (Gollancz, London).

Rhodes, R. 1986, *The Making of the Atomic Bomb* (Simon & Schuster, New York).

Roberts, A. 2015, *The Thing Itself* (Gollancz, London).

Robles, J. A. et al. 2008, 'A Comprehensive Comparison of the Sun to Other Stars: Searching for Self-Selection Effects', *Astrophys. J.* **684**, 691–706.

Robson, J. 2005, *Natural History* (Bantam Books, New York).

Rose, C. and Wright, G. 2004, 'Inscribed Matter as an Energy-Efficient Means of Communications with an Extraterrestrial Civilization', *Nature* **431**, 47–9.

Rosen, J. 2010, *Lawless Universe: Science and the Hunt for Reality* (Johns Hopkins University Press, Baltimore).

Rospars, J.-P. 2011, 'Terrestrial Biological Evolution and its Implication for SETI', *Acta Astronaut.* **67**, 1361–5.

Roy, K. L., Kennedy, R. G., III, and Fields, D. E. 2009, 'Shell Worlds: An Approach to Terraforming Moons, Small Planets and Plutoids', *J. Brit. Interplanet. Soc.* **62**, 32–8.

Roy, K. L., Kennedy, R. G., III, and Fields, D. E. 2014, 'Shell Worlds: The Question of Shell Stability', *J. Brit. Interplanet. Soc.* **67**, 364–8.

Rubtsov, V. V. 1991, 'Criteria of Artificiality in SETI', in J. Heidmann and M. J. Klein (eds.) *Bioastronomy: The Search for Extraterrestial Life*, Lecture Notes in Physics, Vol. 390 (Springer-Verlag, Heidelberg), 306–10.

Rudder Baker, L. 2013, *Naturalism and the First-Person Perspective* (Oxford University Press, Oxford).

Ruddiman, W. F. 2013, 'The Anthropocene', *Annu. Rev. Earth Planet. Sci.* **41**, 45–68.

Ruderman, M. A. 1974, 'Possible Consequences of Nearby Supernova Explosions for Atmospheric Ozone and Terrestrial Life', *Science* **184**, 1079–81.

Rummel, J. D., and Billings, L. 2004, 'Issues in Planetary Protection: Policy, Protocol and Implementation', *Space Policy* **20**, 49–54.

Russell, D. A. 1983, 'Exponential Evolution: Implications for Extraterrestrial Intelligent Life', *Adv. Space Res.* **3**, 95–103.

Russell, D. A. 1995, 'Biodiversity and Time Scales for the Evolution of Extraterrestrial Intelligence', in G. Seth Shostak (ed.) *Progress in the Search for Extraterrestrial Life*, ASP Conference Series, Vol. 74 (Astronomical Society of the Pacific San Francisco), 143–51.

Russell, L. 1999, 'Last Year At Marienbad' (online article, http://www.culturecourt.com/F/NewWave/Marienbad1.htm, last accessed 15 September 2015).

Saberhagen, F. 1992, *Berserker* (Ace, New York).

Saberhagen, F. 1998, *Berserkers: The Beginning* (Baen, Riverdale, NY).

Sagan, C. 1973a, 'On the Detectivity of Advanced Galactic Civilizations', *Icarus* **19**, 350–2.

Sagan, C. (ed.). 1973b, *Communication with Extraterrestrial Intelligence* (MIT Press, Cambridge, MA).

Sagan, C. 1985, *Contact* (Simon & Schuster, New York).

Sagan, C. 1994, *Pale Blue Dot: A Vision of the Human Future in Space* (Random House, New York).

Sagan, C. 1995, *The Demon-Haunted World: Science as a Candle in the Dark* (Random House, New York).

Sagan, C. [1973] 2000, *Carl Sagan's Cosmic Connection: An Extraterrestrial Perspective* (Cambridge University Press, Cambridge).

Sagan, C. and Newman, W. I. 1983, 'The Solipsist Approach to Extraterrestrial Intelligence', *Q. J. Roy. Astron. Soc.* **24**, 113–21.

Sagan, C. and Salpeter, E. E. 1976, 'Particles, Environments, and Possible Ecologies in the Jovian Atmosphere', *Astrophys. J. Suppl. Ser.* **32**, 737–55.

Sagan, C. and Walker, R. G. 1966, 'The Infrared Detectability of Dyson Civilizations', *Astrophys. J.* **144**, 1216–18.

Sainsbury, R. M. 1995, *Paradoxes* (Cambridge University Press, Cambridge).

Sakharov, A. D. 1975, 'Peace, Progress, Human Rights', Nobel Lecture, 11 December 1975 (translation, http://www.nobelprize.org/nobel_prizes/peace/laureates/1975/sakharov-lecture.html, last accessed 1 September 2015).

Salzberg, C., Antony, A., and Sayama, H. 2004, 'Evolutionary Dynamics of Cellular Automata-Based Self-Replicators in Hostile Environments', *Biosystems* **78**, 119–34.

Samuelson, D. N. 1973, '*Childhood's End*: A Median Stage of Adolescence?' *Sci. Fict. Stud.* **1**, 4–17.

Sanchez, J. P. and McInnes, C. R. 2012, 'Assessment on the Feasibility of Future Shepherding of Asteroid Resources', *Acta Astronaut.* **73**, 49–66.

Sandberg, A. 2000, 'The Physics of Information Processing Superobjects: Daily Life Among the Jupiter Brains', *J. Evol. Technol.* **5**, 1–34.

Sandberg, A., Armstrong, S., and Ćirković, M. M. 2016, 'That Is Not Dead Which Can Eternal Lie: The Aestivation Hypothesis for Resolving Fermi's Paradox', *J. Brit. Interplanet. Soc.* **69**, 406–15.

Sandels, R. 1986, 'UFOs, Science Fiction and the Postwar Utopia', *J. Pop. Cult.* **20**, 141–51.

Sato, K., Kodama, H., Sasaki, M., and Maeda, K.-I. 1982, 'Multi-Production of Universes by First-Order Phase Transition of a Vacuum', *Phys. Lett. B* **108**, 103–7.

Scalo, J. and Wheeler, J. C. 2002, 'Astrophysical and Astrobiological Implications of Gamma-Ray Burst Properties', *Astrophys. J.* **566**, 723–37.

Schaller, R. R. 1997, 'Moore's Law: Past, Present and Future', *IEEE Spectrum* **34** (June), 52–9.

Scharf, C. 2014, *The Copernicus Complex: Our Cosmic Significance in a Universe of Planets and Probabilities* (Farrar, Straus and Giroux, New York).

Scharf, C. and Cronin, L. 2016, 'Quantifying the Origins of Life on a Planetary Scale', *Proc. Natl. Acad. Sci. USA* **113**, 8127–32.

Schechter, P. 1976, 'An Analytic Expression for the Luminosity Function for Galaxies', *Astrophys. J.* **203**, 297–306.

Scheffer, L. K. 1994, 'Machine Intelligence, the Cost of Interstellar Travel and Fermi's Paradox', *Q. J. Roy. Astron. Soc.* **35**, 157–75.

Scheffer, L. K. 2011, 'Large Scale Use of Solar Power May Be Visible across Interstellar Distances', in D. A. Vakoch, ed., *Communication with Extraterrestrial Intelligence* (State University of New York Press, Albany), 161–75.

Schindewolf, O. 1962, 'Neokatastrophismus?' *Deut. Geol. Ges. Z. Jahrg.* **114**, 430–45 (in German).

Schroeder, K. 2001, *Ventus* (Tor Books, New York).

Schroeder, K. 2002, *Permanence* (Tor Books, New York).

Schroeder, K. 2005, *Lady of Mazes* (Tor Books, New York).

Schroeder, K. 2014, *Lockstep: A Novel* (Tor Books, New York).

Schulze-Makuch, D. and Darling, D. 2010, *We Are Not Alone: Why We Have Already Found Extraterrestrial Life* (Oneworld, Oxford).

Schwartzman, D. and Middendorf, G. 2011, 'Multiple Paths to Encephalization and Technical Civilizations', *Orig. Life Evol. Biospheres* **41**, 581–5.

Scott, R. (director). 2012, *Prometheus* (DVD, 20th Century Fox Film).

Sergi, V., Galletta, G., Celotti, L., and Turatto, M. 2001, 'Nearby Supernova Explosions and Radiation Effects on the Earth's Environment', in P. Ehrenfreund, O. Angerer, and B. Battrick (eds.) *Proceedings of the First European Workshop on Exo- and Astro-Biology*, ESA SP-496 (ESA Publications Division, Noordwijk), 409–12.

Shanahan, T. 2004, *The Evolution of Darwinism: Selection, Adaptation, and Progress in Evolutionary Biology* (Cambridge University Press, Cambridge).

Shaviv, N. J. 2002a, 'Cosmic Ray Diffusion from the Galactic Spiral Arms, Iron Meteorites, and a Possible Climatic Connection', *Phys. Rev. Lett.* **89**, 051102-1/4.

Shaviv, N. J. 2002b, 'The Spiral Structure of the Milky Way, Cosmic Rays, and Ice Age Epochs on Earth', *New Astron.* **8**, 39–77.

shCherbak, V. I. and Makukov, M. A. 2013, 'The "Wow! Signal" of the Terrestrial Genetic Code', *Icarus* **224**, 228–42.

Shea, M. A. and Smart, D. F. 2000, 'Cosmic Ray Implications for Human Health', *Space Sci. Rev.* **93**, 187–205.

Sheaffer, R. 1995, 'An Examination of Claims That Extraterrestrial Visitors to Earth Are Being Observed', in B. Zuckerman and M. H. Hart (eds.) *Extraterrestrials, Where Are They?* (2nd edition, Cambridge University Press, Cambridge), 20–8.

Shechtman, I. 2006, 'Is the Universe Teeming with Super Civilizations?' *J. Brit. Interplanet. Soc.* **59**, 257–61.

Sheridan, M. A. 2009, *SETI's Scope: How the Search for Extraterrestrial Intelligence Became Disconnected from New Ideas about Extraterrestrials* (Ph.D. dissertation; Drew University, Madison; an adapted version is available at http://www.daviddarling.info/encyclopedia/S/SETI_critical_history_cover.html, last accessed 15 February 2018).

Shermer, M. 2002, *In Darwin's Shadow: The Life and Science of Alfred Russel Wallace* (Oxford University Press, Oxford).

Shklovsky, I. S. 1962, *Universe—Life—Intelligence* (AN SSSR, Moscow, in Russian).

Shklovsky, I. S. 1973, 'Problema Vnezemnykh Tsivilizatsei i ee Filosofskie Aspekti', *Вопросы философии* **2**, 76–93 (in Russian).

Shklovsky, I. S. and Sagan, C. 1966, *Intelligent Life in the Universe* (Holden-Day, San Francisco).

Shostak, S. 2009, *Confessions of an Alien Hunter: A Scientist's Search for Extraterrestrial Intelligence* (National Geographic, Washington, DC).

Shostak, S. 2013, 'Are Transmissions to Space Dangerous?' *Int. J. Astrobiol.* **12**, 17–20.

Shostak, S., Ekers, R., and Vaile, R. 1996, 'A Search for Artificial Signals from the Small Magellanic Cloud', *Astron. J.* **112**, 164–6.

Shostak, S. G. 2003, 'Searching for Sentience: SETI Today', *Int. J. Astrobiol.* **2**, 111–14.

Siddiqi, A. A. 2008, 'Imagining the Cosmos: Utopians, Mystics, and the Popular Culture of Spaceflight in Revolutionary Russia', *Osiris* **23**, 260–88.

Simpson, F. 2017, 'The Longevity of Habitable Planets and the Development of Intelligent Life', *Int. J. Astrobiol.* **16**, 266–70.

Simpson, G. G. 1964, 'The Nonprevalence of Humanoids', *Science* **143**, 769–75.

Simpson, G. G. 1968, 'Evolutionary Effects of Cosmic Radiation', *Science* **162**, 140–1.

Slobodian, R. E. 2015, 'Selling Space Colonization and Immortality: A Psychosocial, Anthropological Critique of the Rush To Colonize Mars', *Acta Astronaut.* **113**, 89–104.

Slysh, V. I. 1985, 'A Search in the Infrared for Astroengineering Activity', in M. D. Papagiannis (ed.) *The Search for Extraterrestrial Life: Recent Developments*, Proceedings of the 112th IAU Symposium (Reidel Publishing Co., Dordrecht), 315–19.

Smart, J. J. C. 2004, 'The Brain in the Vat and the Question of Metaphysical Realism', *Stud. Hist. Philos. Sci. C* **35**, 237–47.

Smart, J. M. 2012, 'The Transcension Hypothesis: Sufficiently Advanced Civilizations Invariably Leave Our Universe, and Implications for METI and SETI', *Acta Astronaut.* **78**, 55–68.

Smith, D. S., Scalo, J., and Wheeler, J. C. 2004, 'Importance of Biologically Active Aurora-Like Ultraviolet Emission: Stochastic Irradiation of Earth and Mars by Flares and Explosions', *Orig. Life Evol. Biospheres* **34**, 513–32.

Sober, E. 1993, *Philosophy of Biology* (Westview Press, Boulder).

Solé, R. V. and Valverde, S. 2001, 'Information Transfer and Phase Transitions in a Model of Internet Traffic', *Physica A* **289**, 595–605.

Sorkin, A. R. 2009, *Too Big to Fail: The Inside Story of How Wall Street and Washington Fought to Save the Financial System—and Themselves* (Viking Press, New York).

Spitoni, E., Matteucci, F., and Sozzetti, A. 2014, 'The Galactic Habitable Zone of the Milky Way and M31 from Chemical Evolution Models with Gas Radial Flows', *Mon. Notices Royal Astron. Soc.* **440**, 2588–98.

Spivey, R. J. 2015, 'A Cosmological Hypothesis Potentially Resolving the Mystery of Extraterrestrial Silence with Falsifiable Implications for Neutrinos', *Phys. Essays* **28**, 254–64.

Stancil, D. D. et al. 2012, 'Demonstration of Communication Using Neutrinos', *Mod. Phys. Lett. A* **27**, article id. 1250077 (10 pp.).

Stapledon, O. 1930, *Last and First Men* (Methuen, London).

Stapledon, O. 1937, *Star Maker* (Methuen, London).

Starling, J. and Forgan, D. H. 2014, 'Virulence as a Model for Interplanetary and Interstellar Colonization: Parasitism or Mutualism?' *Int. J. Astrobiol.* **13**, 45–52.

Steckline, V. S. 1983, 'Zermelo, Boltzmann, and the Recurrence Paradox', *Am. J. Phys.* **51**, 894–7.

Steel, D. 1995, 'SETA and 1991 VG', *Observatory* **115**, 78–83.

Stephenson, D. G. 1979, 'Extraterrestrial Cultures within the Solar System', *Q. J. Roy. Astron. Soc.* **20**, 422–8.

Sterelny, K. 2005, 'Another View of Life', *Stud. Hist. Phil. Biol. Biomed. Sci.* **36**, 585–93.

Stevens, A., Forgan, D., and James, J. O. 2016, 'Observational Signatures of Self-Destructive Civilizations', *Int. J. Astrobiol.* **15**, 333–44.

Stevenson, D. J. 1999, 'Life-Sustaining Planets in Interstellar Space?' *Nature* **400**, 32.

Stewart, J. E. 2010, 'The Meaning of Life in a Developing Universe', *Found. Sci.* **15**, 395–409.

Stewart, J. E. 2012, 'The Future of Life and What It Means for Humanity', *Found. Sci.* **17**, 47–50.

Story, R. 1976, *The Space-Gods Revealed: A Close Look at the Theories of Erich von Däniken* (Harper & Row, New York).

Stozhkov, Y. I. 2003, 'The Role of Cosmic Rays in the Atmospheric Processes', *J. Phys. G* **29**, 913–23.

Strateva, I. et al. 2001, 'Color Separation of Galaxy Types in the Sloan Digital Sky Survey Imaging Data', *Astron. J.* **122**, 1861–74.

Strauss, M. A. et al. 2002, 'Spectroscopic Target Selection in the Sloan Digital Sky Survey: The Main Galaxy Sample', *Astron. J.* **124**, 1810–24.

Stride, S. C. 2001, 'An Instrument-Based Method to Search for Extraterrestrial Interstellar Robotic Probes', *J. Brit. Interplanet. Soc.* **54**, 2–13.

Strigari, L. E., Barnabè, M., Marshall, P. J., and Blandford, R. D. 2012, 'Nomads of the Galaxy', *Mon. Notices Royal Astron. Soc.* **423**, 1856–65.

Stross, C. 2005, *Accelerando* (Orbit Books, London).

Stross, C. 2006, *Missile Gap* (Subterranean Press, Burton).

Stross, C. 2009, *Wireless* (Ace Books, New York).

Stross, C. [1993] 2011, *Scratch Monkey* (NESFA Press, Boston).

Strugatsky, A. and Strugatsky, B. [1985] 1987, *The Time Wanderers* (translated by A. W. Bouis; Richardson & Steirman, New York).

Strugatsky, A. and Strugatsky, B. [1977] 2007, *The Roadside Picnic* (Gollancz, London).

Suffern, K. G. 1977, 'Some Thoughts on Dyson Spheres', *Proc. Astron. Soc. Aust.* **3**, 177–9.

Suthar, F. and McKay, C. P. 2012, 'The Galactic Habitable Zone in Elliptical Galaxies', *Int. J. Astrobiol.* **11**, 157–61.

Svensmark, H. 2012, 'Evidence of Nearby Supernovae Affecting Life on Earth', *Mon. Notices Royal Astron. Soc.* **423**, 1234–53.

Swain, C. M. and Nieli, R. (eds.). 2003, *Contemporary Voices of White Nationalism in America* (Cambridge University Press, Cambridge).

Swan, M. 2012, 'Sensor Mania! The Internet of Things, Wearable Computing, Objective Metrics, and the Quantified Self 2.0', *Sens. Actuator Netw.* **1**, 217–53.

Swan, M. 2013, 'The Quantified Self: Fundamental Disruption in Big Data Science and Biological Discovery', *Big Data* **1**, 85–99.

Swinburn, B. A. et al. 2011, 'The Global Obesity Pandemic: Shaped by Global Drivers and Local Environments', *Lancet* **378**, 804–14.

Swirski, P. 2000, *Between Literature and Science: Poe, Lem, and Explorations in Aesthetics, Cognitive Science, and Literary Knowledge* (McGill-Queen's University Press, Montreal).

Swirski, P. (ed.). 2006, *The Art and Science of Stanislaw Lem* (McGill-Queen's University Press, Montreal).

Tadross, A. L. 2003, 'Metallicity Distribution on the Galactic Disk', *New Astron.* **8**, 737–44.

Taleb, N. N. 2005, *Fooled by Randomness* (2nd edition, Random House, New York).

Taleb, N. N. 2007, *The Black Swan: The Impact of the Highly Improbable* (Random House, New York).

Tarn, W. W. 1951, *The Greeks in Bactria and India* (Cambridge University Press, Cambridge).

Tarter, J. 2001, 'The Search for Extraterrestrial Intelligence (SETI)', *Annu. Rev. Astron. Astrophys.* **39**, 511–48.

Tarter, J. C. et al. 2007, 'A Reappraisal of the Habitability of Planets around M Dwarf Stars', *Astrobiology* **7**, 30–65.

Taylor, A. J. P. 1976, *The Origins of the Second World War* (Hamish Hamilton, London).

Tegmark, M. 1998, 'Is "The Theory of Everything" Merely the Ultimate Ensemble Theory?' *Annals of Physics* **270**, 1–51.

Tegmark, M. 2008, 'The Mathematical Universe', *Found. Phys.* **38**, 101–50.

Teodorani, M. 2014, 'Search for High-Proper Motion Object with Infrared Excess', *Acta Astronaut.* **105**, 547–52.

Terry, K. D. and Tucker, W. H. 1968, 'Biologic Effects of Supernovae', *Science* **159**, 421–3.

Thomas, B. C. 2009, 'Gamma-Ray Bursts as a Threat to Life on Earth', *Int. J. Astrobiol.* **8**, 183–6.

Thomas, B. C. and Honeyman, M. D. 2008, 'Amphibian Nitrate Stress as an Additional Terrestrial Threat from Astrophysical Ionizing Radiation Events?' *Astrobiology* **8**, 1–3.

Thomas, B. C. and Melott, A. L. 2006, 'Gamma-Ray Bursts and Terrestrial Planetary Atmospheres', *New J. Phys.*, **8**, article id. 120 (14 pp.).

Thomas, B. C. et al. 2005a, 'Gamma-Ray Bursts and the Earth: Exploration of Atmospheric, Biological, Climatic, and Biogeochemical Effects', *Astrophys. J.* **634**, 509–33.

Thomas, B. C. et al. 2005b, 'Terrestrial Ozone Depletion due to a Milky Way Gamma-Ray Burst', *Astrophys. J.* **622**, L153–L156.

Thomas, B. C., Melott, A. L., Fields, B. D., and Anthony-Twarog, B. J. 2008, 'Superluminous Supernovae: No Threat from $\eta$ Carinae', *Astrobiology* **8**, 9–16.

Thorsett, S. E. 1995, 'Terrestrial Implications of Cosmological Gamma-Ray Burst Models', *Astrophys. J.* **444**, L53–L55.

Tilgner, C. N. and Heinrichsen, I. 1998, 'A Program To Search for Dyson Spheres with the Infrared Space Observatory', *Acta Astronaut.* **42**, 607–12.

Timofeev, M. Yu., Kardashev, N. S., and Promyslov, V. G. 2000, 'Search of the IRAS Database for Evidence of Dyson Spheres', *Acta Astronaut.* **46**, 655–9.

Tipler, F. J. 1980, 'Extraterrestrial Intelligent Beings do not Exist', *Q. J. Roy. Astron. Soc.* **21**, 267–81.

Tipler, F. J. 1981a, 'A Brief History of the Extraterrestrial Intelligence Concept', *Q. J. Roy. Astron. Soc.* **22**, 133–45.

Tipler, F. J. 1981b, 'Additional Remarks on Extraterrestrial Intelligence', *Q. J. Roy. Astron. Soc.* **22**, 279–92.

Tipler, F. J. 1981c, 'Extraterrestrial Intelligent Beings Do Not Exist', *Physics Today* **34** (4), 9–10.

Tipler, F. J. 1982, 'Anthropic-Principle Arguments against Steady-State Cosmological Theories', *Observatory* **102**, 36–9.

Tipler, F. J. 1986, 'Cosmological Limits on Computation', *Int. J. Theor. Phys.* **25**, 617–61.

Tipler, F. J. 1994, *The Physics of Immortality* (Doubleday, New York).

Tipler, F. J. 2003, 'Intelligent Life in Cosmology', *Int. J. Astrobiol.* **2**, 141–8.

Tirard, S., Morange, M., and Lazcano, A. 2010, 'The Definition of Life: A Brief History of an Elusive Scientific Endeavor', *Astrobiology* **10**, 1003–9.

Tlusty, T. 2010, 'A Colorful Origin for the Genetic Code: Information Theory, Statistical Mechanics and the Emergence of Molecular Codes', *Phys. Life Rev.* **7**, 362–76.

Tolkien, J. R. R. 1983, *The Monsters and the Critics and Other Essays* (Houghton Mifflin, Boston).

Tough, A. 1990, 'A Critical Examination of the Factors that Might Encourage Secrecy', *Acta Astronaut.* **21**, 97–101.

Tough, A. 1998, 'Small Smart Interstellar Probes', *J. Brit. Interplanet. Soc.* **51**, 167–74.

Tremblay, R. E., Hartup, W. W., and Archer, J. 2005, *Developmental Origins of Aggression* (Guilford Press, New York).

Treumann, R. A. 1993, 'Evolution of the Information in the Universe', *Astrophys. Space Sci.* **201**, 135–47.

Trifonov, E. N. 2011, 'Vocabulary of Definitions of Life Suggests a Definition', *J. Biomol. Struct. Dynamics* **29**, 259–66.

Trifonov, E. N. 2012, 'Definition of Life: Navigation through Uncertainties', *J. Biomol. Struct. Dynamics* **29**, 647–50.

Troitskij, V. S. 1989, 'Development of Extraterrestrial Intelligence and Physical Laws', *Acta Astronaut.* **19**, 875–87.

Tryka, K. A., Brown, R. H., Anicich, V., Cruikshank, D. P., and Owen, T. C. 1993, 'Spectroscopic Determination of the Phase Composition and Temperature of Nitrogen Ice on Triton', *Science* **261**, 751–4.

Tsiolkovsky, K. E. 1933, 'The Planets Are Occupied by Living Beings', (in Russian; manuscript in archives of Tsiolkovsky State Museum of the History of Cosmonautics, Kaluga, Russia).

Tsokolov, S. A. 2009, 'Why Is the Definition of Life So Elusive? Epistemological Considerations', *Astrobiology* **9**, 401–12.

Turner, D. 2007, *Making Prehistory: Historical Science and the Scientific Realism Debate* (Cambridge University Press, Cambridge).

Ulvestad, E. 2002, 'Biosemiotic Knowledge: A Prerequisite for Valid Explorations of Extraterrestrial Intelligent Life', *Sign Syst. Stud.* **30**, 283–92.

Uzan, J.-P. 2003, 'The Fundamental Constants and their Variation: Observational and Theoretical Status', *Rev. Mod. Phys.* **75**, 403–55.

Vakoch, D. A. (ed.). 2014, *Archaeology, Anthropology, and Interstellar Communication* (NASA, Washington, DC).

Valdes, F. and Freitas, R. A., Jr. 1980, 'Comparison of Reproducing and Non-reproducing Starprobe Strategies for Galactic Exploration', *J. Brit. Interplanet. Soc.* **33**, 402–8.

Valdes, F. and Freitas, R. A., Jr. 1986, 'A Search for the Tritium Hypefine Line from Nearby Stars', *Icarus* **65**, 152–7.

Valéry, P. 1989, *Monsieur Teste* (translated by J. Mathews; Princeton University Press, Princeton).

van den Bergh, S. 1994, 'Astronomical Catastrophes in Earth History', *Publ. Astron. Soc. Pac.* **106**, 689–95.

Verendel, V. and Häggström, O. 2017, 'Fermi's Paradox, Extraterrestrial Life and the Future of Humanity: A Bayesian Analysis', *Int. J. Astrobiol.* **16**, 14–18.

Verma, S. 2007, *Why Aren't They Here? The Question of Life on Other Words* (Icon Books, Cambridge).

Vico, G. [1725] 1968, *The New Science* (translated by M. H. Fisch and T. G. Bergin; Cornell University Press, Ithaca).

Viewing, D. 1975, 'Directly Interacting Extra-Terrestrial Technological Communities', *J. Brit. Interplanet. Soc.* **28**, 735–44.

Vinge, V. 1986, *Marooned in Realtime* (Bluejay Books, New York).

Vinge, V. 1992, *A Fire Upon the Deep* (Tor Books, New York).

Virgo, N., Fernando, S., Bigge, B., and Husbands, P. 2012, 'Evolvable Physical Self-Replicators', *Artificial Life* **18**, 129–42.

Volk, T. 2002, 'The Gaia Hypothesis: Fact, Theory, and Wishful Thinking', *Climatic Change* **52**, 423–30.

von Bloh, W., Bounama, C., and Franck, S. 2007, 'Dynamic Habitability for Earth-Like Planets in 86 Extrasolar Planetary Systems', *Planet. Space Sci.* **35**, 651–60.

von Hoerner, S. 1975, 'Population Explosion and Interstellar Expansion', *J. Brit. Interplanet. Soc.* **28**, 691–712.

von Hoerner, S. 1978, 'Where Is Everybody?' *Naturwissenschaften* **65**, 553–7.

von Hoerner, S. 1980, 'Manifestations of Advanced Cosmic Civilizations: An Introduction', in M. D. Papagiannis (ed.) *Strategies for the Search for Life in the Universe* (Reidel, Dordrecht), 189–96.

Voros, J. 2014, 'Galactic-Scale Macro-Engineering: Looking for Signs of Other Intelligent Species, as an Exercise in Hope for Our Own', in L. Grinin, D. Baker, E. Quaedackers, and A. Korotayev (eds.) *Teaching and Researching Big History: Exploring a New Scholarly Field*; (Uchitel Publ House, Volgograd) 283–304.

Vukotić, B. 2010, 'The Set of Habitable Planets and Astrobiological Regulation Mechanisms', *Int. J. Astrobiol.* **9**, 81–7.

Vukotić, B. and Ćirković, M. M. 2007, 'On the Timescale Forcing in Astrobiology', *Serb. Astron. J.* **175**, 45–50.

Vukotić, B. and Ćirković, M. M. 2008, 'Neocatastrophism and the Milky Way Astrobiological Landscape', *Serb. Astron. J.* **176**, 71–9.

Vukotić, B. and Ćirković, M. M. 2012, 'Astrobiological Complexity with Probabilistic Cellular Automata', *Orig. Life Evol. Biospheres* **42**, 347–71.

Vukotić, B., Steinhauser, D., Martinez-Aviles, G., Ćirković, M. M., Micic, M., and Schindler, S. 2016, '"Grandeur in this View of Life": N-Body Simulation Models of the Galactic Habitable Zone', *Mon. Notices Royal Astron. Soc.* **459**, 3512–24.

Vulpetti, G. 1999, 'On the Viability of the Interstellar Flight', *Acta Astronaut.* **44**, 769–92.

Wagner, K. et al. 2016, 'Direct Imaging Discovery of a Jovian Exoplanet within a Triple-Star System', *Science* **353**, 673–8.

Wald, R. M. 1974, 'Energy Limits on the Penrose Process', *Astrophys. J.* **191**, 231–4.

Waldrop, M. M. 2011, 'SETI Is Dead, Long Live SETI', *Nature* **475**, 442–4.

Wallace, A. R. 1903, *Man's Place in the Universe; A Study of the Results of Scientific Research in Relation to the Unity or Plurality of Worlds* (Chapman & Hall, London).

Wallace, A. R. 1907, *Is Mars Habitable? A Critical Examination of Professor Percival Lowell's Book 'Mars and Its Canals', With an Alternative Explanation* (Macmillan & Co., London).

Wallis, M. K. and Wickramasinghe, N. C. 2004, 'Interstellar Transfer of Planetary Microbiota', *Mon. Not. R. Astron. Soc.* **348**, 52–61.

Walters, C., Hoover, R. A., and Kotra, R. K. 1980, 'Interstellar Colonization: A New Parameter for the Drake Equation?' *Icarus* **41**, 193–7.

Wandel, A. 2015, 'On the Abundance of Extraterrestrial Life after the Kepler Mission', *Int. J. Astrobiol.* **14**, 511–16.

Wang, T., Sha, R., Dreyfus, R., Leunissen, M. E., Maass, C., Pine, D. J., Chaikin, P. M., and Seeman, N. C. 2011, 'Self-Replication of Information-Bearing Nanoscale Patterns', *Nature* **478**, 225–8.

Ward, P. 2005, *Life as We Do Not Know It* (Viking Press, New York).

Ward, P. D. and Brownlee, D. 2000, *Rare Earth: Why Complex Life Is Uncommon in the Universe* (Springer, New York).

Watts, P. 2006, *Blindsight* (Tor Books, New York).

Webb, S. 2002, *Where is Everybody? Fifty Solutions to the Fermi Paradox* (Copernicus, New York).

Webb, S. 2015, *Where is Everybody? Seventy-Five Solutions to the Fermi Paradox and the Problem of Extraterrestrial Life* (Springer, New York).

Weiler, H. 1998, 'The Fermi Paradox and 1991 VG', *Observatory* **118**, 226.

Weinberg, S. 1972, *Gravitation and Cosmology: Principles and Applications of the General Theory of Relativity* (Wiley-VCH, Hoboken, NJ).

Weinberg, S. 2008, *Cosmology* (Oxford University Press, Oxford).

Weisberg, M. and Muldoon, R. 2009, 'Epistemic Landscapes and the Division of Cognitive Labor', *Philos. Sci.* **76**, 225–52.

Wesson, P. S. 1990, 'Cosmology, Extraterrestrial Intelligence, and a Resolution of the Fermi-Hart Paradox', *Q. J. Roy. Astron. Soc.* **31**, 161–70.

Wesson, P. S. 1991, 'Olbers's Paradox and the Spectral Intensity of the Extragalactic Background Light', *Astrophys. J.* **367**, 399–406.

Wesson, P. S. 2011, 'Panspermia, Past and Present: Astrophysical and Biophysical Conditions for the Dissemination of Life in Space', *Space Sci. Rev.* **156**, 239–52.

Wesson, P. S., Valle, K., and Stabell, R. 1987, 'The Extragalactic Background Light and a Definitive Resolution of Olbers's Paradox', *Astrophys. J.* **317**, 601–6.

Westfahl, G. 1997, 'The Case against Space', *Sci. Fict. Stud.* **24**, 193–206.

Weston, A. 1988, 'Radio Astronomy as Epistemology: Some Philosophical Reflections on the Contemporary Search for Extraterrestrial Intelligence', *Monist* **71**, 88–100.

Whitmire, D. P. and Wright, D. P. 1980, 'Nuclear Waste Spectrum as Evidence of Technological Extraterrestrial Civilizations', *Icarus* **42**, 149–56.

Whitten, R. C., Borucki, W. J., Wolfe, J. H., and Cuzzi, J. 1976, 'Effect of Nearby Supernova Explosions on Atmospheric Ozone', *Nature* **263**, 398–400.

Wilcox, F. M. (director). 1956, *Forbidden Planet* (DVD, Metro-Goldwyn-Mayer Corp.).

Wilczek, F. 1993, 'Weinberg: A Philosopher in Spite of Himself', *Phys. Today* **46**, 59–60.

Wiley, K. B. 2014, 'The Fermi Paradox, Self-Replicating Probes, and the Interstellar Transportation Bandwidth', *h+ Magazine*, 18 November 2014 (online article, http://hplusmagazine.com/2014/11/18/fermi-paradox-self-replicating-probes-interstellar-transportation-bandwidth/, last accessed 15 March 2016).

Wilkinson, D. 2013, *Science, Religion, and the Search for Extraterrestrial Life* (Oxford University Press, Oxford).

Williams, D. M., Kasting, J. F., and Wade, R. A. 1997, 'Habitable Moons around Extrasolar Giant Planets', *Nature* **385**, 234–6.

Williams, L. 2010, 'Irrational Dreams of Space Colonization', *Peace Rev.* **22**, 4–8.

Wilson, P. A. 1994, 'Carter on Anthropic Principle Predictions', *Brit. J. Phil. Sci.* **45**, 241–53.

Wilson, R. A. and Hill, M. J. 1998, *Everything Is Under Control: Conspiracies, Cults and Cover-Ups* (HarperCollins, New York).

Wilson, R. C. 2003, *Blind Lake* (Tor Books, New York).

Wilson, T. L. 1984, 'Bayes' Theorem and the Real SETI Equation', *Q. J. Roy. Astron. Soc.* **25**, 435–48.

Wilson, T. L. 2001, 'The Search for Extraterrestrial Intelligence', *Nature* **409**, 1110–14.

Woosley, S. E. and Bloom, J. S. 2006, 'The Supernova Gamma-Ray Burst Connection', *Ann. Rev. Astron. Astrophys.* **44**, 507–56.

Wright, J. C. 2002, *The Golden Age* (Tor Books, New York).

Wright, J. C. 2003a, *The Phoenix Exultant* (Tor Books, New York).

Wright, J. C. 2003b, *The Golden Transcendence* (Tor Books, New York).

Wright, J. C. 2011, *Count to a Trillion* (Tor Books, New York).

Wright, J. T. 2018, 'Prior Indigenous Technological Species', *Int. J. Astrobiol.* **17**, 96–100.

Wright, J. T., Cartier, K. M. S., Zhao, M., Jontof-Hutter, D., and Ford, E. B. 2016, 'The Search for Extraterrestrial Civilizations with Large Energy Supplies. IV. The Signatures and Information Content of Transiting Megastructures', *Astrophys. J.* **816**, article id. 17 (22 pp.).

Wright, J. T., Griffith, R. L., Sigurdsson, S., Povich, M. S., and Mullan, B. 2014, 'The Ĝ Infrared Search for Extraterrestrial Civilizations with Large Energy Supplies. II. Framework, Strategy, and First Result', *Astrophys. J.* **792**, article id. 27 (12 pp.).

Wright, J. T., Mullan, B., Sigurdsson, S., and Povich, M. S. 2014, 'The Ĝ Infrared Search for Extraterrestrial Civilizations with Large Energy Supplies. I. Background and Justification', *Astrophys. J.* **792**, article id. 26 (16 pp.).

Wright, J. T. and Sigurdsson, S. 2016, 'Families of Plausible Solutions to the Puzzle of Boyajian's Star,' *The Astrophys. J. Lett.* **829**, article id. L3, (12pp.).

Yockey, H. P. 1977, 'A Calculation of the Probability of Spontaneous Biogenesis by Information Theory', *J. Theor. Biol.* **67**, 377–98.

Yoon, J. H., Putman, M. E., Thom, C., Chen, H.-W., and Bryan, G. L. 2012, 'Warm Gas in the Virgo Cluster. I. Distribution of Lyα Absorbers', *Astrophys. J.* **754**, article id. 84 (14 pp.).

York, D. G. et al. 2000, 'The Sloan Digital Sky Survey: Technical Summary', *Astron. J.* **120**, 1579–87.

Young, G. M. 2012, *The Russian Cosmists: The Esoteric Futurism of Nikolai Fedorov and His Followers* (Oxford University Press, Oxford).

Zackrisson, E., Calissendorff, P., Asadi, S., and Nyholm, A. 2015, 'Extragalactic SETI: The Tully–Fisher Relation as a Probe of Dysonian Astroengineering in Disk Galaxies', *Astrophys. J.* **810**, article id. 23 (12 pp.).

Zackrisson, E., Calissendorff, P., González, J., Benson, A., Johansen, A., and Janson, M. 2016, 'Terrestrial Planets across Space and Time', *Astrophys. J.* **833**, article id. 214 (12 pp.).

Zalasiewicz, J., Williams, M., Waters, C. N., Barnosky, A. D., and Haff, P. 2014, 'The Technofossil Record of Humans', *Anthropocene Rev.* **1**, 34–43.

Zamyatin, Y. [1924] 1993, *We* (translated by C. Brown; Penguin Books, New York).

Zank, G. P. and Frisch, P. C. 1999, 'Consequences of a Change in the Galactic Environment of the Sun', *Astrophys. J.* **518**, 965–73.

Zubrin, R. 1999, *Entering Space: Creating a Spacefaring Civilization* (Jeremy P. Tarcher/ Putnam, New York).

Zuckerman, B. 1985, 'Stellar Evolution: Motivation for Mass Interstellar Migrations', *Q. J. Roy. Astron. Soc.* **26**, 56–9.

Zuckerman, B. and Hart, M. H. (eds.). 1995, *Extraterrestrials. Where Are They?* (Cambridge University Press, Cambridge).

# INDEX